Air Pollution XXVII

WITPRESS

WIT Press publishes leading books in Science and Technology.
Visit our website for the current list of titles.
www.witpress.com

WITeLibrary

Home of the Transactions of the Wessex Institute.
Papers contained in this volume are archived in the WIT eLibrary in volume 236 of WIT Transactions on Ecology and the Environment (ISSN 1743-3541).
The WIT eLibrary provides the international scientific community with immediate and permanent access to individual papers presented at WIT conferences.
Visit the WIT eLibrary at www.witpress.com.

TWENTY SEVENTH INTERNATIONAL CONFERENCE ON MODELLING, MONITORING AND MANAGEMENT OF AIR POLLUTION

Air Pollution 2019

Conference Chairmen

G. Passerini
Marche Polytechnic University, Italy
Member of WIT Board of Directors

C. Borrego
University of Aveiro, Portugal

J. Longhurst
University of the West of England, UK

M. Lopes
University of Aveiro, Portugal

J. Barnes
University of the West of England, UK

International Scientific Advisory Committee

Organised by

Wessex Institute, UK

University of Aveiro, Portugal

University of The West of England, UK

Sponsored by

WIT Transactions on Ecology and the Environment

International Journal of Environmental Impacts

WIT Transactions

Wessex Institute
Ashurst Lodge, Ashurst
Southampton SO40 7AA, UK

Senior Editors

Editorial Board

Air Pollution XXVII

Editors

G. Passerini
Marche Polytechnic University, Italy
Member of WIT Board of Directors

C. Borrego
University of Aveiro, Portugal

J. Longhurst
University of the West of England, UK

M. Lopes
University of Aveiro, Portugal

J. Barnes
University of the West of England, UK

WIT*PRESS* Southampton, Boston

Editors:

G. Passerini
Marche Polytechnic University, Italy
Member of WIT Board of Directors

C. Borrego
University of Aveiro, Portugal

J. Longhurst
University of the West of England, UK

M. Lopes
University of Aveiro, Portugal

J. Barnes
University of the West of England, UK

Published by

WIT Press
Ashurst Lodge, Ashurst, Southampton, SO40 7AA, UK
Tel: 44 (0) 238 029 3223; Fax: 44 (0) 238 029 2853
E-Mail: witpress@witpress.com
http://www.witpress.com

For USA, Canada and Mexico

Computational Mechanics International Inc
25 Bridge Street, Billerica, MA 01821, USA
Tel: 978 667 5841; Fax: 978 667 7582
E-Mail: infousa@witpress.com
http://www.witpress.com

British Library Cataloguing-in-Publication Data

A Catalogue record for this book is available
from the British Library

ISBN: 978-1-78466-343-8
eISBN: 978-1-78466-344-5
ISSN: 1746-448X (print)
ISSN: 1743-3541 (on-line)

The texts of the papers in this volume were set individually by the authors or under their supervision. Only minor corrections to the text may have been carried out by the publisher.

Preface

Air Pollution 2019 is the 27th Annual Meeting in the successful series of international conferences organised by the Wessex Institute, the University of Aveiro and the University of the West of England Bristol concerned with advances in Modelling, Monitoring and Management of Air Pollution.

The series started in Monterrey (1993) and continued in Barcelona (1994), Halkidiki (1995), Toulouse (1996), Bologna (1997), Genova (1998), San Francisco (1999), Cambridge (2000), Ancona (2001), Segovia (2002), Catania (2003), Rhodes (2004), Cordoba (2005), New Forest (2006), Algarve (2007), Skiathos (2008), Tallinn (2009), Kos (2010), Malta (2011), A Coruña (2012), Siena (2013), Split (2014),Valencia (2015), Crete (2016), Cadiz (2017) and most recently convened at the Parthenope University in Naples for a well-attended and stimulating meeting discussing a wide range of air pollution topics.

This important conference brings together contributions from scientists from around the world to present recent work on various aspects of air pollution phenomena. The meetings have discussed and considered many important air pollution issues and the international nature of the attendees has ensured that the conference findings and conclusions enjoy a wide and rapid dissemination amongst the air pollution science and policy communities. Air pollution sources and impacts remain one of the most challenging problems facing the international community. The series has demonstrated the widespread nature of the air pollution phenomena and has contributed to the evolving understanding of the science and policy contexts of air pollution management.

The conference continues to meet the demands of a discerning conference audience through the quality of the science and policy presented at the meetings, the publication formats and the interesting conference venues. This series has discussed important air pollution issues at an international, national and local level and by virtue of the international composition of the delegates has brought to the discussion a unique suite of perspectives. Notable in each of the conferences in this series has been the opportunity to foster scientific exchange between participants. Each meeting has provided a further opportunity for identifying new areas of air pollution science demanding collaborative investigation.

The conference series has consistently acknowledged that science remains the key to identifying the nature and scale of air pollution sources and impacts and reaffirmed that science is essential in the formulation of policy relevant information for regulatory decision-making. The conference series also acknowledged, at a very early stage, that science alone will not improve a polluted atmosphere. The scientific knowledge derived from well-designed studies needs to be allied with further technical and economic studies in order to ensure cost effective and efficient mitigation. In turn, the science, technology and economic outcomes are necessary but not sufficient. Increasingly, the conference series has recognised that the outcome of such research need to be contextualised within well formulated communication strategies that help policy makers and citizens to understand and appreciate the risks and rewards arising from air pollution management. Consequently, the series has enjoyed a wide range of high quality presentations that develop the fundamental science of air pollution and an equally impressive range of presentations that places these new developments within the frame of mitigation and management of air pollution. The peer reviewed nature of the conference volumes enables policy makers to use the new findings with confidence to formulate sustainable decisions and to build public acceptance and understanding of the nature and scale of the air pollution problem.

The 27th meeting in 2019 takes place at a time when the problems of air pollution are manifesting themselves across the world. The science to explain air pollution phenomena at different spatial and temporal scales has continued to advance throughout the life of this series. The impacts of air pollution on health, ecosystems and the built environment are better understood, as is the contribution of different sources to observed phenomena.

However, air pollution is not an accident; it results from the decisions others take to shape land use, develop transport infrastructure and to grow the economy. Air pollution results from the interactions of these processes and the way the practices of citizens and businesses are conditioned by the systems in which they operate. The outcome is a risk to, and often an impact on, public health and the environment. Different groups in society have different experiences of pollution, in general, the richest in society emit the most pollution and are exposed the least whilst the poorest in society emit the least pollution and are exposed the most. In many parts of the world public concern about the scale and consequence of air pollution exposure is rising, as exemplified by the BreathLife campaign of WHO, UNEP and others.

What cannot be avoided is the enormous scale, complexity and impact of the global problem of air pollution. The WHO estimates that 4.2 million people die prematurely each year as a result of exposure to ambient air pollution, principally fine particulates and nitrogen dioxide, and that 91% of the world's population lives in places where air quality exceeds WHO guidelines. The European Court of Auditors describe air pollution as the biggest environmental health risk in the European Union with some 400 000 premature deaths per year in the EU and hundreds of billions of euros in health-related external costs attributable to air pollution. In the United Kingdom, estimates suggest the annual premature death toll associated with exposure to fine particles and nitrogen dioxide is in the range of 28 000 to 36 000 people. Public Health England estimated the 2017 cost of air pollution to the NHS and social care in England as £157 million, whilst the House of Commons Joint Select Committee on Improving Air Quality estimated air pollution to cost the UK some £20 billion per annum in health care costs and lost productivity. Furthermore, there is increasing recognition that air pollution must be viewed within the wider context of urban stressors (e.g. noise, heat and lack of access to green space) that are collectively and cumulatively affecting life expectancy and quality of life.

Recent medical research points to a range of potential health impacts arising from exposure to air pollution across the human life course. Much needs to be done to limit the impacts on health and the environment and to create effective policy and regulatory instruments. Where good policy and legislation exists implementation and enforcement has rarely been followed through with the full force of the law. This may be because politicians are concerned about the effect such policies and regulations may have on the electorate and investment. Whatever the reason for national inaction, city governance has started to respond to concern about air pollution with decisions by Paris, Mexico, Athens, Madrid and Rome, amongst others, to restrict certain vehicle types within their cities. This may be the beginning of a more coherent municipal response to such issues with London's introduction of an Ultra Low Emission Zone, a further encouraging sign of municipal action. However, city authorities on their own cannot deal with all of the complexities and challenges of air pollution. To do so they need carefully designed and well-integrated national and international policy and regulatory actions that share the burden of managing air pollution at the appropriate temporal and spatial scale.

Part of the forthcoming challenge is that many of the easy wins with air quality policy have been secured and the causes of the remaining challenges are deeply entrenched and resistant to change. Policy development and its effective implementation is becoming tougher as the direct costs, financial and political, of improving air quality increase. Yet the public's concern is increasing. This challenge is one the air quality community must recognise, confront and overcome. It will require recognition that air quality is a multifaceted, emergent problem, necessitating the integration of social science and natural science perspective to inform policy and to avoid unintended consequences. A systems approach will be needed to derive appropriate solutions in the long and short term. Key to success will be the integration of effective air quality management into land use, transport, industrial and economic planning processes.

Air quality can be seen as an emergent property arising from anthropogenic activities and resulting from policy decisions from differing governmental spheres, often disconnected and asynchronous. True management of air pollution can be only be achieved through a systematic understanding of the often multiple causes, feedback loops and impacts. Ultimately, air pollution is a choice society makes through its collective and individual behaviours and collective practices. However the consequences of those behaviours and choices will play out in many different ways with those who are least able to exercise choice having air pollution concentrations imposed upon them. However, history shows that concerted, collective and sustained action can lead to dramatic improvements in air quality.

There remains much to do and the Modelling, Monitoring and Management of Air Pollution conference series continues to play an important role in providing an opportunity for an international audience to discuss both long-standing and emergent issues in air pollution science and policy.

The papers selected for presentation and published in this volume are part of the Transactions on Ecology and the Environment. They have been archived online in the WIT eLibrary (www.witpress.com/elibrary) where they are easily and permanently available to the international scientific community. These collected papers provide an invaluable record of the development of science and policy pertaining to air pollution.

The Editors are grateful to the University of Aveiro for having hosted the meeting in their excellent facilities. This gave the participants the occasion to visit one of Portugal's hidden gems, an inspiring city with ancient roots, and an impressive Art Nouveau architecture in a water setting that gives rise to Aveiro being called the Venice of Portugal.

The Editors wish to thank the authors for their contributions to the conference and to acknowledge the assistance of the eminent members of the International Scientific Advisory Committee for their support for the conference and in particular for their peer reviewing of the manuscripts.

The Editors
Aveiro, Portugal
June 2019

Contents

AIR QUALITY COMMUNITY ACTION NETWORK

MICHAEL R. OGLETREE & GREGG W. THOMAS
Environmental Quality Division, Department of Public Health and Environment,
City and County of Denver, USA

ABSTRACT

In Denver, 11% of public school children have asthma and some schools have asthma rates as high as 29%, leading to increased school absenteeism. Denver's high levels of air pollution exacerbate the problem. To address this inequity, Denver, in partnership with Denver Public Schools (DPS), is creating a citywide air quality monitoring network to provide real-time air quality data – utilizing low-cost cutting-edge air pollution sensor technology, redeveloped with solar, battery storage, and data connectivity to make it useful for widescale deployment and replicable in any municipality. To date, no US municipality has a city-led, community-based air-monitoring program. While the air sensors are foundational to the project, the heart of the solution is the collaborative, culturally-appropriate and scientifically-validated approach to programming. Each participating school (with asthma rates above the median) will receive a sensor, air quality dashboard and programming. Programming options are appropriate for elementary, middle or high school and include STEM curriculum, anti-idling campaign, event plan for extremely high air pollution days (e.g., wildfires), built environment safety study to change traffic patterns near schools, and school-based challenges to incentivize new ideas for driving behaviour change and reducing pollution. The dashboard will display real time data and suggested behaviour changes, while the backend data platform will create insights for air quality patterns near each school – leading to policy and institutional changes for the City and DPS – as well as generate automated alerts for stakeholders.
Keywords: sensors, PM, schools, community, innovation, dashboard, human centred design.

1 INTRODUCTION

As one of the fastest-growing US cities, Denver experiences significant construction and traffic congestion, worsening the air quality – the 14th worst among major US cities [1]. Only 53% of residents realize the impacts of poor air quality [2], including that children are more susceptible to its effects, such as decreased lung function and missed school days. While multiple factors influence exposure to air pollution, schools are an ideal intervention point for sensor deployment, education and empowerment. Denver will provide real-time, hyper-local air quality data measured at Denver schools, supported by evidence-based, culturally responsive programming to empower communities to limit exposure and reduce pollution.

A winner of the 2018 US Bloomberg Mayor Challenge, Denver's idea has received $1,000,000 to implement the program at scale in Denver Public Schools (DPS), the city's primary public school district with 207 schools and 92,331 students [3]. Working closely with DPS, the City and County of Denver will scale the project from its 3 pilot schools in 2018, to 40 over the next three years.

2 US AIR QUALITY REGULATORY STRUCTURE

Within the United States, air quality regulatory structure is defined by the Clean Air Act (CAA) of 1970. The CAA is a United States federal law designed to control air pollution on a national level [4]. It was first established in 1970 and has continued to evolve with its most recent major amendment in 1990. The CAA authorised the development of comprehensive federal and state regulations limiting emissions from stationary as well as mobile sources. It led to the creation the National Ambient Air Quality Standards (NAAQS), which sets limits

WIT Transactions on Ecology and the Environment, Vol 236, © 2019 WIT Press
www.witpress.com, ISSN 1743-3541 (on-line)
doi:10.2495/AIR190011

for six major pollutants: sulphur dioxide, particulate matter, carbon monoxide, ozone, nitrogen dioxide, and lead.

Under the CAA the federal agency responsible for implementation and oversight is the United States Environmental Protection Agency (USEPA). The USEPA sets national limits on pollutants and approves state, tribal and local agency plans for reducing pollution from sources to meet those limits. Individual state, tribal, and local agencies may have stricter regulations but not more lenient. State, tribal, and local agencies take the lead on developing solutions to meet pollution limits as they have a better understanding of conditions such as local industry, geography, housing, and travel patterns, as well as other influencing factors. They also take the lead on monitoring air quality in their respective jurisdictions.

Specifics can vary from state to state in terms of the number of local and tribal agencies, as well as which local agencies are responsible for which facilities. Additionally, it is common for state and local agencies to work cooperatively to oversee facilities and other sources in a specific region. In Denver Colorado, the local agency, the Denver Department of Public Health and Environment (DDPHE) does some local inspection and monitoring. DDPHE is overseen by the state agency, the Colorado Department of Public Health and Environment (CDPHE) which is overseen by the federal agency, the USEPA. There is also an additional agency which is the lead on regional planning, the Regional Air Quality Council (RAQC) (Fig. 1).

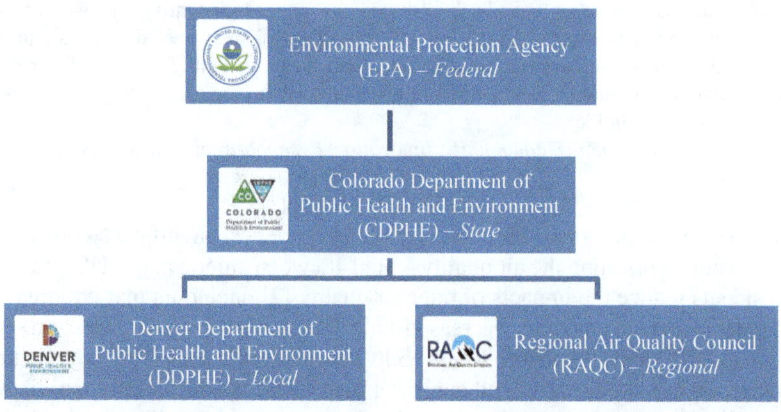

Figure 1: Organizational chart of federal, state, and local air agencies in Denver, Colorado.

3 THE MAYORS CHALLENGE

On 26 June 2017, Bloomberg Philanthropies announced the US Mayor Challenge. The fourth iteration of the challenge, the second of which was focused in the United States, the Mayors Challenge is an ideas competition, challenging cities to find innovative solutions to solving urgent problems within their municipalities [5]. The 2018 Mayors Challenge was open to US cities larger than 30,000. 324 cities applied for the challenge and on 20 February 2018, 35 cities, including Denver, were selected as Champion Cities. These cities received $100,000 each, with a timeline of 6 months to pilot their idea and show that they could deliver on their projects. At the end of the pilot, the 35 cities then submitted a second application. On 20 October 2018, nine cities were then selected as winners, each receiving $1,000,000 to implement their projects at scale in their cities. Selection criteria for winning cities included: vision, impact, implementation, and transferability. In addition to the funding, each city in the pilot phase were assigned an innovation coach and an implementation coach. These coaches helped the cities refine their ideas and provided guidance and training in new and innovative methods such as human centred design. The participation and collaboration with the implementation coaches will continue through the three-year project assisting through in-person work sessions, regular project conference calls, and replication events.

4 PILOT TESTING

During the six-month pilot phase, Denver tested key components of the idea with three pilot schools. The methodology incorporated iterative testing with key stakeholders while co-creating solutions using a human centred design approach. Components tested included: stakeholder identification and buy-in, influencing behaviour change, dashboard/information sharing, and sensor/data platform technology.

4.1 Stakeholder identification and buy-in

This was a key component in the project because of the fact that in the Denver Public School system each individual school is given the freedom to select programs and projects in which it participates. The City and County of Denver had already been collaborating with the school district on the idea, but it was necessary to received buy-in and approval from principals on a school by school basis.

Tested at three public schools, the initial approach was direct outreach to school principals through phone calls and email introductions. The team provided an overview of programs and asked principals if the phone call or email approach was effective and received feedback on how it could be more effective. The Denver team refined their approach and developed a strategy for future school initial engagement.

This strategy involved approaching engaged nurses and science teacher at events they would regularly participate as part of their specific roles. The team attended workshops and built relationships with nurses working in the schools. Those individuals would then champion the idea to the principal of their schools leading to in-person meetings and approval of participation.

4.2 Influencing behaviour change

Behavioural change experiments were conducted with parents to test various idling reduction methods at three schools. Volunteers from the city as well as community members and parents monitored idling at pilot schools during morning drop-off and afternoon pick-up. Vehicle counts as well as idling duration were recorded. One week of baseline, as well as one week for each intervention methodology were tracked. The first method tested was anti-idling signage. This method differed from signage previously present at schools in that the signage was facing the direction of oncoming traffic as well as near eye level with drivers (Fig. 2). The second test was an anti-idling pledge sent home with students in weekly folders. The third intervention was intended to use a district media platform to relay messages daily to the school communities. This last intervention was not tested as there were some hurdles in allowing testing though the district portal that the team was not able to overcome before school let out for summer.

Many lessons were learned through this testing. Manually tracking idling at a school of 300+ students is very difficult with only 3–4 volunteers per school. The detail of notes varied from volunteer to volunteer which resulted in inconsistent data sets. During the period of sending pledges home, it was found there was often a disconnect with the person who picked

Figure 2: Dashboard mock-ups created in parent focus group.

up the student from school and the person who went through the weekly folder and received the pledge. While the Denver team did not see expected results through this test, they found the experience to be very insightful. Being present on the ground at schools, gave a more detailed understanding of the situation during pick-up and drop-off times. It was shown that each school is unique. There are some basic strategies that can be implemented to help ease some of the challenges, however, a customized plan developed with each school will have the biggest impact.

4.3 Dashboard information sharing

Information sharing with school communities on near-real time air quality can be a challenge. The approach taken was to conduct iterative focus groups with the parents, teachers, and nurses with the aim of developing a dashboard relaying the information that was useful in a way that was understood by the community. The team started with hand drawn low fidelity prototypes, progressing to hand built dashboards mock-ups (Fig. 3), finally to digital dashboards displaying local data from sensors. Different air quality scales, colours, and ranges were also tested in the focus groups.

As in other tests, the team learned a great deal from having the focus groups and hearing directly from stakeholders. The specific outcome was a dashboard design that will be implemented in schools in 2019. The platform which hosts the dashboard allows for continued refinement and customization for each school.

4.4 Sensor and data platform technology

The air quality sensor technology used in this project was co-developed with a local aerospace engineering company for the specific use case of deployment at schools within the project. There were limitations of off the shelf instrumentation, with custom modification

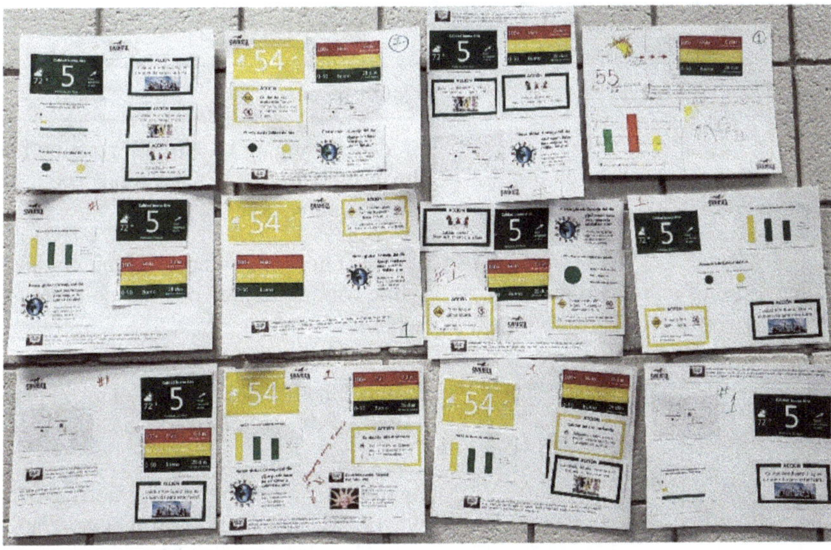

Figure 3: Dashboard mock-ups created in parent focus group.

costly and impractical for scaling up to a city-wide network. By co-designing and developing the units, the Denver team was able to limit the costs while also achieving the level of data resolution and accuracy needed for the project. Testing of the sensors was done through extended collocations at state agency sites equipped with instrumentation that met the standards of the federal agency. Sensors were deployed at three different sites with two different types of instruments and one triplicate collocation for sensor to sensor variability comparison. The results showed very high correlation ($r^2 > 0.9$) of sensors to both types of high quality instruments and extremely high correlations (r2 > 0.95) between sensors.

In addition to development of the air quality sensors, the Denver team also worked with a developer to create a platform for management of air quality sensor data. This platform can ingest data from a variety of sensors as well as state run stations. By ingesting both instruments in the same platform it allows for near-real time analytics of the entire network as well as quality assure collocated sensors continuously. Having permanent sensors located at state sites, the platform will have the ability to dynamically generate correction algorithms which would then be pushed to the sensor network allowing for on-going automated calibration. In addition, the platform manages the dashboard interfaces as well as alerting thresholds and automated network reporting, allowing for ease of network management with minimal oversite.

5 MITIGATION METHODOLOGY AND MEASUREMENT

The strategies for mitigating exposure and reducing nearby sources of emission are split into three categories: individual, school community, and city government. The first two categories have exposure reduction as well as pollution reduction sub-categories, while city government is limited to pollution reduction strategies (Table 1). Research was conducted during the pilot phase of the project to identify available programs and strategies for inclusion. While many options existed, we found only limited evaluation of the effectiveness. As the team continues to implement the project at scale, programs will be evaluated to measure the impact and effectiveness of each.

Table 1: Air pollution mitigation strategy categories with examples.

Category	Example
Individual – exposure reduction	Students using air quality information from the dashboard to make decisions about their outdoor activities.
Individual – pollution reduction	Parents choosing to turn off their engines while waiting to pick up their children from school.
School community – exposure reduction	Schools having policies and plans in place for days with pollution above safe levels.
School community – pollution reduction	Schools having reduced emissions events where they swap out fossil fuel generators with electric power banks.
City government – pollution reduction	Prioritizing construction activities near schools to times when schools are not in session.

5.1 Program selection and implementation

The team shared the first iteration of programmatic options with DPS as well as the three pilot schools. It was suggested that the list be developed into a "menu" of options that schools would be able to select from. This would provide schools with flexibility to choose programs that met their needs. Schools with limited time could select simple options such as the installation of anti-idling signs, while other schools would be able to take on bigger projects such as walking school busses or changing recess schedules. By giving schools the ability to opt into programs that suited their needs, they were provided options versus simply implementing a program that may not be well suited for a particular school.

5.2 Measuring impact

As programs are selected by schools, measuring the impact of each will allow the narrowing of options as the project continues to scale. The foundations for measuring this impact are the air quality sensors deployed at schools in addition to health metrics collected at each school. Through a data sharing agreement with the school district, the City of Denver has aggregated baseline asthma indicators including inhaler usage and absenteeism, as well as several others. As programs are implemented, the air quality and health data will be cross analysed to identify which have the most impact on the health of students.

6 CONCLUSION

The Air Quality Community Action Network (AQ-CAN) has evolved since 2017 when first submitted as an idea to the Bloomberg Mayors Challenge. Denver has piloted and evolved the project through direct interaction and feedback from stakeholders. It has continued to learn and will follow the same methodology moving forward. Scaling up to 40 schools will undoubtedly come with additional unforeseen challenges. However, the team is confident in its ability to overcome obstacles and make a positive impact on the respiratory health and air quality of Denver's residents.

ACKNOWLEDGEMENTS

The City and County of Denver would like to thank Bloomberg Philanthropies for their continued support of the project and belief in the idea. Mayor Michael Hancock for supporting the effort in Denver. The Core Team: Liz Babcock, Christy Haas-Howard, Joni Rix, Bob McDonald, and Dave Edinger. The innovation coaches: Mayo Nissen and Minza Zahid. The Denver Smart City Implementation Team. All the volunteers from both the City and County of Denver as well as those within Denver Public Schools, parents, teachers, and nurses. Pilot school principals who gave time to provide feedback: Gilberto Munoz, Antionette Hudson, and Kirsten Frassanito. All those people who went out of their way to help during the pilot phase of the Bloomberg Mayors Challenge.

REFERENCES

[1] American Lung Association, State of the Air 2018. www.lung.org/assets/documents/healthy-air/state-of-the-air/sota-2018-full.pdf. Accessed on: 2 Mar. 2019.

[2] National Research Center & ICMA, The National Citizen Survey, Denver, CO, Technical Appendices. www.denvergov.org/content/dam/denvergov/Portals/344/documents/Financial_Reports/Citizen%20Surveys/2018/The%20NCS%20Technical%20Appendices-Denver%202017.pdf. Accessed on: 5 Mar. 2019.

[3] Denver Public Schools, Facts and Figures, DPS by the Numbers. www.dpsk12.org/about-dps/facts-figures/#1473890264817-1aa2ce27-4615. Accessed on: 8 Mar. 2019.

[4] U.S. Environmental Protection Agency, Office of Air Quality Planning and Standards, The Plain English Guide To The Clean Air Act. www.epa.gov/sites/production/files/2015-08/documents/peg.pdf. Accessed on: 6 Mar. 2019.

[5] Mayors Challenge, Competition Overview 2018 Bloomberg Philanthropies. https://mayorschallenge.bloomberg.org/competition-overview/. Accessed on: 9 Mar. 2019.

AIR QUALITY INDEX AS A TOOL FOR MONITORING ENVIRONMENTAL DEGRADATION AND HEALTH IMPLICATIONS

M. SANMI AWOPETU & J. OLUGBENGA ARIBISALA
Department of Civil Engineering, Ekiti State University, Nigeria

ABSTRACT

Considering the importance of air to human existence, air pollution is a critical issue that requires collective effort for its prevention and control. The anthropogenic activities keyed into a crystal responsibility which resulted in environmental dilapidation and ruin. One of the tools that can be used for such a campaign is Air Quality Index (AQI). This study carried out an air quality survey of some air pollutants in Ado-Ekiti, Ekiti State Nigeria with the view to develop the AQI. The AQI was based on the concentrations of four pollutants. The index is calculated from the concentrations of the following pollutants: $PM_{2.5}$, PM_{10}, CO and SO2. The air quality sample was taking in July 2017 (rainy season) and January 2018 (dry season) for a period of one week in each season. Seven sampling points across two environmental zones in the study area, namely commercial and residential (high-income and low-income areas), were considered, resulting in 42 samples for each of the 4 air pollutants, totaling 392 samples. The results show that the AQI was generally lower during the rainy season than the dry season, and that the AQI in the high-income residential areas was most favorable, ranging from good to unhealthy for sensitive group. The worst scenario was recorded in the commercial environmental zones, which ranged from moderate to hazardous. There is a need for constant and continuous monitoring of air quality for development of AQI, which in turn will enable clear communication of how clean or unhealthy the air in the study area is; it will usher in environmental degradation, dilapidation and will ruin the awareness campaign.

Keywords: air, quality, index, pollution, Ado Ekiti, environmental, degradation, pollutants.

1 INTRODUCTION

Air pollution can be described as a leading environmental problem associated with urban areas around the world. A range of monitoring programs have been put in place to determine the quality of air by generating huge amount of data on concentration of each air pollutant in different parts of the world. The large data often do not convey the air quality status to the scientific community, government officials, policy makers, and in particular to the general public in a simple and straightforward manner. This problem is addressed by determining the Air Quality Index (AQI) of a given area.

Human existence in comparison with its environment is more and more getting threatened sequel to air pollution occasioned majorly by human coupled with natural activities. Earth is getting warmer, the ozone layer is getting depleted, and acid rain is being experienced, all as a result of air pollution (Awopetu [1], Awopetu [2]). Nigeria is facing myriad of environmental issues which include but not limited to air pollution, improper waste disposal, water pollution, climate change and global warming that are of great concern to environmentalists (Awopetu [2]). The effects of this environmental degradation are detrimental to human, plant and animal existence. The anthropogenic activities keyed into a crystal responsibility which resulted in environmental dilapidation and ruin. In order for this trend to change there has to be environmental degradation, dilapidation and ruin campaign which will usher in awareness. One of the tools that can be used for such campaign is AQI. The AQI is a reporting system; an important tool of risk communication.

WIT Transactions on Ecology and the Environment, Vol 236, © 2019 WIT Press
www.witpress.com, ISSN 1743-3541 (on-line)
doi:10.2495/AIR190021

Consciousness of everyday levels of air pollution is vital to the citizens, particularly for those who are suffering from illnesses occasioned by exposure to air pollution. Achievement of a nation to advance air quality is not unconnected with the cooperation and support of its citizens who are knowledgeable about local and national air pollution problems and about the progress of mitigation efforts (San Salvador [3]). Therefore, an uncomplicated and effective communication of air quality is very important. The theory of an AQI that transforms weighted values of individual air pollution related parameters such as Particulate matter (PM_{10}, $PM_{2.5}$), Carbon monoxide (CO) and Sulfur dioxide (SO_2) into a single number or set of numbers is widely used for air quality communication. In what follows is the explanation of individual pollutants that constituted the AQI

Carbon monoxide (CO) is a colorless, odorless gas created when a fuel is burned or from incomplete combustion of hydrocarbons in gasoline-powered engines such as generator, this is common especially in developing countries. It is practically impossible to detect the presence of CO through senses in an environment since CO has no smell or taste. It is worthy of note that there are reported cases of breathlessness, restlessness and unconsciousness following inhalation of fumes produced by an electric generator that was put in a confined area [4], [5]. As reported by (Aliyu and Ibrahim [6]) was a case of CO poisoning resulted in loss of consciousness as seen in a family of six children who slept in an overcrowded room, polluted with burning charcoal which was meant to generate heat for warmth.

According to Nordqvist [7]. Hemoglobin is the molecule in red blood cells that carries oxygen from the lungs to tissues all over the body, and it brings carbon dioxide (CO_2) back from the tissues. CO binds to hemoglobin over 200 times more easily than oxygen does, so if CO is present, oxygen will not be able to find space to get into the hemoglobin. This is because the space is occupied with CO. As a result, parts of the body will be starved of oxygen, and the affected parts will die. The human body needs oxygen, but it has no use for CO.

Sulfur dioxide (SO_2) belongs to the family of sulfur oxide (SOx) gases. These gases are formed when fuel containing sulfur (mainly coal, gasoline and fuel oil) is burned and during metal smelting and other industrial processes as well as in the oxidation of naturally occurring sulfur gases, as in volcanic eruptions. High concentrations of SO_2 are associated with multiple health and environmental effects. Short-term exposure to airborne SO_2 has been associated with various adverse health effects [8], [9]. Multiple human clinical studies, epidemiological studies, and toxicological studies support a causal relationship between short-term exposure to airborne SO_2 and respiratory morbidity. Sulfur dioxide also causes acid rain which can damage or kill trees and crops.

PM_{10} is particulate matter 10 micrometers or less in diameter, $PM_{2.5}$ is particulate matter 2.5 micrometers or less in diameter. $PM_{2.5}$ is generally described as fine particles. The components of particulate matter (PM) include finely divided solids or liquids such as dust, fly ash, soot, smoke, aerosols, fumes, mists and condensing vapors that can be suspended in the air for extended periods of time. The smaller the particles, the deeper they can penetrate into the respiratory system and the more hazardous they are to breathe. The $PM_{2.5}$ is more dangerous since they are so small and light, fine particles tend to stay longer in the air than heavier particles.

PM is also known to trigger or worsen chronic disease such as asthma, heart attack, bronchitis and other respiratory problems. Exposure to such particles can affect both lungs and heart. Numerous scientific studies have linked particle pollution exposure to a variety of problems, including; decreased lung function, increased respiratory symptoms, such as irritation of the airways, coughing or difficulty breathing.

Several studies from across the world have documented the many ways in which air pollution can affect people's health, including but not limited to making it difficult to breathe for those with asthma or other respiratory diseases, regularly sending the young and old to hospital or causing them to miss school or work, and contributing to early death from heart and lung disease (Health Effects Institute [10]). Therefore, an AQI is helpful for: (i) public to know air quality in a simplified mode, (ii) politicians to know they cannot completely shield away from effects of air pollution and then invoke quick actions, (iii) a decision maker to be aware of the trend of air pollution events and to map out corrective pollution control strategies.

2 MATERIAL AND METHODS

2.1 Study area

Ado Ekiti is a city in southwest Nigeria, the state capital and headquarters of the Ekiti State. It is also known as Ado. It has a population of above 424, 340. The people of Ado Ekiti are mainly of the Ekiti sub-ethnic group of the Yorubas. Ado Ekiti has four tertiary educational institutions namely: Ekiti State University, Afe Babalola University and The Federal Polytechnic Ado Ekiti and Ekiti State School of Nursing and Midwifery. It also plays host to two local television and three radio stations; NTA Ado Ekiti, Ekiti State Television (ESBS), Ekiti FM, Voice FM and Progress FM Ado Ekiti. Various commercial banks and enterprises operate in Ado Ekiti. Ado Ekiti also have ninety-four (94) hotels and more than fifty (50) petrol stations all running on generating sets as source of electricity between two to twenty-four hours per day.

The town lies between the latitude $7^0\ 33^1$ and $7^0\ 42^1$ North of the equator and the longitude $5^0\ 11^1$ and $5^0\ 20^1$ East on a low land surrounded by several isolated hills and inselbergs, (Oyedele [11]). Geologically, the region lies entirely within the pre-Cambrian basement complex rock group, which underlies much of Ekiti State (Awosusi and Jegede [12]). The temperature of this area is almost uniform throughout the year; with little deviation from the mean annual temperature of 27°C. February and March are the hottest 28°C and 29°C respectively, while June with temperature of 25°C is the coolest (Adebayo [13]). The mean annual rainfall is 1,367 mm with a low co-efficient variation of about 10% and 117 raining days in year 2017. Rainfall is highly seasonal with well-marked wet and dry season. The wet season lasts from April to October, with a break in August.

2.2 Research method

2.2.1 Sampling

Air sampling collection and analysis is required in order to quantify the air pollutants in the study area. To obtain valid data considering the fact that measuring air pollution is a complex task and requires due care and diligence, the following issues were put into consideration: (i) appropriateness of the sample points; (ii) how representative will the sample be in time and space; and (iii) how appropriate is the sampling equipment, analysis and calibration techniques.

Hand-held portable Aeroqual series 500 ambient air quality sampling equipment was used to measure $PM_{2.5}$, PM_{10}, CO and SO_2. The air quality sample was taking in July 2017 (rainy season) and January 2018 (dry season) for a period of one week in each season. All sampling locations were sampled at different times of the day (morning, afternoon and evening).

Morning readings were taken between 8am–11am, afternoon readings between 12pm–3pm and evening readings were taken between 4pm–7pm.

Seven sampling points across two environmental zones in the study area namely; commercial and residential (high income and low-income areas) were considered, resulting in 42 samples for each of the 4 air pollutant totaling 168 samples. Air monitoring was carried out in seven core sites which are as follows:

i) Old Garage: (this is characterized by retail shops, market, high vehicle and pedestrian traffic, it also serves as transfer point for minibuses and taxi linking other towns, urban, peri-urban and rural destinations);
ii) GRA: (represented high economic status residential area with low vehicular and pedestrian traffic volume);
iii) Ajilosun: (represented medium economic status residential area where majority of the residents either use kerosene or cooking gas for cooking);
iv) Dalimore Junction: (this serves as an important commuter route within ado Ekiti which represented heavy-traffic sites);
v) Odo Ado: Odo Ado-Ekiti (represent rural background area);
vi) Fajuyi Park: (represented civil engineering construction activity area); and
vii) Ilokun: (represented low economic status residential area where the houses are built of mud bricks without plastering and the floors were not paved or cemented. A lot of firewood burning activities were taking place).

The purpose of the AQI is to help people understand what local air quality means to their health. To make it easier to understand, USEPA has developed an AQI that is used to report air quality which is divided into six categories indicating increasing levels of health concern (Table 1). An AQI value over 300 represents hazardous air quality and below 50 the air quality is good, (US EPA, [14]).

Table 1: AQI. *(Source: US EPA [14].)*

AQI Value	AQI Color	AQI Color
0–50	Good	Green
51–100	Moderate	Yellow
101–150	Unhealthy for sensitive groups	Orange
151–200	Unhealthy	Red
201–300	Very unhealthy	Purple
301–500	Hazardous	Maroon

The AQI is a yardstick that runs from 0 to 500. The higher the AQI value, the greater the level of air pollution and the greater the health concern. For example, an AQI value of 50 represents good air quality with little or no potential to affect public health, while an AQI value over 300 represents air quality so hazardous that everyone may experience serious effect (Table 2).

Table 2: AQI health messages. *(Source: US EPA [14].)*

AQI Value	Health Message	AQI Color
0–50	None	Green
51–100	Unusually sensitive people should reduce prolonged or heavy exertion	Yellow
101–150	Sensitive groups should reduce prolonged or heavy exertion	Orange
151–200	Sensitive groups should avoid prolonged or heavy exertion; general public should reduce prolonged or heavy exertion	Red
201–300	Sensitive groups should avoid all physical activity outdoors; general public should avoid prolonged or heavy exertion	Purple
301–500	Everyone should avoid all physical activity outdoors	Maroon

2.2.2 Computing the AQI

Fourteen days (Monday to Sunday) monitors record concentrations of four pollutants at seven locations across the study area were taken. These raw measurements were converted into a separate AQI value for each day using standard equations (eqn (1)) developed by EPA. The AQI values are recorded as the AQI value for that day. The AQI is a piecewise linear function of the pollutant concentration. At the boundary between AQI categories (Table 3), there is a discontinuous jump of one AQI unit. To convert from concentration to AQI, eqn (1) is used [15].

$$I = \frac{I_{high} - I_{low}}{C_{high} - C_{low}} (C - C_{low}) + I_{low}, \tag{1}$$

where I = AQI; C = pollutant concentration; C_{low} = concentration breakpoint, that is ≤ C; C_{high} = concentration breakpoint that is ≥ C; I_{low} = index breakpoint corresponding to C_{low}; I_{high} = index breakpoint corresponding to C_{high}.

Table 3: Breakpoints for the AQI. *(Source: US EPA [15].)*

S/N	Breakpoints				Equal these AQIs	Category
1	$PM_{2.5}$ (μg/m^2)	PM_{10} (μg/m^2)	CO (ppm)	SO_2 (ppm)	AQI	
2	0.0–15.4	0–54	0.0–4.4	0.000–0.034	0–50	Good
3	15.5–40.4	55–154	4.5–9.4	0.035–0.144	51–100	Moderate
4	40.5–65.4	155–254	9.5–12.4	0.145–0.224	101–150	Unhealthy for sensitive group
5	65.5–150.4	255–354	12.5–15.4	0.225–0.304	151–200	Unhealthy
6	150.5–250.4	355–424	15.5–30.4	0.305–0.604	201–300	Very unhealthy
7	250.5–350.4	425–504	30.5–40.4	0.605–0.804	301–400	Hazardous
8	350.5–500.4	505–604	0.000–0.034	0.805–1.004	401–500	Hazardous

3 RESULTS AND DISCUSSION

It is seen from Tables 4 to 17 that the AQI ranges between good to hazardous during the raining season (Tables 4 to 10) and moderate to hazardous during the dry season (Tables 11 to 17). During the raining season, GRA (Table 5) had the best AQI which ranged between good and moderate while Old Garage (Table 4) had the worst AQI which ranged between moderate to hazardous. The friendly and acceptable AQI in GRA can be linked to the economic status of the people living in the area. GRA represents high economic status residential area with low vehicular and pedestrian traffic volume. The area also had a very low commercial activity with lots of green vegetations. Old Garage on the other hand is characterized with retail shops, market, high vehicle and pedestrian traffic; it also serves as transfer point for minibuses and taxi linking other towns, urban, peri-urban and rural destinations. It is the hub of commercial activities in the study area. Most of the retail shops in the Old Garage used generator for electricity supply. Woods, charcoal and kerosene stoves were used for cooking activities. All these anthropogenic activities are responsible for the worst AQI in the study area.

It was generally observed that AQI was worst in the dry season when compared with raining season, except Fajuyi (Table 9) which recorded hazardous in two days (Monday and Tuesday). GRA still recorded the friendliest AQI during the dry season, the AQI ranges between moderate to unhealthy for sensitive groups while Old Garage recorded the worst AQI (hazardous for seven days) during the same period. Hazardous levels of health concern, which are AQI values over 300, trigger health warnings of emergency conditions. The entire population is even more likely to be affected by serious health effects. Fajuyi (Table 16) also recorded hazardous on a Saturday while other days are mostly very unhealthy meaning everyone may experience more serious health effects.

Table 4: Daily AQI for Old Garage in Ado Ekiti during the raining season.

Day of the week	$PM_{2.5}$ µg/m^3	PM_{10} µg/m^3	CO ppm	SO_2 ppm	AQI	Colors	Levels of health concerns
Monday	148	636	139	516	360		Hazardous
Tuesday	154	668	201	457	370		Hazardous
Wednesday	85	94	117	119	104		Unhealthy for sensitive group
Thursday	157	626	171	654	402		Hazardous
Friday	72	430	142	393	259		Unhealthy
Saturday	102	118	25	118	91		Moderate
Sunday	108	451	301	206	267		Unhealthy

Table 5: Daily AQI for GRA in Ado Ekiti during the raining season.

Day of the week	$PM_{2.5}$ µg/m^3	PM_{10} µg/m^3	CO ppm	SO_2 ppm	AQI	Colors	Levels of health concerns
Monday	59	45	23	68	49		Good
Tuesday	57	46	9	68	45		Good
Wednesday	59	45	17	102	56		Moderate
Thursday	67	45	12	119	61		Moderate
Friday	66	57	20	119	65		Moderate
Saturday	61	54	12	65	48		Good
Sunday	54	49	18	10	33		Good

Table 6: Daily AQI for Ajilosun in Ado Ekiti during the raining season.

Day of the week	$PM_{2.5}$ µg/m^3	PM_{10} µg/m^3	CO ppm	SO_2 ppm	AQI	Colors	Levels of health concerns
Monday	85	128	28	83	81		Moderate
Tuesday	87	125	37	65	78		Moderate
Wednesday	88	136	35	68	82		Moderate
Thursday	82	147	68	117	103		Unhealthy for sensitive group
Friday	73	133	65	179	112		Unhealthy for sensitive group
Saturday	73	94	77	189	109		Unhealthy for sensitive group
Sunday	63	60	65	54	61		Moderate

Table 7: Daily AQI for Dalimore in Ado Ekiti during the raining season.

Day of the week	$PM_{2.5}$ µg/m^3	PM_{10} µg/m^3	CO ppm	SO_2 ppm	AQI	Colors	Levels of health concerns
Monday	122	118	90	29	72		Moderate
Tuesday	119	122	110	73	85		Moderate
Wednesday	112	117	111	123	92		Moderate
Thursday	117	110	144	135	101		Unhealthy for sensitive group
Friday	107	91	176	139	103		Unhealthy for sensitive group
Saturday	99	87	201	127	103		Unhealthy for sensitive group
Sunday	81	67	198	198	109		Unhealthy for sensitive group

Table 8: Daily AQI for Odo Ado in Ado Ekiti during the raining season.

Day of the week	$PM_{2.5}$ µg/m^3	PM_{10} µg/m^3	CO ppm	SO_2 ppm	AQI	Colors	Levels of health concerns
Monday	81	87	121	101	97		Moderate
Tuesday	84	98	51	122	89		Moderate
Wednesday	124	160	78	82	111		Moderate
Thursday	157	193	87	74	128		Unhealthy for sensitive group
Friday	154	184	55	56	112		Moderate
Saturday	111	352	61	97	155		Unhealthy for sensitive group
Sunday	66	315	117	95	148		Unhealthy for sensitive group

Table 9: Daily AQI for Fajuyi in Ado Ekiti during the raining season.

Day of the week	$PM_{2.5}$ µg/m^3	PM_{10} µg/m^3	CO ppm	SO_2 ppm	AQI	Colors	Levels of health concerns
Monday	190	979	79	29	319		Hazardous
Tuesday	191	979	98	73	335		Hazardous
Wednesday	171	523	109	123	231		Very unhealthy
Thursday	183	448	112	135	219		Very unhealthy
Friday	181	624	127	139	268		Very unhealthy
Saturday	173	457	122	127	220		Very unhealthy
Sunday	155	500	163	198	254		Very unhealthy

Table 10: Daily AQI for Ilokun in Ado Ekiti during the raining season.

Day of the week	$PM_{2.5}$ µg/m^3	PM_{10} µg/m^3	CO ppm	SO_2 ppm	AQI	Colors	Levels of health concerns
Monday	67	61	3	300	108		Moderate
Tuesday	66	57	4	520	161		Unhealthy
Wednesday	64	56	1	505	156		Unhealthy
Thursday	69	58	1	473	150		Unhealthy for sensitive group
Friday	59	55	1	269	96		Moderate
Saturday	62	48	9	269	97		Moderate
Sunday	67	45	12	119	61		Moderate

Table 11: Daily AQI for Old Garage in Ado Ekiti during the dry season.

Day of the week	$PM_{2.5}$ µg/m^3	PM_{10} µg/m^3	CO ppm	SO_2 ppm	AQI	Colors	Levels of health concerns
Monday	166	821	185	654	457		Hazardous
Tuesday	167	842	206	599	453		Hazardous
Wednesday	156	1032	154	511	463		Hazardous
Thursday	179	819	205	534	434		Hazardous
Friday	191	553	157	463	341		Hazardous
Saturday	158	640	83	656	384		Hazardous
Sunday	175	606	233	513	382		Hazardous

Table 12: Daily AQI for GRA in Ado Ekiti during the dry season.

Day of the week	$PM_{2.5}$ µg/m^3	PM_{10} µg/m^3	CO ppm	SO_2 ppm	AQI	Colors	Levels of health concerns
Monday	63	52	55	237	102		Unhealthy for sensitive group
Tuesday	59	53	50	237	100		Moderate
Wednesday	63	56	54	248	105		Unhealthy for sensitive group
Thursday	58	56	49	355	129		Unhealthy for sensitive group
Friday	58	58	53	308	119		Unhealthy for sensitive group
Saturday	52	54	36	267	102		Unhealthy for sensitive group
Sunday	54	51	38	251	99		Moderate

Table 13: Daily AQI for Ajilosun in Ado Ekiti during the dry season.

Day of the week	$PM_{2.5}$ µg/m^3	PM_{10} µg/m^3	CO ppm	SO_2 ppm	AQI	Colors	Levels of health concerns
Monday	106	375	57	45	146		Unhealthy for sensitive group
Tuesday	107	392	63	65	157		Unhealthy
Wednesday	77	178	124	82	115		Unhealthy for sensitive group
Thursday	105	386	150	21	165		Unhealthy
Friday	90	291	128	188	174		Unhealthy
Saturday	112	384	69	14	145		Unhealthy for sensitive group
Sunday	176	292	124	32	156		Unhealthy

Table 14: Daily AQI for Odo Ado in Ado Ekiti during the dry season.

Day of the week	$PM_{2.5}$ µg/m³	PM_{10} µg/m³	CO ppm	SO_2 ppm	AQI	Colors	Levels of health concerns
Monday	66	84	172	453	194		Unhealthy
Tuesday	93	253	149	250	186		Unhealthy
Wednesday	83	200	146	578	252		Very unhealthy
Thursday	71	236	189	403	225		Very unhealthy
Friday	68	195	213	285	190		Unhealthy
Saturday	103	303	212	413	258		Very unhealthy
Sunday	115	92	145	386	185		Unhealthy

Table 15: Daily AQI for Dalimore in Ado Ekiti during the dry season.

Day of the week	$PM_{2.5}$ µg/m³	PM_{10} µg/m³	CO ppm	SO_2 ppm	AQI	Colors	Levels of health concerns
Monday	137	140	154	385	204		Very unhealthy
Tuesday	143	136	148	350	194		Unhealthy
Wednesday	127	104	55	579	216		Very unhealthy
Thursday	148	171	179	398	224		Very unhealthy
Friday	142	209	93	373	204		Very unhealthy
Saturday	151	208	150	218	182		Unhealthy
Sunday	153	154	139	330	194		Unhealthy

Table 16: Daily AQI for Fajuyi in Ado Ekiti during the dry season.

Day of the week	$PM_{2.5}$ µg/m³	PM_{10} µg/m³	CO ppm	SO_2 ppm	AQI	Colors	Levels of health concerns
Monday	138	621	79	283	280		Very unhealthy
Tuesday	137	354	112	67	167		Unhealthy
Wednesday	105	412	163	100	195		Unhealthy
Thursday	139	614	98	173	256		Very unhealthy
Friday	150	494	127	191	241		Very unhealthy
Saturday	140	780	207	88	304		Hazardous
Sunday	131	153	129	460	218		Very unhealthy

Table 17: Daily AQI for Ilokun in Ado Ekiti during the dry season.

Day of the week	$PM_{2.5}$ µg/m³	PM_{10} µg/m³	CO ppm	SO_2 ppm	AQI	Colors	Levels of health concerns
Monday	80	62	6	563	178		Unhealthy
Tuesday	69	55	5	549	170		Unhealthy
Wednesday	73	46	8	355	121		Unhealthy for sensitive group
Thursday	81	54	16	642	198		Unhealthy
Friday	67	41	0	307	104		Unhealthy for sensitive group
Saturday	69	53	11	320	113		Unhealthy for sensitive group
Sunday	108	63	9	255	109		Unhealthy for sensitive group

Tables 18 and 19 shows the weekly AQI for the study area during raining and dry season respectively. The overall AQI index in GRA for both seasons still remains most favorable, followed by that of Ilokun while Old Garage and Fajuyi still recorded the worst AQI scenario.

Table 18: Weekly AQI for Ado Ekiti during the raining.

Location	$PM_{2.5}$ µg/m^3	PM_{10} µg/m^3	CO ppm	SO_2 ppm	AQI	Colors	Levels of health concerns
Old Garage	118	432	157	352	265		Very unhealthy
GRA	60	48	16	79	51		Moderate
Ajilosun	79	118	54	108	89		Moderate
Dalimore	108	102	147	118	119		Unhealthy for sensitive group
Odo Ado	111	198	81	90	120		Unhealthy for sensitive group
Fajuyi	178	644	116	118	264		Very unhealthy
Ilokun	65	54	4	350	118		Unhealthy for sensitive group

Table 19: Weekly AQI for Ado Ekiti during the dry season.

Location	$PM_{2.5}$ µg/m^3	PM_{10} µg/m^3	CO ppm	SO_2 ppm	AQI	Colors	Levels of health concerns
Old Garage	170	759	175	561	416		Hazardous
GRA	58	54	48	272	108		Unhealthy for sensitive group
Ajilosun	111	328	102	64	151		Unhealthy
Dalimore	86	195	175	396	213		Very unhealthy
Odo Ado	143	160	131	376	203		Very unhealthy
Fajuyi	134	490	131	194	237		Very unhealthy
Ilokun	78	53	8	427	142		Unhealthy for sensitive group

4 CONCLUSIONS AND RECOMMENDATIONS

This study presents the AQI for Ado Ekiti in Nigeria. As shown; the inhabitants of the study area are perpetually exposed to a diversity of pollutants as indicated by the measured AQI concentration. The main conclusion drawn is that, Ado Ekiti air is polluted and the level of human exposure to the pollutants requires further investigation to ascertain the health effect among the populace. The conclusion is supported by a number of epidemiological studies on the effects of air pollutants such as $PM_{2.5}$, PM_{10}, CO, and SO_2 on human health. It was discovered that most of the air pollutants sampled for the development of AQI was disgustingly higher than the World health organization (WHO) standard thereby posing great risk to the public health in particular and the environment in general. The federal, state and local government is doing nothing to mitigate the air pollutant in the study area. There is a need for continuous of air pollutants, regular development of AQI and public education on the adverse effects of air pollution on both health and environment. As it were, air pollution and its attendant consequences in the urban area under study should be made public. Steps that could be taken for air pollution mitigation at individual and domestic level should be clearly spelt out. Considering the AQI in most of the study locations, government intervention in terms of air quality regulation and mitigation is much needed.

REFERENCES

[1] Awopetu M.S., Effect of air pollution on rain water: A case study of Ado-Ekiti, Nigeria. *International Journal of Advanced Engineering Research and Science*, **5**(8), pp. 19–24, 2018.

[2] Awopetu, M.S., Effect of air pollution on dew water: A case study of Ado-Ekiti. *Nigeria American Journal of Engineering Research*, **7**(8), pp. 88–93, 2018.

[3] US EPA, Air Quality Index (AQI) *Air Quality Communication Workshop*, San Salvador, El Salvador, 2012. www.epa.gov/sites/production/files/2014-05/documents/zell-aqi.pdf Accessed on: 19 Aug. 2018.

[4] Afolayan, J.M., Edomwonyi, N.P. & Esangbedo, S.E., Carbon monoxide poisoning in a Nigerian home: case reports. *Nigerian Postgraduate Medical Journal*, **21**(2), pp. 199–202, 2014.

[5] Seleye-Fubara, D., Etebu, E.N. & Athanasius, B., Pathology of deaths from carbon monoxide poisoning in Port Harcourt: an autopsy study of 75 cases. *Nigerian Journal of Medicine*, **20**(3), pp. 337–340, 2011.

[6] Aliyu, I. & Ibrahim, Z.F., Accidental carbon monoxide poisoning in a family of six: Diagnosis and treatment challenges in a resource limited setting. *Journal of Medical Investigation and Practice*, **9**, 130–131, 2014.

[7] Nordqvist, C., Carbon monoxide (CO), the silent killer. *Medical News Today*, 11 November 2017.

[8] US EPA., *Supplement to the Second Addendum (1986) to Air Quality Criteria for Particulate Matter and Sulfur Oxides (1982): Assessment of New Findings on Sulfur Dioxide and Acute Exposure Health Effects in Asthmatic Individuals*. EPA/600/FP-93/002, Washington, DC, 1994.

[9] ATSDR (Agency for Toxic Substances and Disease Registry), *Toxicological Profile for Sulfur Dioxide*. Atlanta, 1998. www.atsdr.cdc.gov/ToxProfiles/tp116.pdf.

[10] Health Effects Institute, State of global air 2018. Special report, Health Effect Institute: Boston, MA, 2018.

[11] Oyedele, E.A.A. & Olayinka, A.I., Statistical evaluation of groundwater potential of Ado-Ekiti, South West Nigeria. *Transnational Journal of Science & Technology*, **2**(6), pp. 110–127, 2012.

[12] Awosusi, O.O. & Jegede A.O., Challenges of sustainability and urban development: A case study of Ado-Ekiti, Ekiti State, Nigeria. *International Education Research*, **1**(1), 22–29, 2013.

[13] Adebayo, W.O., Weather and climate. Ado-Ekiti region. *A Geographical Analysis and Master Plan*, ed. F.S. Ebisemiju, Alpha Prints: Lagos, pp. 11–14, 1993.

[14] US EPA (2014) AQI: A guide to air quality and your health. www3.epa.gov/airnow/aqi_brochure_02_14.pdf.

[15] US EPA (2016) Technical assistance document for the reporting of daily air quality the air quality index (AQI). www3.epa.gov/airnow/aqi-technical-assistance-document-may2016.pdf.

LONG-TERM TRENDS OF TROPOSPHERIC OZONE IN THE CZECH REPUBLIC, 1993–2018

MILAN VANA & JAROSLAV PEKAREK
Czech Hydrometeorological Institute, Košetice Observatory, Czech Republic

ABSTRACT

The regular measurement of tropospheric ozone in the Czech Republic began within the National Air Pollution Monitoring Network operated by the Czech Hydrometeorological Institute in 1993. The long-term trend study is based on the data from the stations with the longest homogeneous data series: Košetice (EMEP, GAW, and ACTRIS) and Svratouch (EMEP), Praha-Libuš (suburban area of the capital Prague) and Churáňov (mountain site). Non-parametric Mann–Kendall method was used for trend evaluation. Slightly decreasing trends in mean annual concentrations were found at mountain and background level in the whole period as well as in the first part (1993–2005). After 2006 no trend was found. The warm period (April–September) displays similar patterns as the whole year at regional and mountain level. In contrast, no trend was found in the whole period at EMEP stations and slightly decreasing tendency at mountain stations in the cold period. The suburban level is characterized by a slightly increasing trend. The difference between mean annual concentrations at regional and suburban stations was smaller in last years. The annual variation is characterized by maxima at the end of spring. According to the current EU air quality legislation, the target value for protection of human health is exceeded when the 8-hour running mean is higher than 120 $\mu g.m^{-3}$ 25 times on average for 3 years. The limit was exceeded at all stations during the period 1993–2005, but at the same time significant drop of high ozone episodes was found. After 2006, the 3-year mean fluctuated around the target limit at all stations. Assessment of the ozone impacts on ecosystems using the AOT40 index suggests that the critical level was exceeded for long periods, not only in the regional areas but almost over the whole of the Czech Republic. In the last decade, the values varied round the requested limit.

Keywords: tropospheric ozone, long-term trends, critical levels, background scale, Czech Republic.

1 INTRODUCTION

Tropospheric ozone does not have a source in the atmosphere. This is a secondary substance formed by a number of complicated nonlinear photochemical reactions, which were described in detail, e.g. in [1]. Tropospheric ozone precursors include nitrogen oxides (NO_x) and non-methane volatile organic compounds (NMVOC), while methane (CH_4) and carbon monoxide (CO) play a role on a background scale. The photolysis of NO_2 by radiation with a wavelength of 280–430 nm is an important reaction, forming NO and atomic oxygen. Ozone molecules are formed by the reaction of atomic and molecular oxygen in the presence of a catalyst. Simultaneously, ozone is titrated with nitrogen oxide, NO, with the formation of NO_2 and O_2. If ozone is replaced by radicals in this reaction, its concentration increases in the atmosphere. The OH radical plays an especially important role in this reaction.

In the formation of tropospheric ozone from precursors, the absolute amounts of precursors is not important, but rather their mutual ratios [2], [3]. In areas where the regime is limited by NO_x, characterized by relatively low concentrations of NO_x and high concentrations of VOC, the ozone concentrations increase with increasing NO_x concentrations, but only minimally with increasing VOC concentrations. On the other hand, in areas with a regime limited by VOC, the ozone concentrations decrease with increasing NO_x concentrations and the ozone concentrations increase with increasing VOC concentrations. Areas with a high NO_x/VOC ratio are typically polluted areas around the centers of large cities. The dependence of ozone formation on the initial concentrations of

WIT Transactions on Ecology and the Environment, Vol 236, © 2019 WIT Press
www.witpress.com, ISSN 1743-3541 (on-line)
doi:10.2495/AIR190031

VOC and NO_x is frequently expressed by ozone isopleth diagrams, which depict the maximum attained ozone concentration as a function of the initial NO_x and VOC concentrations [4]. Not only the concentrations of precursors, but also meteorological conditions, play an important role in the ozone formation [5]. The ozone concentrations increase with increasing ultraviolet radiation and temperature, but decrease with increasing relative atmospheric humidity. High concentrations are often connected with prolonged anticyclone situations. In addition to the above-described photochemical mechanisms, the concentrations of ozone can also increase in episodes as a result of penetration of stratospheric ozone into the troposphere and also during thunderstorms. Recently, there has also been an increase in the importance of long-range transmission of ozone in the air streams in the northern hemisphere to Europe and North America from source areas in south-east Asia. Tropospheric ozone is removed from the atmosphere by reaction with NO and dry deposition. Ozone can be transported in the troposphere over long distances, accumulate and reach concentrations far from its place of origin [6].

NO_x are formed in all combustion processes. NMVOC are emitted from a number of anthropogenic sources (transport, manipulation with petroleum and its derivatives, refineries, the use of coatings and solvents, etc.), and also natural sources (e.g. biogenic emissions from vegetation). In the Czech Republic, NOx emissions dropped by 80% in the period 1990–2016 and by 55% after 2000. The total NO_x emissions are negatively affected by a considerable increase in emissions from mobile sources. The share of NO_x emissions from mobile sources increased from 27.0% in 1990 to 55% by nowadays. In the whole European Union (EU), NOx emissions dropped EU by 58% between 1990 and 2016 and by 42% after the year 2000. The VOCs emissions in the Czech Republic dropped by 63% in the period 1990–2016 and 30% after 2000. The figures for the whole of the EU were 62% in 1990–2016 and 40% after 2000 [7].

The main effect of ozone on the human organisms is irritative. It irritates the conjunctiva, nasal mucous membranes and bronchi. Short-term studies show that ozone concentrations may have adverse effects on pulmonary function leading to lung inflammation and respiratory problems [8]. Higher concentrations cause irritation-induced narrowing of the respiratory tract and breathlessness. More sensitive to ozone are people with chronic obstructive diseases of the lungs and asthma patients. Higher concentrations of ozone are purportedly associated with increased mortality during the day. Tropospheric ozone damages vegetation, impairs plant growth and decreases crop yields; it can damage forest ecosystems and reduce biodiversity [8].

Measurement of tropospheric ozone concentrations, in view of its key role in chemical and physical processes in the troposphere, is one of the compulsory parts of international air quality monitoring programmes at regional level. In the Czech Republic, the measurements began within the CHMI's Automatic Monitoring of Ambient Air Pollution network in 1993. This study is based on the data from the stations with longest data series: Košetice and Svratouch are the background stations, involved in EMEP programme, Praha-Libuš represents the suburban area of the capital Prague and Churáňov is a typical mountain site [9]. The geographical co-ordinates of the stations are indicated in Table 1. The objective of the study is to evaluate the trends of tropospheric ozone concentrations at the regional scale in the Czech Republic in the period 1993–2018. The period under review was divided into two identical sub-periods 1993–2005 and 2006–2018. The motivation for dividing of the whole period was an effort towards better understanding of influence of environmental measures for reduction of ozone precursor´s emission as well as the changing climatological conditions. The similar approach was used also in pan-European assessment [12].

Table 1: Geographical co-ordinates of selected stations.

	Altitude (m a.s.l.)	Longitude	Latitude
Košetice	534 m	49°35′N	15°05′E
Praha-Libuš	301 m	50°00′N	14°27′E
Svratouch	737 m	49°73′N	16°03′E
Churáňov	1,118 m	49°04′N	13°37′E

The nonparametric Mann–Kendall method [10] was used for statistical evaluation of tropospheric ozone trends significance. This test divides the trend significance to four categories:

***	if trend at $\alpha = 0{,}001$ level of significance;
**	if trend at $\alpha = 0{,}01$ level of significance;
*	if trend at $\alpha = 0{,}5$ level of significance;
+	if trend at $\alpha = 0{,}1$ level of significance.

Test Z characterize the existence and significance of the trend. A positive (negative) value of Z indicates an upward (downward) trend. Sen′s slope estimate Q represents estimator for true slope of the trend per year.

2 TREND OF TROPOSPHERIC OZONE CONCENTRATIONS

Fig. 1 shows general tendency of mean annual ozone concentrations at selected stations, while the trends in warm (April–September) and cold (October–March) periods are specified in Figs 2 and 3. The results of trend analysis are shown in Table 2. and the mean annual concentrations in Table 3.

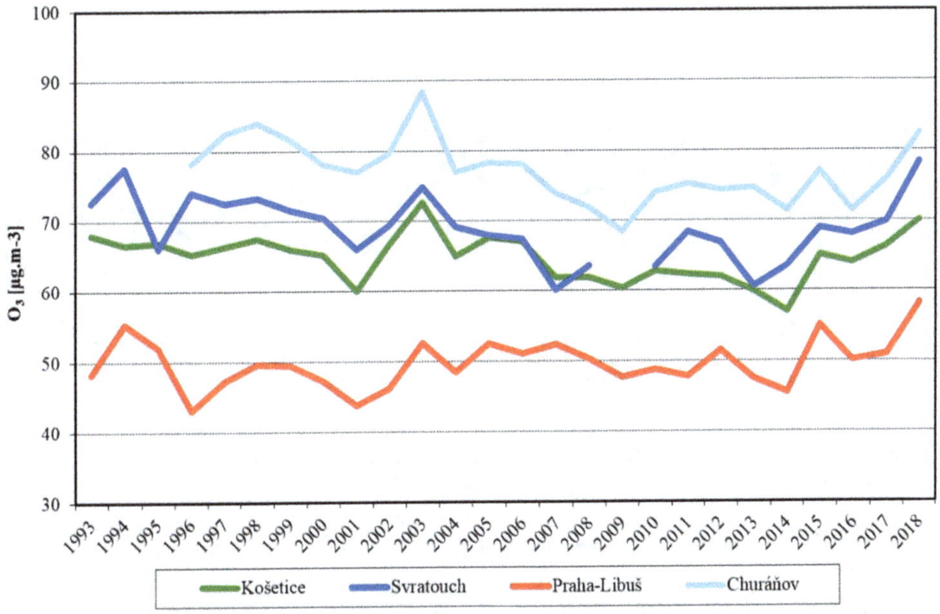

Figure 1: Mean annual tropospheric ozone concentrations at stations Košetice, Svratouch, Praha-Libuš, Churáňov (1993–2018).

WIT Transactions on Ecology and the Environment, Vol 236, © 2019 WIT Press
www.witpress.com, ISSN 1743-3541 (on-line)

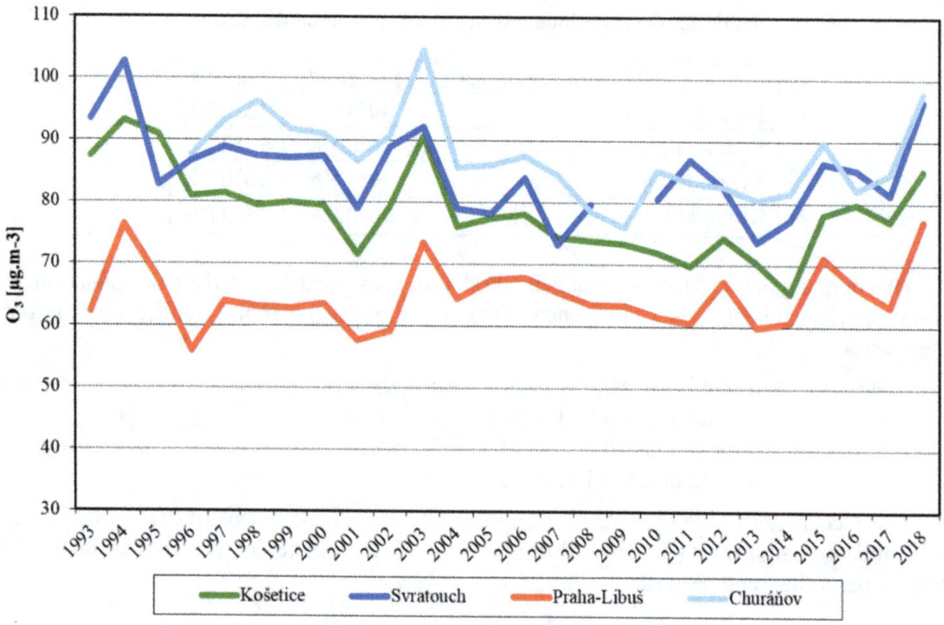

Figure 2: Mean annual tropospheric ozone concentrations in the warm period at stations Košetice, Svratouch, Praha-Libuš, Churáňov (1993–2018).

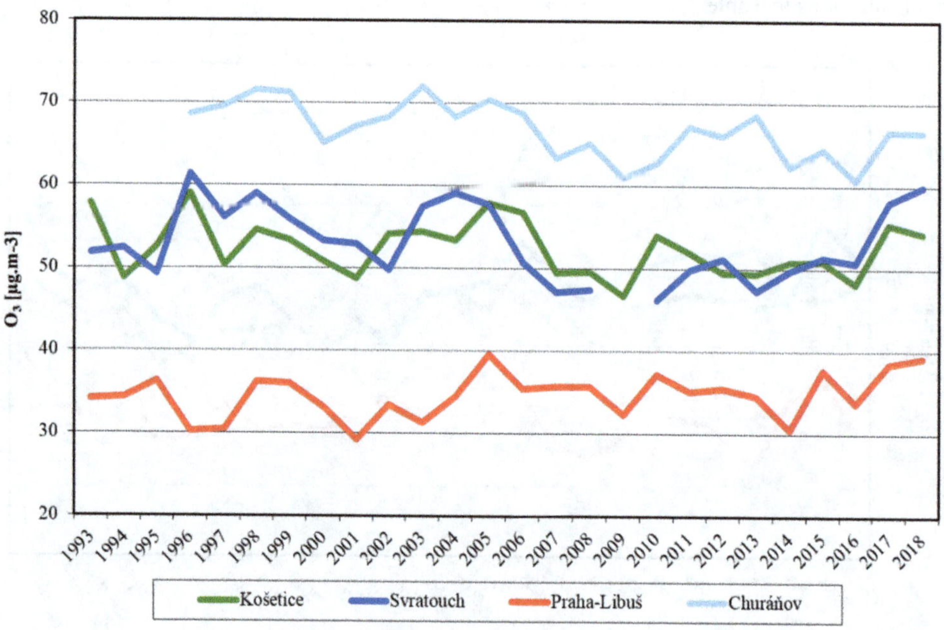

Figure 3: Mean annual tropospheric ozone concentrations in the cold period at stations Košetice, Svratouch, Praha-Libuš, Churáňov (1993–2018).

Table 2: Trend analysis of tropospheric ozone trends from the Czech stations Košetice, Svratouch, Churáňov and Libuš.

	Košetice			Svratouch			Churáňov			Libuš		
Period	Q	Z	Sig.	Q	Z	Sig.	Q	Z	Sig.	Q	Z	Sig.
Whole												
1993–2018	-0.2	-2.1	*	-0.4	-2.3	*	-0.4	-2.2	*	0.1	1.1	
1993–2005	-0.2	-0.8		-0.7	-2.3	*	-0.3	-0.5		0.0	0.0	
2006–2018	0.5	1.4		0.9	2.1	*	0.2	0.8		0.1	0.3	
Warm												
1993–2018	-0.6	-3.2	**	-0.4	-1.9	+	-0.4	-2.4	*	0.1	0.6	
1993–2005	-0.8	-2.6	**	-1.0	-1.5		-0.5	-1.3		0.0	0.1	
2006–2018	0.3	0.6		0.9	1.5		0.3	0.4		0.0	0.0	
Cold												
1993–2018	-0.1	-0.9		-0.1	-0.3		-0.3	-2.5	*	0.1	1.6	
1993–2005	0.2	0.6		0.5	1.3		0.1	0.2		0.1	0.2	
2006–2018	0.1	0.3		0.9	2.8	**	-0.1	-0.2		0.2	0.4	

Table 3: Mean annual tropospheric ozone concentrations at the Czech stations Košetice, Svratouch, Churáňov and Libuš ($\mu g.m^{-3}$).

	Košetice	Svratouch	Churáňov	Libuš
Whole period				
1993–2018	64.6	68.2	77.8	49.7
1993–2005	66.4	71.3	80.4	48.8
2006–2018	62.8	65.1	75.7	50.6
Warm period				
1993–2018	78.4	84.1	88.0	64.9
1993–2005	82.1	87.4	91.4	64.3
2006–2018	74.7	80.7	85.4	65.5
Cold period				
1993–2018	52.4	51.9	67.5	34.7
1993–2005	53.5	54.6	69.2	33.8
2006–2018	51.3	49.2	66.1	35.6

Slightly significant downward trend was found at background stations and the mountain station in the whole period under review as well as in the first period 1993–2005. On contrary, after 2005, no trend was found, the mean annual concentrations displayed upward tendency. An exception was the year 2003 with extremely hot summer, when the concentrations were the highest in the whole period under review. Long-term average values at EMEP stations Kosetice (65 $\mu g.m^{-3}$) and Svratouch (68 $\mu g.m^{-3}$) belong together with similarly located German and Austrian stations among the highest in the whole EMEP network [11]. Most stations of EMEP network reported a downward trend, however, because of the large inter-annual variability of ozone, the downward trends are statistically significant at only 30–50% of the sites [12].

Warm period displays similar patterns as the whole year at background and mountain scale. The situation in the cold period is quite different. No trend was found in the whole period at background stations and slightly decreasing tendency at mountain station In the period 2006–2018, increasing trend was found (at Svratouch even statistically significant). The difference between the mountain site and background level was visibly higher in the cold period. Suburban station is characterized by slightly increasing tendency, which is not statistically significant. The difference between mean annual concentrations at background and suburban stations was smaller during the period under review from 20 μg.m^{-3} in the 90 to 12 μg.m^{-3} in last five years.

Tropospheric ozone concentrations in the Czech Republic have a marked annual variation with maxima at the end of spring (Fig. 4). In June and sometimes also in July there is a decrease caused by the onset of the so-called “continental monsoon”, which brings increased cloud cover and a drop in solar radiation. We then register a second maximum in July and August (especially in very hot summers). In the very cool year 1996 we recorded a maximum concentration as early as April. Concentrations at mountain station are higher all year round. The annual variation of tropospheric ozone concentrations at the background stations is comparable with similarly located Central European EMEP stations. EMEP Assessment report [12] indicates an increase in winter and springtime mean ozone in EMEP domain, which is generally attributed to changes in baseline ozone (both intercontinental transport and stratosphere–troposphere exchange) and also local effects such as the longer lifetime of ozone because of reduced availability of NO (reduced titration).

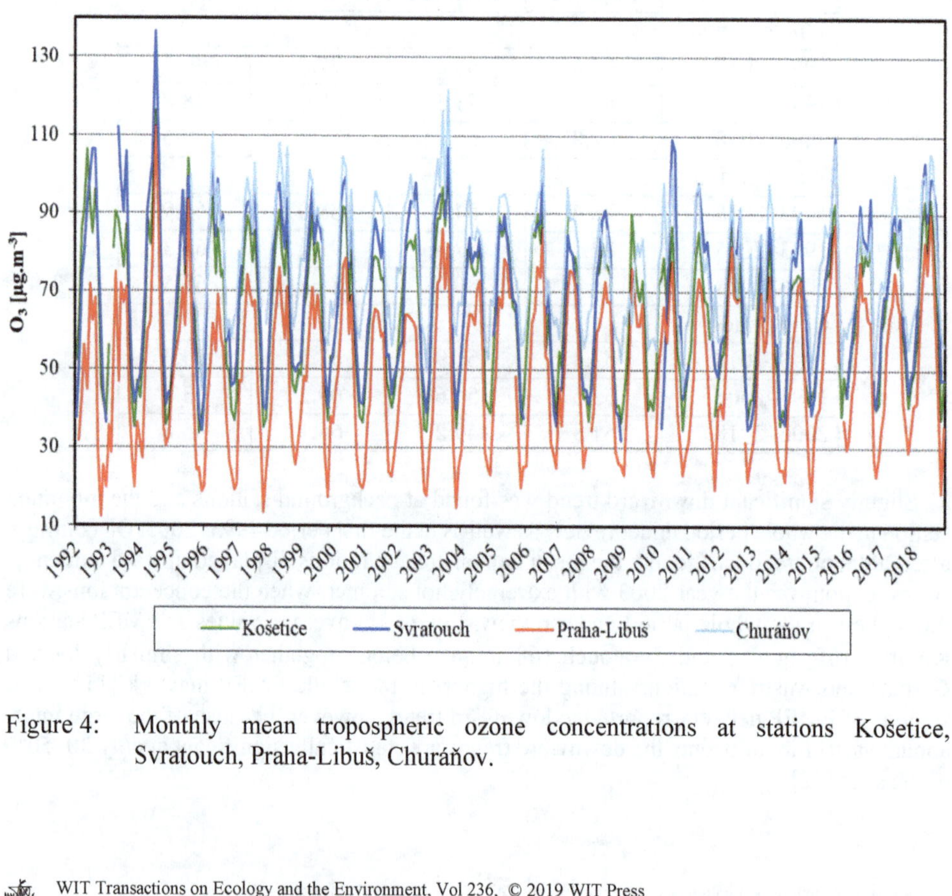

Figure 4: Monthly mean tropospheric ozone concentrations at stations Košetice, Svratouch, Praha-Libuš, Churáňov.

Fig. 5 clearly shows the diurnal variation of tropospheric ozone concentrations with a minimum at around 6 am and a maximum around 2 pm. The diurnal variation at the mountain station is less significant; mean annual concentrations in the nigh time are about 75 μg.m^{-3}. The more marked daily variation at the Košetice observatory than at Svratouch is clearly influenced by the lower elevation. The difference is particularly noticeable at night, when the Košetice values fall to an average of 10 μg.m^{-3} less than in Svratouch, while during the day the average concentrations are equal. At Prague suburb, the daily variation in more significant, the concentrations are lower by 10 μg.m^{-3} during the daylight and by 20 μg.m^{-3} in the night in comparison with background level [13].

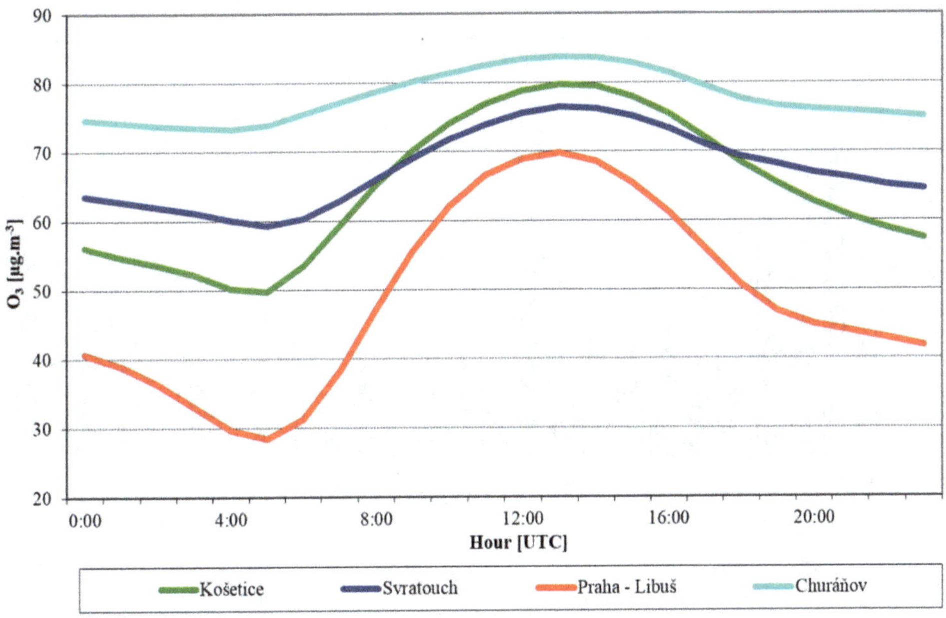

Figure 5: Diurnal variation of tropospheric ozone concentrations at stations Košetice, Svratouch, Praha-Libuš and Churáňov.

4 CRITICAL LEVELS

To assess the effects of tropospheric ozone on human health and on vegetation, international and national bodies have gradually introduced a whole system of ambient air pollution limits or critical values during the period under review. In this study, a comparison with current Czech and European ambient pollution limits is used. Target value of tropospheric ozone for protection of human health is exceeded when 8-hour running mean is higher than 120 μg.m^{-3} 25 times in average for 3 years. The alert threshold for protection of human health is exceeded when the concentration is at least once per day higher than 180 μg.m^{-3}. The limit for protection of ecosystems and vegetation is based on AOT40 index. AOT40 is the sum of differences between the hourly concentration higher than 80 μg.m^{-3} (=40 ppb) and the value 80 μg.m^{-3} in the given period by using only hourly values measured between 8:00 and 20:00 CET. The target value is surpassed when the AOT40 index, calculated between May and July is higher than 18 000 μg.m^{-3}.h in average for 5 years [14].

The target limit for human health protection was exceeded at all selected stations almost during the whole period 1993–2005, only at suburb station Prague-Libuš the three-year mean values varied round the target limit in the period 1998–2002 (Fig. 6). Generally, significant drop was found analyzing the whole period at all types of stations. The year 2003 brought the hot summer comparable with early nineties. Although the emissions of ozone precursors declined in the Central Europe, the number of days in which the limit was exceeded in 2003 reached the level of the beginning of the nineties. In the period 2006–2018, significant drop of high ozone episodes continued. The three-year mean fluctuated around the target limit at all stations. The summer 2018 was ones of the hottest in the history of climatological measurement in the Czech Republic and the number of high ozone episodes increased again.

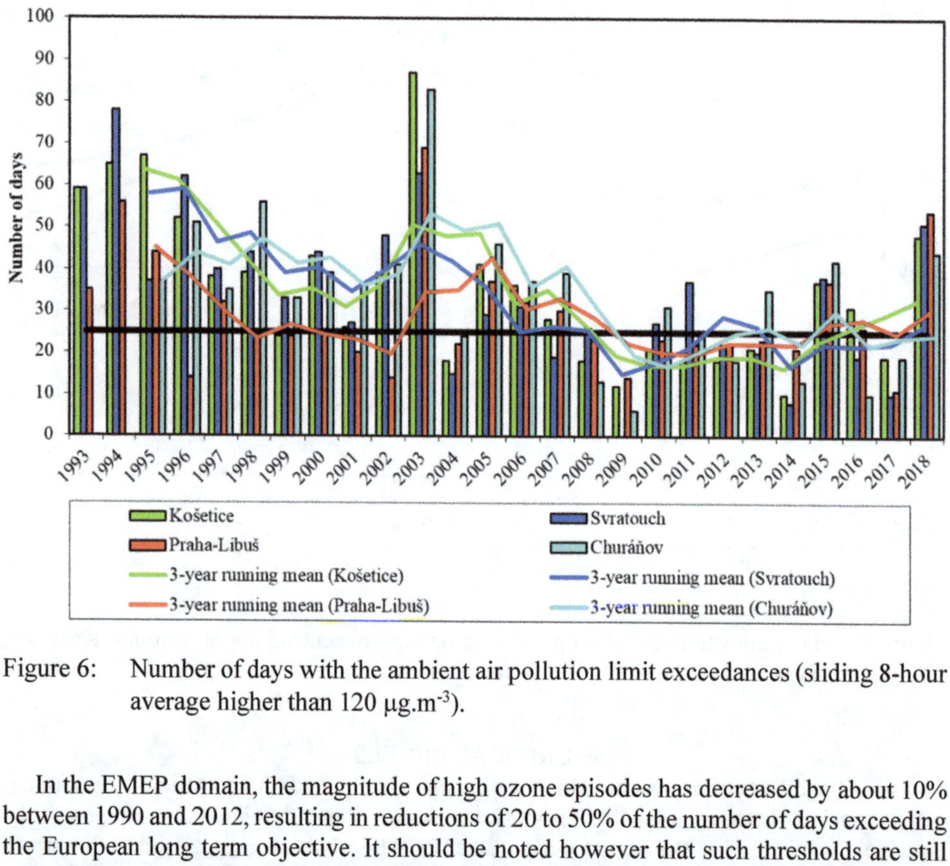

Figure 6: Number of days with the ambient air pollution limit exceedances (sliding 8-hour average higher than 120 $\mu g.m^{-3}$).

In the EMEP domain, the magnitude of high ozone episodes has decreased by about 10% between 1990 and 2012, resulting in reductions of 20 to 50% of the number of days exceeding the European long term objective. It should be noted however that such thresholds are still exceeded at a majority of stations, thereby demonstrating both the efficiency of control measures undertaken over the past 20 years, and the need for further action.

Fig. 7 shows very significant drop of high ozone episodes, when the alert threshold (180 $\mu g.m^{-3}$) was exceeded. Such episodes were quite frequent in the very beginning of the data series (first half of the nineties). Generally, the episodes were recorded more often at the suburban station. By far the highest amount of the episodes with alert threshold exceedances was recorded in the year 1994. The summer of 1994 was extremely warm and, at the same time, the emissions of ozone precursors were still higher than in the rest of the period under review. After 2000, the episodes of alert threshold exceedances occurred very rarely, namely in the years with very hot summers (2003, 2015, 2018).

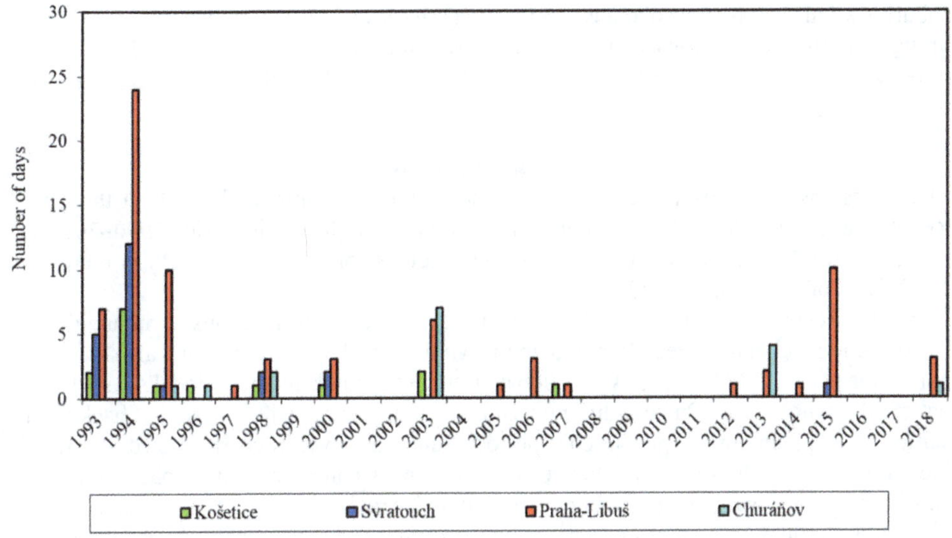

Figure 7: Number of days of the ozone alert threshold exceedance (180 μg.m-3).

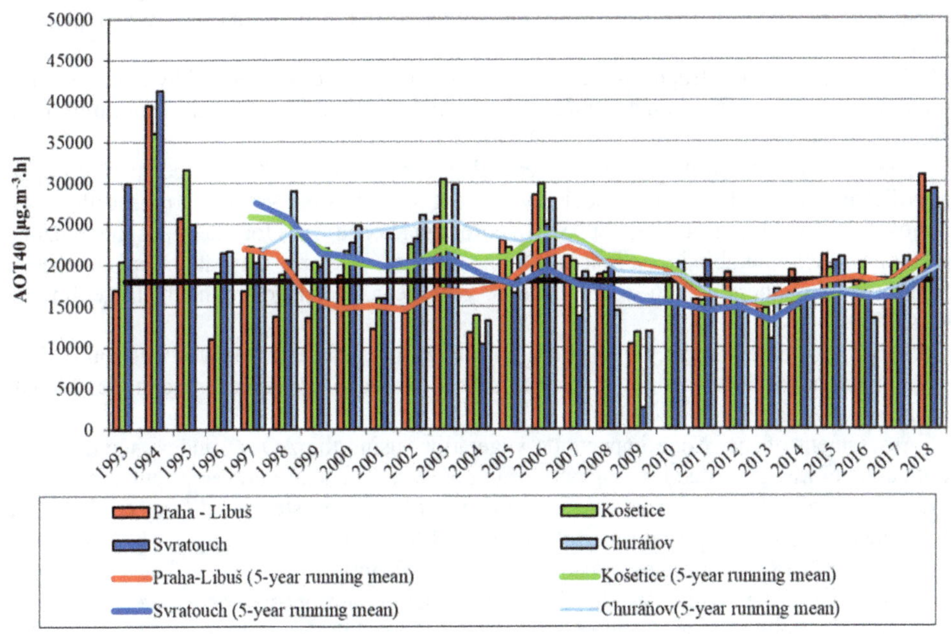

Figure 8: Trend of AOT40 index.

The evaluation of the ozone impacts on vegetation implies that, not only in rural areas, but practically throughout the Czech Republic, critical levels were significantly surpassed in the nineties (Fig. 8). In 1994 the limit was exceeded twice at both background stations. In the second half of the nineties, the values dropped significantly at all stations. The values at suburb station Praha-Libuš declined under the target level. The warmer period 2003–2006 brought short-term increase of AOT40 index above the target limit. In the last decade, AOT index varied round the requested level round 18,000 μg.m^{-3}.h. In the very hot summer 2018 the values exceeded the limit significantly.

5 CONCLUSIONS

The emissions of tropospheric ozone precursors dropped significantly both in the Czech Republic and in whole the EU. NOx emissions dropped by 80% in the period 1990–2016 and by 55% after 2000 in the Czech Republic. The figures for VOCs were 63% in the period 1990–2016 and 30% after 2000.

Slightly decreasing trend was found in mean annual concentrations at mountain and background level of the Czech Republic in the whole period under review and also in the first part of the period (1993–2005). After 2006 no trend was found. Warm period displays similar patterns as whole year. On contrary, no trend was found in the whole period at background stations and mountain site in the cold period. Suburban level is characterized by slightly increasing trend. The difference between mean annual concentrations at background and suburban stations was smaller during the period under review.

Tropospheric ozone concentrations at the Czech stations have a marked annual variation with maxima at the end of spring. In June and sometimes also in July there is a decrease caused by the onset of the so-called "continental monsoon", which brings increased cloud cover and a drop in solar radiation. We then register a second maximum in July and August (especially in very hot summers). Concentrations at mountain station are higher all year round.

Diurnal variation of tropospheric ozone concentrations is characterized by a minimum at around 6 am and a maximum around 2 pm. At suburb station, the diurnal variation in more significant.

The number of episodes with target value for human health exceedances dropped significantly during the period under review, but annual target values surpassed the limit almost in the whole period 1993–2005. Meteorological conditions for the tropospheric ozone creation were favourable in the first half of the nineties, when the majority of episodes were registered. In the period 2006–2018, the three-year mean fluctuated around the target limit at all stations.

Very significant drop of the episodes with alert threshold exceedances was found. The highest amount of the episodes with alert threshold exceedances was recorded in the first half of the nineties, especially in 1994.

Assessment of the ozone impacts on vegetation, using the AOT40 index, suggests that critical level was exceeded for long periods not only in the background areas but almost over the whole Czech Republic; this causes damage to forest ecosystems and farm crops. In the last decade, the values varied round the requested level round 18,000 μg.m^{-3}.h. In the very hot summer 2018 the values exceeded the limit significantly.

In the end of the second decade of the new millennium, tropospheric ozone remains one of the most serious environmental problems of the Central European region; however certain prognoses from the beginning of the nineties, which predicted further continuous growth in tropospheric ozone concentrations in next decades, have not been fulfilled.

ACKNOWLEDGEMENTS

The research leading to these results has received funding from the project for support of national research infrastructure ACTRIS – participation of the Czech Republic (ACTRIS-CZ – LM2015037) – Ministry of Education, Youth and Sports of the Czech Republic.

The research leading to these results has received funding from the ACTRIS-2 Integrating Activities (IA), European Union's Horizon 2020 research and innovation programme (grant agreement No 654109).

REFERENCES

[1] Seinfeld, J.H. & Pandis, S.N., *Atmospheric Chemistry and Physics: From Air Pollution to Climate Change*, John Wiley & Sons: New York, 2006.

[2] Sillman, S., Logan, J.A. & Wofsy, S.C., 1990. The sensitivity of ozone to nitrogen oxides and hydrocarbons in regional ozone episodes. *Journal of Geophysical Research: Atmospheres*, **9**, pp. 1837–1851, 1990.

[3] Fiala, J. & Závodský, D., Compendium of air quality protection. Part 2. Chemical aspects of polluted air – tropospheric ozone. *Appendix to the Journal Ochrana ovzduší (Air Quality Protection)*, 2002.

[4] Moldanová, J., Gas-phase chemistry. *Atmosphere and Climate. Topic Problems of Ambient Air Protection,* M. Braniš & I. Hůnová, eds, Karolinum: Prague, 2009.

[5] Colbeck, I. & Mackenzie, A.R., *Air Pollution by Photochemical Oxidants. Air Quality Monographs*, Vol. 1, Elsevier: Amsterdam, 1994.

[6] Brookes, D. et al., Air Pollution in the UK 2012. Department for Environment, Food and Rural Affairs: London, 2013. http://uk-air.defra.gov.uk/library/annualreport/viewonline?year=2012_issue_1.

[7] EEA, European Union emission inventory report 1990–2016 under the UNECE convention on long-range transboundary air pollution (LRTAP). EA Report No. 6/2018, 2018. DOI: 10.2800/571876.

[8] EEA, Air quality in Europe – 2013 report. EEA technical report 9/2013. EEA: Copenhagen, 2013. www.eea.europa.eu/publications/air-quality-in-europe-2013.

[9] *Air Quality in the Czech Republic 1994–2018*, ČHMÚ: Prague.

[10] Salmi, T., Anttila, P., Maattaa, A. & Ruoho-Airola, T., Detecting trends of annual values of atmospheric pollutants by Mann–Kendall test and Sen's slope estimates. Finnish Meteorological Institute, 2002.

[11] *Ozone Measurements 1990–2017*, NILU: Kjeller.

[12] Air pollution trends in the EMEP region between 1990 and 2012. Joint Report of : EMEP Task Force on Measurements and Modelling (TFMM), Chemical Co-ordinating Centre (CCC), Meteorological Synthesizing Centre-East (MSC-E), Meteorological Synthesizing Centre-West (MSC-W). EMEP/CCC Report 1/2016.

[13] Váňa, M. & Pekárek, J., Long-term trends of surface ozone trends in the Czech Republic. *Košetice Observatory – 25 years*, ČHMÚ: Prague, 2013.

[14] EC, Directive 2008/50/EC of the European Parliament and of the Council of 21 May 2008 on ambient air quality and cleaner air for Europe, 2008. http://eurlex.europa.eu/LexUriServ/LexUriServ.do?uri=OJ:L:2008:152:0001:0044:CS:PDF.

AIR POLLUTION, CITIZEN DATA COLLECTIVES AND COMMUNICATION AGENDA SETTING IN COLOMBIA

JUAN-CARLOS VALENCIA & OSCAR FONSECA
Department of Communication, Universidad Javeriana, Colombia

ABSTRACT

Air pollution levels in the cities of Bogotá and Medellín (Colombia) are now reaching dangerous levels, as measured by the few government monitoring stations available and operative. So far, government actions have focused on enforcing private car and industrial emission limits, prohibiting the circulation of vehicles for a few days during particularly polluted periods and vague promises of switching a percentage of public transport vehicles to alternative power. Official measurements are increasingly being reported in the national and local media, as well as through social media, but citizen distrust of their quality has mounted. Citizen collectives are starting to work designing low-cost mobile sensors, monitoring pollution in some areas and in public transport, sharing their georeferenced data over the internet and trying to raise awareness of the dangers of pollution and the necessity of radical actions to deal with the problem. This paper describes the current actions of some of these citizen collectives and their results in setting the media and public agenda on air pollution problems.

Keywords: social movements, citizen collectives, air pollution, data activism, Colombia, citizen sensing, participatory monitoring, environmental justice.

1 INTRODUCTION

Air pollution is a mounting problem in cities from London to Mumbai, from Detroit to Bogotá. An estimated 4.5 billion people around the world are currently exposed to particulate matter (PM) levels above the concentration that the World Health Organization (WHO) considers safe [1]. It is one of the most visible and worrying effects of what some authors call the Anthropocene or the Capitalocene [2]. Air pollution intensification in cities around the world demonstrates that a hegemonic, yet not universal civilizatory model, has reached a breaking point and that its technological solutions cannot really solve the problems it has created and continues to create [3]. The overproduction and widespread use of fossil-fuel powered vehicles, the accelerated growth of urban populations, the expansion and nomadic character of largely environmentally-unfriendly industries, the expansion of power generation plants and the ideas that unlimited consumption and constant economic growth are the ultimate goals of public institutions and private corporations have resulted in a deepening, sadly irreversible environmental crisis. Only highly coordinated, big budgeted, radical actions in places like China have somewhat reversed this trend, but even their air pollution levels are still above WHO's recommended values [4].

There is growing concern among Social Science academics and sectors of public opinion around the world that market forces and democratic institutions are not managing to confront this dangerous crisis: regulations, policies, international agreements, declarations of intent are not resulting in effective solutions and environmental indicators worldwide continue to deteriorate. We are already experiencing the limits to growth [5]. Despair at inaction has led to recent protests in Great Britain such as the so-called Extinction Rebellion. Perhaps our only hope is that, as it has happened in the past, citizens will take direct action, promote real although hard solutions, start civilizatory paradigm transitions and adopt more realistic humans-as-part-of nature lifestyles. This paper comes out of the early stages of participatory research conducted in Colombia, South America with citizen collectives that worried about air-quality deterioration in the cities of Bogotá and Medellin, ineffective or insufficient public

WIT Transactions on Ecology and the Environment, Vol 236, © 2019 WIT Press
www.witpress.com, ISSN 1743-3541 (on-line)
doi:10.2495/AIR190041

policies, corruption and irresponsible, short-term corporate decisions, have decided to take a hands on, radical approach. They are designing and building low-cost air-quality sensors, installing them around their under-monitored cities, using mobile monitoring equipment and circulating their data through social media platforms, raising awareness about the problem across all sectors of society and confronting public institutions, officials and corporations with sufficiently accurate, urgent information.

2 AIR POLLUTION IN BOGOTÁ AND MEDELLIN

Bogotá is a city of 1775 km² and more than 8,300,000 inhabitants. Medellín's size is 1,152 km² with a population of 3,731,000. Both cities have been important light-industry hubs in Colombia for the past half century. Industries are basically food, chemical, plastics, metallurgy and construction materials. They use carbon, heavy oils and natural gas as fuel for machinery and release particulate matter to the atmosphere. Both cities are located high on the Andes in fairly narrow valleys surrounded by tall mountains, a geography that makes heavy particles and gases difficult to disperse. Private cars and motorcycles, trucks, taxis and public transport vehicles have increased substantially in both cities, especially after neoliberal reforms introduced in the early 1990s. Power generation in Colombia relies mostly on hydroelectric power, although there are some carbon thermoelectric plants running. Farmers displaced from the countryside because of violence have dramatically increased the population of Colombian cities in the past four decades. Most live in precarious settlements in the outskirts of cities, many in unstable terrains on hills and mountains, connected to the flat land through unsealed roads.

The authorities estimate that there could be nearly one million motorcycles, 60,000 taxis, nearly 20.000 public transport vehicles, nearly one million private cars and about 25,000 trucks circulation in Bogotá [6]. They mostly use gasoline and diesel as fuel with a small percentage using natural gas. A subway system was built in Medellin in the 1980s but the one planned for Bogotá has been postponed for decades because of the barriers mounted by special interests. Bogotá introduced Transmilenio, an extensive Bus Rapid Transit System, more than twenty years ago but the buses use Diesel [7]. Citizen and experts' requests that the buses use natural gas went unheard. Legislation to control pollutant emissions has been introduced since the 1970s and the first air quality measurements were conducted in the late 1960s. Fixed and mobile sensors were installed and replaced when they served their lifespan. The Red de Monitoreo de Calidad del Aire de Bogotá (RMCAB) was created and in the mid-1990s it was planning to set up 32 monitoring stations throughout the city that would measure PST, PM_{10}, O_3, CO and NO_x [8]. Two decades later just 13 high-quality fixed sensors were installed and running but there are worrying measurement blackouts and areas of Bogotá that are still unmeasured. The national government introduced mandatory private vehicle emission measurements early in the 21st century and has tightened the thresholds continuously. There have been improvements in the quality of fuels available to the public. But trucks, public transport vehicles, industries and construction works continue to pollute the air and their impact seems to increase year after year. Half of the nearly 2,000 industries registered in Bogotá release high impact emissions.

PM_{10} is the most significant air pollutant in Bogotá [9]. Annual average PM_{10} levels for Bogotá range from 9.89 µg/m³ to 160 µg/m³ with peaks in the months of March and April. PM_{10} mean is 37.5 µg/m³, below the Colombian annual standard of 50 µg/m³ but twice as high as the WHO guideline of 20 µg/m³ [10]. The sources are fugitive dust, road dust, metal processing, industrial emissions but mostly, about 50%, vehicular traffic.

$PM_{2.5}$ levels fluctuate between 45 and 15 µg/m³ depending on the location of the measurement station and the time of the year. Average annual value is 24.3 µg/m³, barely below the maximum allowed in the national guideline [11]. PST in Bogotá always tops the maximum recommended level of 100 μg/m^3. A worrying fact is that the most affected areas in Bogotá (Puente Aranda, Fontibón, Tunjuelito and Kennedy) are precisely those where lower-income citizens live, and they are the majority of the population [12]. Occasional restrictions on the circulation of private vehicles and heavy trucks when contamination levels reach alarming levels result in even higher $PM_{2.5}$ and PM_{10} measurements because industries and small businesses resort to older, smaller vehicles and public transport circulation increases. Recent research has determined that despite unsubstantiated claims to the contrary by the Mayor of Bogotá, Enrique Peñalosa, particulate levels inside the buses of Transmilenio, the Bus Rapid Transit System are dangerously high, and they are even higher in some bus stations in the system [7].

Medellin has a smaller population than Bogotá but air pollution problems have become more serious because population density is higher, the valley area is smaller, the surrounding mountains are taller, industries operate throughout the city, not concentrated in some specific sections, and the concentration of cars and motorcycles per citizen is much higher. A vehicle census from 2014 reported 1,234,946 cars and motorcycles moving through Medellin. The Sistema de Alerta Temprana de Medellín y el Aburrá (SIATA) has 20 high-precision sensors monitoring air quality in Medellin [13]. 58% of $PM_{2.5}$ particles in Medellín come from old trucks built in the 1990s or earlier and used to transport merchandise through the city. Most of them are owned by small entrepreneurs with no other sources of income so the authorities have faced stiff opposition to prohibit them and force their replacement.

Annual average PM_{10} was 61 μg/m^3 with the most polluted months being April (83 μg/m^3) and May (72 μg/m^3), while for the remaining months, average emissions exceeded 60 μg/m^3 (except October with 49 μg/m^3). Annual $PM_{2.5}$ average is 26.9 μg/m^3 above the national guideline. 64.6% of this particulate material is related to vehicles while 8.5% is associated with industrial processes. Ozone annual average is 31 μg/m^3 [14].

$PM_{2.5}$ and PM_{10} emissions are substantial in Downtown Medellín, where there is a vast number of people. It has been estimated that around 1 million people intermittently walk in this area every day. Unofficial data suggests that around 9,000 informal street vendors work in Downtown Medellín and therefore are exposed to substantial air pollution [15]. But as recently as March 2019, 15 of the 20 monitoring stations in the city reported air pollution values above the maximum permitted. Bicycle circulation restrictions had to be applied as pollution levels were considered too unsafe.

Medical health expenses associated with environmental issues as calculated by the Colombian Department of Planning in 2015 top 2.6% of the national GDP and result in more than 10,000 annual deaths. Between 2011 and 2014 there were 2,877,892 patients treated in Bogotá's hospitals due to illnesses related to poor air quality [16].

3 THE DISTRUST OF THE AUTHORITIES AND THE EMERGENCE OF CITIZEN SENSING

British journalist George Monbiot [17] argues that had governments, corporations and ordinary citizens put as much effort into preventing environmental catastrophe as they have spent on making excuses for inaction, they would have solved it by now. Powerful economic and political interests have blocked radical action and ridiculed civil society demands and initiatives around the world. The media, with a few exceptions, is disinterested or even actively hostile. When broadcasters cover these issues, they carefully avoid any mention of power, talking about air pollution as if it were a random occurrence driven by mysterious, passive forces. They mostly retransmit official communications that propose microscopic, temporary fixes for vast structural problems. In Colombia, and other Global South locations, the myth of development has been reworded [18] and used to delegitimize proposals to solve the air pollution problem and confront the crisis of the Anthropocene. Those who govern nations, manage corporations and shape public discourse cannot be trusted with the preservation of life on Earth. As Monbiot writes, there is no benign authority preserving us from harm. No one is coming to save us.

Additionally, when they exist, institutional monitoring systems around the world have drawbacks [19]: the instruments are large in size, weight and cost and are difficult to calibrate. Their placement in urban areas is impacted by human activities (e.g., construction).

But citizens and social movements are stepping up to the front. Researchers are talking about rising global do-it-yourself (DIY) activism and tactical media use [20]–[23]. Activists are not only raising public awareness of environmental problems but are increasingly taking a hands-on approach, researching the issues, discovering voids in authorities discourses and actions, designing and building tools to collect data, circulating information through social media, pressuring traditional media to focus on topics that matter and constantly scrutinizing institutions, officials and corporations.

Citizens and social movements are buying or building low-cost air pollution sensors to monitor their living environment by themselves, and they are doing it in supposedly democratic societies like the Netherlands, totalitarian regimes like China and poor countries like Colombia. These collectives are implementing community-driven sensing infrastructures in many cities around the world. They calibrate their sensors with data from the official monitoring stations and use them in under-monitored areas or even build mobile devices that they take on the road to gather data about their usual transportation routes. They use open hardware and software, upload data to open servers, add georeferenced information and try to spread their measurements and findings to relevant social actors and the general public. These Citizen Science practices are challenging, they do need professional knowledge and support to be able to retrieve reliable data, but as our research is starting to show, maker movements are starting to spread and are impacting traditional media and public sector agendas in Colombia. Sometimes they seem invisible in our daunting, saturated mediascape of today but they are learning to process big data, to create metadata and to use it in more accessible, everyday narratives that common people can relate to.

4 CITIZEN SENSING IN BOGOTA AND MEDELLIN

In this paper we focus on the work of two citizen collectives, CanAirIO (Canard in Spanish) in Bogotá, and UNLOQUER in Medellín.

4.1 App CanAirIO

The CanAirIO App is a citizen science initiative for the monitoring of air quality in Bogotá using PM 2.5 particulate material sensors. The sensors are low cost (Fig. 1) and are built by citizens in open workshops. There, the problem of air pollution is explained and the sensors are assembled collaboratively.

Daniel Bernal is an electronics engineer. Three years ago he started to ask himself "How could we trust an air monitoring network that is insufficient with only 13 stations in a city as big as Bogotá? In addition, the web page that shows the data is often out of work and the data it shows is incoherent". He came across an article on the Internet that argued that "if you do not feel satisfied about how the government measures air quality, do it yourself". It was an initiative taking place in Germany. The article explained how to assemble an air quality sensor. He brought some sensors from China and started working on building a reliable device to measure air quality. His early measurements detected that air pollution was especially troubling inside the Transmilenio bus stations.

Antonio Vanegas is another member of the CanAirIO App collective. He is a software engineer and develops applications for Windows. He became interested in the topic of air pollution during a trip to the Mediterranean island of Malta. The capital of the island does not have a metro system so public transport relies on a Bus Rapid Transit System similar to Bogotá's Transmilenio. But in Malta, bus emissions were not perceptible. There was no significant air pollution. Antonio Vanegas also spent some time in Berlin and followed the debates on the criteria to choose new public transport buses. He understood that the proposals included buses with much cleaner technologies than those chosen by the government of Bogotá. After returning to Colombia he researched the RMCAB air quality public monitoring network and considered it insufficient. He also noticed the data outages. He saw a TV interview of Daniel Bernal and contacted him. Antonio Vanegas and Daniel Bernal started working together in late 2016 and with time more people started to join them.

The members of the CanAirIO App citizen collective believe that the air pollution information provided by the RMCAB is not fully reliable. Daniel Bernal argues that the RMCAB sensors are installed at a height of between 6 and 10 m, so therefore "the devices do measure in the lower atmosphere but not normal breathing height". The collective decided to create their own low-cost, portable sensors, "but with a comparable measurement efficiency as those of the high-cost sensors". Early prototypes were modified after some tests. They managed to contrast their low-cost sensor measurements with those of a high-cost sensor from the lab of one of the universities in Bogotá. They ended up developing a much better sensor, one with a deviation of only 3.4%. The components of the sensor cost about $120,000 pesos (US$37). The core of the device is a Honeywell particulate material sensor that activates when it receives air from a small fan. The device can be connected to a mobile phone via Bluetooth and through the CanAirIO App air pollution data is uploaded to the internet creating a collaborative georeferenced map of air quality in Bogotá (Fig. 2). The app is available on Google Play.

Figure 1: The CanAirIO sensor [24]. *(Source: Medium©.)*

The CanAirIO App collective wants more citizens to build and use their sensors. So far in 2019 they have conducted two open workshops that help participants build their own sensors from scratch. The workshops last two hours (Fig. 3).

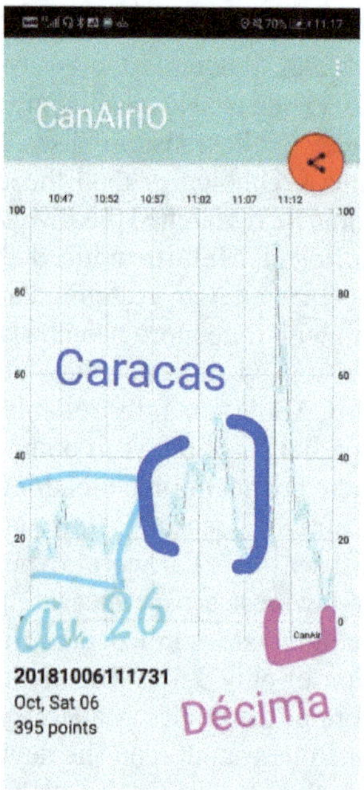

Figure 2: Air pollution data visualization on the CanAirIO app [25]. *(Source: Medium©.)*

Figure 3: Common citizens of Bogotá showing their newly built CanAirIO sensors [25]. *(Source: Angelica Zambrano©.)*

Currently there are more than thirty of these citizen sensors in Bogotá. The goal is to have between 100 and 200 citizens with sensors reporting to form a real and collaborative map of air quality in Bogotá. Although the application is a great initiative that has managed to sensitize many citizens to the serious condition of air pollution in Bogotá, there is still much to be improved in terms of data visualization. The air pollution map could be redesigned to make it easier to understand and follow by citizens with no prior knowledge of the problem. Up to now, the map can only be viewed by those who have the app installed on their smartphones. The collective consists of technology experts with little knowledge of community management and social communication.

4.2 Unloquer

They are a team of technology enthusiasts comprising professionals and amateurs that have been developing citizen science initiatives in Medellín since 2009. They have a hackerspace near downtown Medellín and sometimes convene citizens to explain their projects.

"About two years ago, with the interest of analyzing data, we wanted to see the condition of air quality in the city, but when we went looking for information, we did not find much available. So we decided to collect it ourselves" says Julián Giraldo, an electronics engineer. The problem became more acute with the passing of time. The local government has had to declare public emergencies prohibiting vehicle circulation in Medellín for days, at least 5 times in the past two years. Temporary tighter industrial emission regulations have been proposed but they have been criticized by corporations in the city. So Unloquer started developing a low-cost mobile sensor, to monitor air quality in Medellín.

The current team is made up of 15 members. They researched citizen sensing initiatives around the world and after testing different prototypes came up with a satisfactory, portable, very small $PM_{2.5}$ sensor (Fig. 4). The sensors take into account humidity and temperature conditions. They can be taken on walking tours, on a bicycle, a motorcycle, a car or in any other vehicle.

The devices transmit data through WI-FI connections. They use a GPS to trace routes and add a timestamp to the measurements. Ten devices of this first version were manufactured. Today there are at least 30 Unloquer designed sensors installed in Medellín. The total cost of manufacturing the sensor is $300,000 Colombian pesos ($95 USD). Both the hardware and software designed are open source and documentation is available in the collective web site.

Figure 4: Unloquer air pollution sensors [26]. *(Source: Unloquer©.)*

The project has not only developed a citizen network of mobile sensors to monitor Medellín's air quality, but also created a server system to store and facilitate the visualization of the data, which is readily available to citizens in the city (Fig. 5).

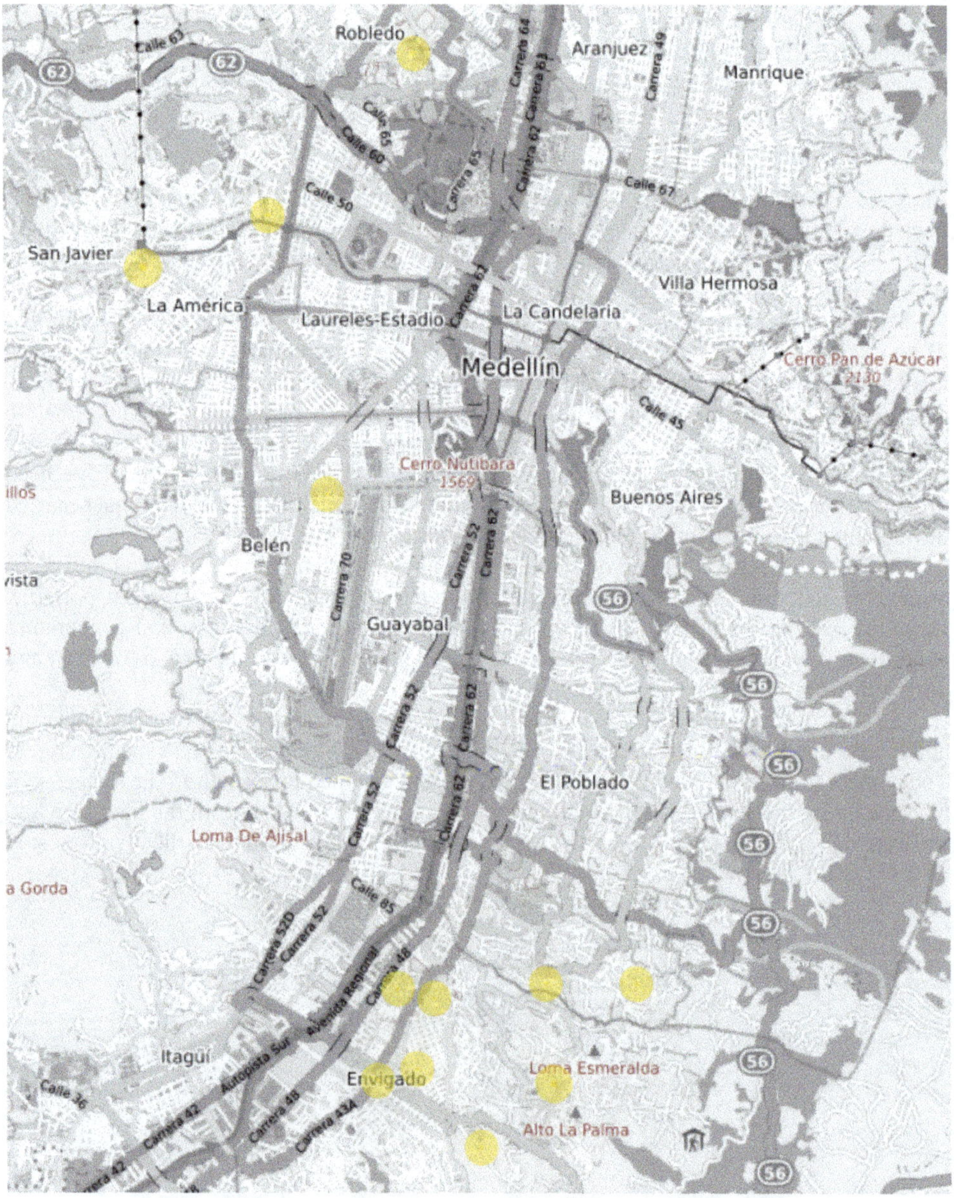

Figure 5: Air pollution map of Medellín obtained from citizen sensors [27]. *(Source: Unloquer©.)*

5 FINAL REMARKS

There are critics of participatory sensing, both in the academic and in the public sectors. Distrust of the accuracy of the early Arduino devices, fears that citizen data could be manipulated by political and economic interests, the absence of audit mechanisms are common complaints. On the other hand, citizen sensing rises new challenges [28], such as protecting user private data and motivating and sustaining participation. Daniel Bernal considers that "the role of the authorities cannot be replaced. Participatory sensing data cannot be fully compared with government data, but they can be correlated and used to discover trends". Citizen sensing alone is not enough but could be a useful tool for activists to pressure governments and corporations to respond to social demands and finally take decisive actions to curb air pollution.

Pressure from Bogotá's citizen collectives using data obtained through participatory sensing and amplified by concerned sectors of traditional media resulted in the local government reconsidering its initial decision of renewing the Transmilenio fleet using only Diesel buses. Now, the local government will replace a total of 1,133 buses and they will rely on Diesel Euro V technology (59%) and Natural Gas (41%). The Mesa Técnica para la Calidad del Aire de Bogotá (MECAB), an umbrella organization that includes citizen collectives, academics and NGO's has managed to become a valid source for traditional media when reporting air pollution issues. The city government has had to interact with this organization as well.

Citizen collectives in Medellin, some of them using data from participatory sensing managed to force the local government to replace fifty buses from the public transport fleet with electric vehicles.

Definitive actions are still to be taken to solve the serious condition of air in Bogotá and Medellin, but citizen sensing collectives are managing to impact public institutions, corporate and general public agendas.

REFERENCES

[1] Ebenstein, A., Fan, M., Greenstone, M., He, G. & Zhou, M., New evidence on the impact of sustained exposure to air pollution on life expectancy from China's Huai River Policy. *PNAS*, **114**(39), pp. 10384–10389, 2017.

[2] Haraway, D., Anthropocene, Capitalocene, Plantationocene, Chthulucene: Making Kin. *Environmental Humanities*, **6**, pp. 159–165, 2015.

[3] De Sousa-Santos, B., Renovar la teoría crítica y reinventar la emancipación social, CLACSO: Buenos Aires, pp. 15, 2006.

[4] Greenstone, M. & Scharza, P., Is China Winning its War on Pollution?; Air Quality Life Index Update March 2018. https://epic.uchicago.edu/sites/default/files/UCH-EPIC-AQLI_Update_8pager_v04_Singles_Hi%20%282%29.pdf. Accessed on: 26 Apr. 2019.

[5] Meadows, D., (ed), *The Limits to Growth; A Report for the Club of Rome's Project on the Predicament of Mankind*, Universe Books: New York, 1972.

[6] Alcadia Mayor de Bogotá, Plan Decenal de Descontaminación del Aire para Bogotá, Alcadia Mayor de Bogotá: Bogotá, pp. 43–46, 2010.

[7] Morales-Betancourt, R., Galvis, B., Rincón-Riveros, J., Rincón-Caro, M., Rodriguez-Valencia, A. & Sarmiento, O., Personal exposure to air pollutants in a Bus Rapid Transit System: Impact of fleet age and emission standard. *Atmospheric Environment*, **202**, pp. 117–127, 2019.

[8] Alcadia Mayor de Bogotá., Plan Decenal de Descontaminación del Aire para Bogotá, Alcadia Mayor de Bogotá: Bogotá, pp. 61–68, 2010.

[9] Ramírez, O., Sánchez, A., Amato, F., Catacolí, R., Rojas, N. & De la Rosa, J., Chemical composition and source apportionment of PM_{10} at an urban background site in a high-altitude Latin American megacity (Bogotá, Colombia). *Environmental Pollution*, **233**, pp. 142–155, 2018.

[10] Alcadia Mayor de Bogotá, Plan Decenal de Descontaminación del Aire para Bogotá, Alcadia Mayor de Bogotá: Bogotá, pp. 69, 2010.

[11] Alcadia Mayor de Bogotá, Plan Decenal de Descontaminación del Aire para Bogotá, Alcadia Mayor de Bogotá: Bogotá, pp. 72, 2010.

[12] García, D., Calidad del aire y políticas públicas en Bogotá: una historia de injusticia ambiental. Ideas Verdes. *Análisis Político*, **14**, pp. 1–22, 2018.

[13] Gaviria, C., Benavides, P. & Tangarife, C., Contaminación por material particulado ($PM_{2.5}$ y PM_{10}) y consultas por enfermedades respiratorias en Medellín (2008–2009). *Revista de la Facultad Nacional de Salud Pública*, **29**(3), pp. 241–250, 2011.

[14] Posada, E., Gómez, M. & Almanza, J., Análisis comparativo y modelación de las situaciones de calidad del aire en una muestra de ciudades del mundo. Comparación con el caso de Medellín. *Revista Politécnica*, **13**(25), pp. 9–29, 2017.

[15] Gaviria, C. & Martínez, D., Air Pollution and the Willingness to Pay of Exposed Individuals in Downtown Medellín, Colombia. *Lecturas de Economía*, **80**, pp. 153–182, 2014.

[16] Rodríguez-Villamizar, L., Rojas-Roa, N., Blanco-Becerra, L., Herrera-Galindo, V. & Fernández-Niño, J., Short-Term Effects of Air Pollution on Respiratory and Circulatory Morbidity in Colombia 2011–2014: A Multi-City, Time-Series Analysis. International *Journal of Environmental Research and Public Health*, **15**(8), 2018.

[17] Monbiot, G. Only rebellion will prevent an ecological apocalypse. *The Guardian*. 15 April 2019. www.theguardian.com/commentisfree/2019/apr/15/rebellion-prevent-ecological-apocalypse-civil-disobedience. Accessed on: 26 Apr. 2019.

[18] Esteva, G., Development. The Development Dictionary. A Guide to Knowledge as Power, ed, J. Sachs, Zed Books: London and New York, pp. 1–23, 2010.

[19] Yi, W., Lo, K., Mak, T., Leung, K., Leung, Y. & Meng, M., A Survey of Wireless Sensor Network Based Air Pollution Monitoring Systems. *Sensors*, **15**(12), pp. 31392–31427, 2015.

[20] Jiang, Q. et al., Citizen Sensing for Improved Urban Environmental Monitoring. *Journal of Sensors*, pp.1-9, 2016.

[21] Matsuzawa, S., Citizen Environmental Activism in China: Legitimacy, Alliances, and Rights-based Discourses. *ASIANetwork Exchange*, **19**(2), pp. 81–91, 2012.

[22] Hasenfratz, D., Saukh, O., Sturzenegger, S. & Thiele, L., *Participatory Air Pollution Monitoring Using Smartphones*. 2nd International Workshop on Mobile Sensing, pp. 1–5, 2012.

[23] Hua Xi, J., Communicating the Right to Know: Social Media in the Do-It-Yourself Air Quality Testing Campaign in Chinese Cities. *International Journal of Communication*, 8, pp. 1374–1393, 2014.

[24] canair.io., *Citizen network for air quality monitoring*. http://canair.io. Accessed on: 2 May 2019.

[25] Pàez, J. D. R., En 6 horas charladitas se puede reportar la calidad del aire en Bogotá. *Medium*, 2018. https://medium.com/@jdreyespaez/en-6-horas-charladitas-se-puede-reportar-calidad-de-aire-en-bogot%C3%A1-88ca8d5bb617. Accessed on: 2 May 2019.

[26] Hackerspace Medellin, Air quality agents. http://wiki.unloquer.org/personas/brolin/proyectos/agentes_calidad_aire. Accessed on: 2 May 2019.

[27] https://bit.ly/2vkDBUi. Accessed on: 2 May 2019.
[28] Mendez, D., Colorado, J., Rodriguez, L., Chacon, A., & Hernandez, M., Monitoring Air Pollution by Combining a Static Structure with a Participatory Sensing Approach: Design and Performance Evaluation. *International Journal of Sustainable Development*, **13**(4): 638–652, 2018.

www.witpress.com, ISSN 1743-3541 (on-line)

AN IOT PLATFORM FOR INDOOR AIR QUALITY MONITORING USING THE WEB OF THINGS

DANIEL IBASETA[1], JULIO MOLLEDA[2], FIDEL DÍEZ[1] & JUAN C. GRANDA[2]
[1]CTIC Technological Centre, Spain
[2]Department of Computer Science and Engineering, University of Oviedo, Spain

ABSTRACT

Platforms populating the Internet of Things (IoT) use dedicated software closely coupled with proprietary hardware, devices and interfaces, which creates silos and a lack of interoperability. The Web of Things (WoT) is a paradigm that incentives the use of web standards to interconnect all kinds of devices and defines an application layer for IoT applications. Multiple organizations and consortiums are pursuing the definition of architectures and standards to deliver interoperability to the IoT application layer. Air quality monitoring is a field in which IoT has a great role to play as it is based on different kinds of sensors and devices which monitor air pollution. Quite a few wireless sensor networks have been proposed in the literature to deal with this monitoring process. In this paper, we propose a low-cost, indoor air quality monitoring platform following the recommendations of the World Wide Web Consortium (W3C) about WoT. The platform is built based on a Web of Things capable of exposing its own Thing Description with 15 to more than 2000 resources, depending on the underlying hardware and the application protocol selected. These resources can serve requests providing measurements of the attached sensors, perform actions on the environment and/or generate events based on these measurements. Although the system is proposed for ambient monitoring, the software architecture developed in this work can be adapted to many embedded applications in the IoT.

Keywords: Internet of Things (IoT), Web of Things (WoT), application layer for IoT, air quality monitoring, ambient air monitoring.

1 INTRODUCTION

The Internet of Things (IoT) is a paradigm combining embedded systems and low-power wireless communication to provide physical objects with Internet connectivity [1]. Services offered by the IoT are the core of smart environments, such as smart homes, smart cities, or smart industries among others. One of the greatest challenges of the IoT is the interoperability of the devices used to create the network of physical objects.

A large number of IoT solutions make use of dedicated software applications coupled with proprietary hardware creating silo solutions that limit the interoperability across different platforms. To overcome this limitation, an application layer is required to communicate among different platforms. The Web of Things (WoT) is a paradigm that focusses on addressing this problem, improving the interoperability and usability of the IoT [2], using open, well-known Web standards to create an application layer for IoT applications. Using these standards, the cyber world and the physical world can communicate through an interoperable infrastructure. The key architectural ideas and technologies enabling the WoT are surveyed in Zeng at al. [3].

One of the fields that can take advantage of the use of IoT is air quality monitoring (AQM). Air is 99.9% nitrogen, oxygen, water vapour, and inert gases. Monitoring the quality of air requires measuring the concentration of several pollutants, such as carbon monoxide (CO), sulphur dioxide (SO_2), nitrogen dioxide (NO_2), and ozone (O_3). In Europe, the threshold levels specified for these pollutants are published in the National Emission Ceilings Directive [4]. These pollutants are released into the air by several human activities and natural sources, harming human health and the environment. Examples of these activities and sources include

www.witpress.com, ISSN 1743-3541 (on-line)
doi:10.2495/AIR190051

burning of fossil fuels in electricity generation, transport, industry and households; industrial processes and solvent use, for example in the chemical and mining industries; agriculture; waste treatment; natural sources, including volcanic eruptions, windblown dust, sea-salt spray and emissions of volatile organic compounds from plants.

As in many other fields, most of the IoT devices and services developed for AQM depend on particular platforms or technologies, some of them proprietary, making it difficult to develop a widely accessible application composing all the devices and services. In this paper we propose a low-cost, indoor air quality monitoring (IAQM) platform following the recommendations of the Word Wide Web Consortium (W3C), based on our previous work in this field [5]. The main benefit of this platform is the interoperability of its sensors with other IoT-based AQM platforms, since they are able to provide a description of their features, making it possible to retrieve measurements whatever the manufacturer. The sensors are accessible worldwide through a standard Web browser, or any other application designed to communicate through HTTP or MQTT. Using this type of sensor, a worldwide accessible AQM platform could be conceived allowing, for instance, remote monitoring of geographically separated environments using only open Web standards.

2 WEB TECHNOLOGIES FOR PHYSICAL OBJECTS

Physical objects were connected through Web technologies to create smart environments worldwide more than two decades ago [6]. One of the first reported works about bridging the Web and the physical world was the Cooltown project by HP Labs [7], which explored an infrastructure to support Web presence for people, places and things. Web servers were integrated into physical objects that were accessible through URLs; on top of the infrastructure, Internet connectivity was used. More recently, the IoT was considered part of the Internet [8], avoiding limitations around host-to-host communications, and focusing on the publishing and retrieval of information, which is the most common use of the Internet. Other pioneer solutions [9]–[11] also demonstrated that it was possible to integrate physical objects directly into the Web, or through the use of a gateway, such as Hwang at al. [12]. Modern solutions bridge heterogeneous Web services from the Internet into the IoT network by creating proxies. Thus, clients access transparently to IoT devices and Web services in the network, as in Jin and Kim [13].

The WoT applies Web solutions to access and retrieve information as well as to provide services by physical objects in the IoT. In IoT solutions, each physical object has an associate digital entity called a "Thing"; in WoT each physical object is expressed as a "Web Thing" [14]. A Web Thing commonly exposes metadata in HTML or JSON representation, provides an API to gain access to its properties, and defines an OWL-based semantic description. Web Things may be integrated in the Web in three different ways [15]: (i) hosted by a Web Server embedded into physical objects; (ii) hosted by a Web Server embedded in a gateway device; or (iii) hosted by a Web Server allocated in a cloud service.

Simple static or dynamic Web pages can be used to abstract functionalities of physical objects in WoT; however, reusable Web services are the preferred choice. A Web service is a service designed for machine-to-machine (M2M) communication through the Web. The most commonly used architecture for Web services is the Web Services Architecture (WSA) defined by the W3C [16]. Two categories of Web services are identified: WS-* style and REST style. In the WS-* style, Web services may expose an arbitrary set of operations, and use HTTP as the transportation medium to provide Remote Procedure Calls (RPC). RESTful services, the name given to Web services designed following REST architecture style [17], manipulate representations of Web resources using a uniform set of stateless operations, where every service is seen as a resource, and each resource is identified by a Uniform Resource Identifier (URI).

In the WoT, physical and virtual entities are abstracted as Web resources. In addition, use of RESTful services is the preferred choice because of their low level of complexity and loose-coupling stateless interaction [2], [3]. Services implementing a RESTful design provide the following features: (i) identification of resources via URI; (ii) uniform interfaces to access the resources, using four HTTP methods: GET, POST, PUT and DELETE; (iii) representation of resources that can be accessed in different formats, such as HTML, JSON or XML, and are self-descriptive as they contain the complete context; and (iv) stateless interaction: the server and clients do not maintain session state as the HTTP request contains all the information about the resource and the message.

2.1 IoT and WoT architectures and standards

Many organizations and consortiums are seeking to define architectures and standards to bring interoperability to the application layer of the IoT for a broad range of applications and industries. They aim to describe, among other scenarios, how to access IoT devices from Web browsers, how to bridge physical objects to the Web, or how to control distributed devices as they are discovered.

Some of the proposals focused on using Web technologies to counter IoT interoperability issues are the following. IEEE proposed a Standard for an Architectural Framework for the Internet of Things [18]; ISO/IEC released a reference architecture for the IoT [19]; ITU-T proposed an architecture where WoT brokers bridge the Web and the physical objects, using HTTP and REST services [20]; OCF developed an open source implementation and a certification program allowing heterogeneous devices to communicate [21]; OGC released a suite of standards to create Web-based interoperable and scalable networks of heterogeneous systems [22]; oneM2M develop architectures to provide interoperability in M2M and IoT solutions [23]; the OpenFog Consortium released an open reference architecture for fog computing [24]; and the W3C proposed recommendations and an architecture for the Web of Things [14].

2.2 W3C WoT

In this work we follow the recommendations of the W3C to overcome interoperability issues in IAQM platforms. In 2016, the W3C launched the Web of Things Working Group (WoT-WG) to propose recommendations and develop standards for the Web of Things. The four main objectives of this group are: (i) counter the fragmentation of the IoT; (ii) reduce the cost of development of IoT solutions; (iii) lessen the risks to both investors and customers; and (iv) foster exponential growth in the market for IoT devices and services. These recommendations are based on scripting languages like JavaScript, data encodings like JSON, and protocols like HTTP and WebSockets, defining also multiple architectures for which such a set of recommendations will be suitable. The W3C Web of Things is devised as the application layer of the IoT, interconnecting existing IoT platforms and complementing available standards.

Following these objectives, the WoT-WG identified four technological building blocks as the key to realizing the WoT [14]: (i) Thing, the abstraction of a physical or virtual entity represented in IoT applications; (ii) Thing Description (TD), the structured data that augments a Thing providing metadata about itself, its interactions, data model, communication and security; (iii) binding templates, the collection of communication metadata that explains how to interact with different IoT platforms; and (iv) scripting API, an optional building block that eases the development of IoT applications by providing a runtime system for IoT applications similar to a Web browser.

3 INDOOR AIR QUALITY MONITORING USING WIRELESS SENSOR NETWORKS AND IOT

Air quality monitoring is a common IoT application. In Postolache et al. [25], a sensor network for indoor and outdoor air quality monitoring using hardwired and wireless air quality sensors is proposed. This system implements a neural network to provide temperature and humidity compensated gas concentration values. The air quality of different locations can be monitored and published on the Web. The sensors express physical magnitudes through voltage levels that are sent to a network controller and Web server for TCP/IP communication to a PC or Web publishing.

A wireless sensor network based on Libelium nodes is used in Bhattacharya et al. [26] to compute an Air Quality Index (AQI). Each node is composed of a gas sensor board and the Waspmote processing board. The AQI computed is published in a Context Aware Framework to command Heating, Ventilating and Air Conditioning (HVAC) systems. This system is able to control HVAC devices based on building occupancy estimated from CO_2 measurements, and is also able to raise an alarm when the AQI is higher than a given threshold.

A wireless sensor network was also used in Yu et al. [27] to monitor temperature, relative humidity and CO_2. This system addressed the problem of firmware update through special purpose messages in the communication protocol among servers, gateways and sensor nodes. Later, the system was evolved to incorporate analysis and prediction capabilities [28].

An air quality monitoring wireless sensor network based on the Arduino open-source development platform is proposed in Abraham and Li [29]. This system uses XBee modules to provide mesh capabilities based on ZigBee specification, and low-cost micro gas sensors able of measuring six air quality parameters. An estimation method for sensor calibration and measurement conversion was also proposed for this system.

A real-time IAQM system using also wireless sensor networks with the ability to measure the concentrations of six gases in addition to temperature and humidity is proposed in Benammar et al. [30]. This solution emphasizes the use of a gateway in processing collected air quality data and its reliable dissemination to end-users through a Web server. The system uses the Raspberry Pi 2 model B single-board computer for the gateway, the open-source IoT platform Emoncms for the Web server, and Libelium sensor nodes (composed of calibrated sensors, the Gas Pro Sensor Board interface and the Waspmote processing board).

IAQ solutions can also monitor working environments, as in Marques and Pitarma [31], where a real-time system collects an IAQ index, temperature, relative humidity and barometric pressure in a laboratory. This system integrates a Web server and a mobile application to provide historical readings, analysis and push notifications to alert users in a timely manner.

4 INDOOR AIR QUALITY MONITORING PLATFORM USING WOT

All the wireless monitoring systems reviewed above can provide air quality measurements of indoor environments. However, these solutions are designed as standalone systems with restricted and/or proprietary communication protocols creating silos in the IoT. Even though all these systems communicate through wireless processing boards, gateways and Web servers, they cannot to communicate with each other since they do not speak the same language. The sensors of a given system are only able to communicate with the network hardware of the same system.

In this work we follow the W3C recommendations to design an IAQM platform based on current Web standards to overcome interoperability issues. This platform uses low-cost nodes abstracted as Web Things, one of the building blocks of the WoT architecture. Any sensor of this platform can be accessible worldwide from any standard Web browser using several URIs. For instance, requesting a temperature reading from a given sensor would require accessing http://WoTthing/temperature, given that WoTthing is defined in some DNS server and links to the root directory of the Web server of the sensor.

The IAQM platform proposed in this work implements the WoT architecture shown in Fig. 1. This architecture, which is a partial implementation of the W3C WoT recommendation [14], is built based on four blocks: Thing, Thing Description, WoT scripting API, and WoT protocol bindings. In addition, a system API provides access to various sensors. The software components of the proposed platform are developed in MicroPython [32], a software implementation of the Python 3 programming language optimized to run on microcontrollers. Therefore, the processing board for this platform must provide MicroPython support.

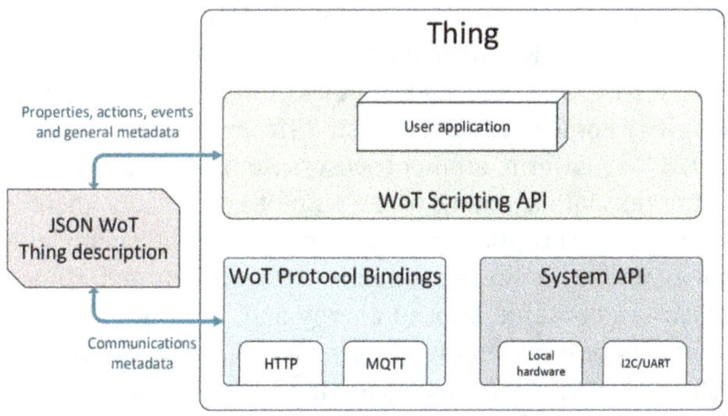

Figure 1: WoT architecture concept.

The scripting API in the proposed IAQM platform exposes Things using two different protocols: HTTP or MQTT.

The exposed description is formatted as required in the W3C WoT recommendations. A Thing can have multiple resources, classified in three groups:

- Properties: A property represents an internal state of a Thing as a value that can be accessed, and sometimes modified, through the corresponding URI.
- Actions: An action is a function that the Thing is capable of performing. Actions can change the internal state of the Thing, manipulate input data or even actuate in the physical world.
- Events: An event is a signal triggered by a change in an internal state or a physical interaction with the Thing. Values that trigger these events may not be exposed as properties.

The interactions are listed and described in the Thing Description, so the larger the number of resources, the larger the JSON file. The WoT protocol binding layer in the proposed IAQM platform manages the handlers and adapters necessary to communicate using the HTTP and MQTT protocols. The HTTP and the MQTT protocol bindings are implemented using the PicoWeb [33] and the AsyncMQTT [34] libraries, respectively. The main benefit of this platform is asynchrony, as it is built upon the uasyncio library [35], a port of the Python asyncio, which enables Micropython to run different processes in parallel. This allows monitoring different data sources at the same time and maintaining the MQTT or HTTP server online without using threading and avoiding all the complications it implies.

The core of each Thing in the IAQM platform is the processing board. The processing board must provide support for the Thing to communicate with a wide variety of sensors. In addition, it must provide wireless connectivity. The processing board must be selected taking into account the requirements imposed by the application software and firmware that will be deployed in the system.

The Things in the proposed IAQM platform can be run using either of two selected processing boards. Firstly, the DOIT ESP32 DevKit V1, which is designed for mobile, wearable electronics and IoT applications. This processing board features an ESP32-WROOM-32 MCU with two CPU cores (low-power Xtensa 32-bit LX6) that can be individually controlled, CPU clock frequency adjustable from 80 MHz to 240 MHz, 520 KB of on-chip SRAM memory, and 4 MB of flash memory. It also provides Wi-Fi, Bluetooth and BLE wireless interfaces. Secondly, the Pycom GPy processing board can also run in the platform. This processing board uses a dual core ESP32 microcontroller, provides more memory (520 KB of on-chip SRAM memory, 4 MB SRAM memory, and 8 MB of flash memory), as well as Wi-Fi, BLE and LTE wireless interfaces. Both processing boards run native Micropython, with some hardware related differences.

In the proposed IAQM platform, ambient sensors (temperature, humidity, gas) are chosen taking power consumption into account. In the case of gas sensors, there are two measuring techniques: (i) heating an area of the sensor and measuring the chemical reaction in the air; and (ii) measuring particles in the air directly through infrared (IR) sensors. Resistive heating-based gas sensors consume a lot of energy and, thus, greatly reduce the lifetime of battery-powered sensor nodes. In contrast, IR sensors consume less energy since they do not have to heat the sensor. The sensors selected are the MH-Z16 infrared sensor, for measuring concentrations of CO_2, and the fully calibrated SHT25 sensor, for measuring temperature and relative humidity.

The devices in the IAQM platform implement Wi-Fi as the main connection technology. Although it is quite a high consumption communication technology, the latest implementations provide a reasonably low consumption profile (about 100 milliamps while connected) and tools to reduce connection time, as well as deep sleep modes. Another advantage of using Wi-Fi is that it is widely implemented and available, making it simpler to deploy a network of IAQM devices, not needing to implement a custom network and allowing the re-use of existing network infrastructure.

The sensor network will be able to access the internet if the Wi-Fi network has internet connection. A gateway is not required (but it is recommended for security reasons) to access the exposed resources as the IAQM things are designed as standalone servers. The network follows a star topology as seen in Fig. 2.

Air quality monitoring platforms manage data acquired by sensors and provide information that can be analysed, for instance, to optimize the operation of an HVAC system. Usually, such analysis involves processing historical and real-time data. Processing historical data, or data at rest, is a time-consuming task that can be done in batch mode and there is no need for "always on" infrastructure. On the other hand, processing real-time data, or data in motion, usually requires a stream or real-time method running on a low latency infrastructure. Predictive analytics can be used to make decisions supported on the knowledge inferred by the analysis, such as automatically adapt and control the HVAC system. Based on the WoT model used in this platform, Things can consume each other using WoT methods and automatically make decisions based on that data.

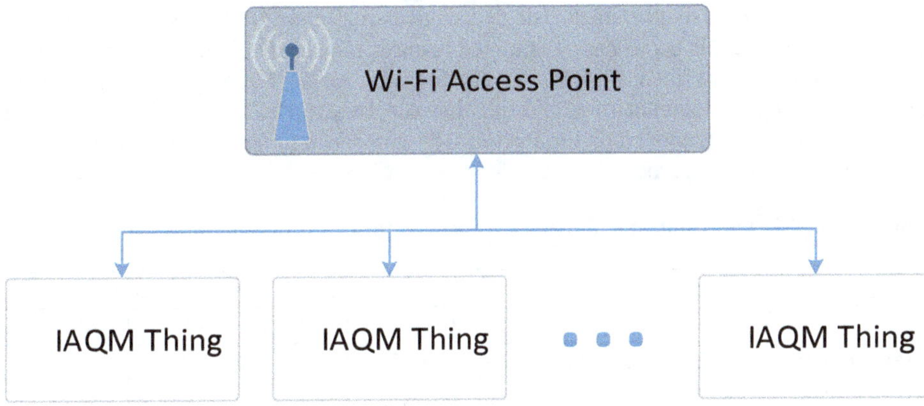

Figure 2: IAQM network topology.

5 EXPERIMENTAL RESULTS

In this section, the proposed WoT architecture and the IAQM platform are evaluated. Systems with the two previously mentioned boards (DOIT ESP32 DevKit V1 and Pycom GPy) will be used. The experiments will be driven using a Lenovo Thinkpad W550s laptop equipped with an Intel Core i7 and 12 GB RAM. A D-Link DIR-655 2.4 GHz Wi-Fi access point is also used to communicate the devices with the laptop.

Two types of experiments have been designed. Firstly, performance tests, in which the capacities of both the hardware and the software components are tested to determine the responsiveness and stability of the platform. Secondly, measurement tests, to check the device while measuring real ambient variables.

5.1 Performance test

The evaluation of the Web of Things is based on load and concurrency tests. Each experiment was run 1000 times. In the HTTP implementation, HTTP REST GET requests are used to access the properties of the Web Thing. In order to make those requests, the ApacheBench tool [36] was used to test the concurrency and load tests. In MQTT, each request consists of a message to a topic and a reply in a response topic which contains the requested property value. MQTT also requires an MQTT broker which is deployed within the same laptop used for the tests.

The load experiment consists of a large number of requests: 1000 sequential requests, with an increasing number of properties exposed. This test was performed for both HTTP and MQTT, although, the concurrency tests are only made for the HTTP implementation, as MQTT is a queuing protocol and the requests are served sequentially, independently of how they are generated.

The concurrency tests create multiple requests at the same time, evaluating how the Thing responses. The Thing holds, for any number of concurrent requests, a constant number of properties exposed in both devices: 45. This value is selected according to DOIT ESP32 board, leaving a margin at its upper limit, since it is the less powerful of the two.

Table 1 shows a brief comparison of the maximum capacities of both processing boards. As can be seen, the Pycom GPy is more powerful than the DOIT ESP32 because it has more RAM, can expose more properties and hold bigger Thing Descriptions, reporting also proportionally better response times. The experiments run in the DOIT ESP32 board return poor results because this board has constrained features. The available RAM memory to run these experiments (less than 500KB) is very low due to the Micropython environment, therefore limiting the amount of properties that can be exposed by the Web Things. In contrast, Web Things deployed on the Pycom GPy, that has 8 times more RAM, are able to expose thousands of properties.

Table 1: IAQM platform load and concurrency tests for two different processing boards.

TESTS	DOIT ESP32		Pycom GPy	
	HTTP	MQTT	HTTP	MQTT
Maximum number of exposed resources	52	15	1,000	2,000
Mean property response time with max. number of resources exposed	334 ms	361 ms	505 ms	388 ms
Mean Thing Description response time with max. number of resources exposed	384 ms	355 ms	836 ms	44,474 ms
Maximum concurrent requests	80	N/A	100	N/A

Fig. 3 shows the average response time exposing the resources through HTTP, using the two different types of processing boards. As can be seen, the response time for properties, which is the most common feature to be used, is always under 500 ms. Although it is a bit high, it depends greatly on the quality of the Wi-Fi access point. A ping to the devices using the same network results in 4 ms on average.

Fig. 4 shows the average response time exposing the resources through MQTT, using both types of processing boards. Using this protocol, the average response time is higher because two topics must be used and the MQTT protocol is not designed to transport large data files, as it is required, for example, to retrieve the TD. It also shows an exponential growth of the TD, because of the amount of resources exposed, that causes the response time to increase gradually as the TD grows in size.

Fig. 5 shows how the devices handle the concurrency of requests. There are two aspects to take in account when analysing these results: (i) Micropython, as well as native Python, is executed in a single thread; therefore, the requests have to be processed one by one in any case; however; (ii) the accumulation of requests creates a different behaviour because there is no time lost between them.

5.2 Measurement tests

To evaluate the stability and reliability of the system in long term operations, a custom software that exposes three resources, each one of them accessible through HTTP GET requests, was developed. The software makes use of the Micropython WoT runtime implementation to expose the properties in a microcontroller (DOIT ESP32 in this case), and another software component, in a remote computer, requests a measurement every 10 mins, and stores it in a Sqlite Data Base.

Fig. 6 shows a six-day period of data gathered with this system. The data loss during this period is lower than to 0.01% and is caused mostly by Wi-Fi disconnections.

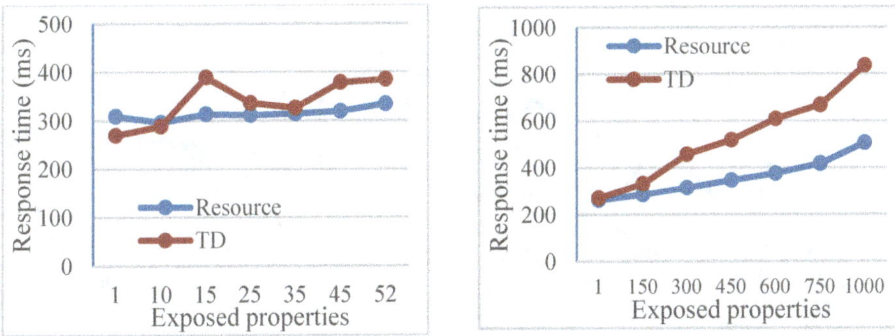

(a) (b)

Figure 3: Average response time with HTTP using (a) DOIT ESP32; and (b) PyCom GPy.

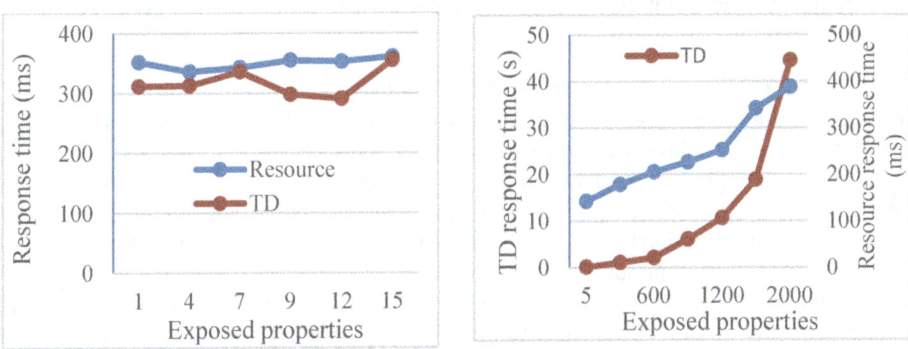

(a) (b)

Figure 4: Average response time with MQTT using (a) DOIT ESP32; and (b) PyCom GPy.

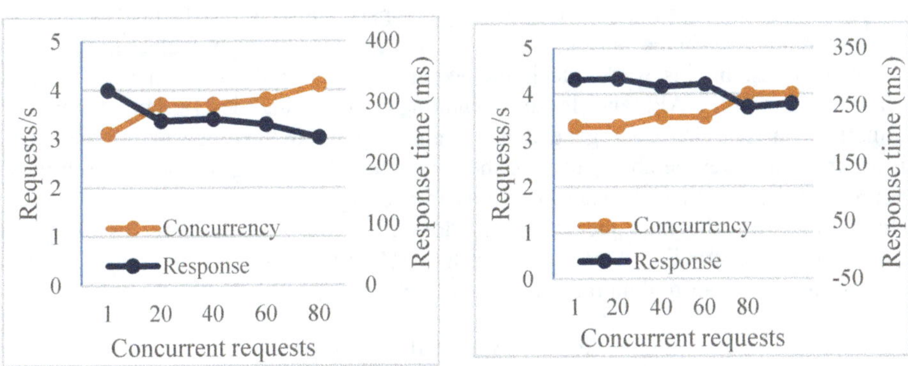

(a) (b)

Figure 5: Average number of requests processed and response time with HTTP running on (a) DOIT ESP32; and (b) PyCom GPy.

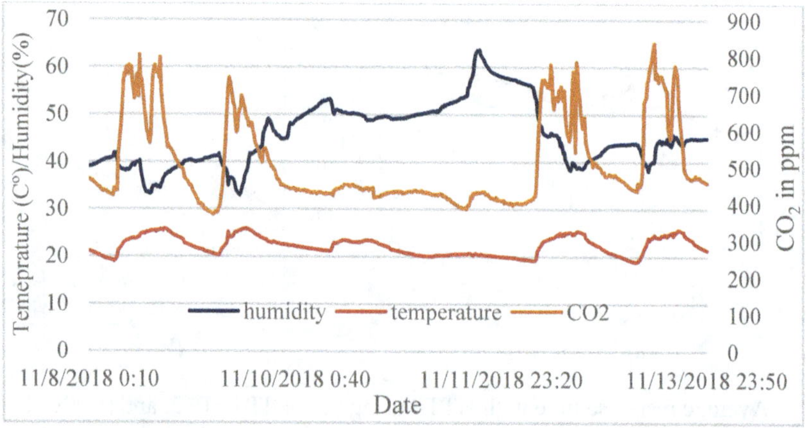

Figure 6: CO_2, temperature and humidity acquired with MH-Z16 and SHT25 sensors.

6 CONCLUSIONS

Air quality monitoring is a key instrument to determine air pollution issues. It is also one of the fields where IoT plays a great role. However, most of the IoT devices and services developed in this field depend on particular platforms or technologies, some of them proprietary, making it difficult to develop a widely accessible application. Using Web standards, the fragmentation between different sensors has been countered, making them able to be accessed through common Web protocols.

In this work, we deal with indoor air quality monitoring and measuring by following the recommendations of the W3C WoT Working Group to design, develop and deploy an ambient sensor network which can be accessed by any standard web browser. The implementation of two protocols, HTTP and MQTT, enables the solution to be more flexible for the desired scenario. The use of WoT standards abstracts access to the useful data from the acquisition process, making it easier to collect, store and process it.

This work is built on our previous work [5], where the W3C WoT standard proposal implementation in an indoor air quality sensor was in an early stage, including few of the required features. In this work, the implementation added and enhanced more building blocks: the Scripting API and Protocol bindings; the Thing description is generated automatically and has more autogenerated metadata required in the proposal. In addition, the performance and interoperability tests of the air quality monitoring sensor were run in two different boards, as well as the interoperability measuring tests.

In future work, the implementation of additional security and Thing discovery will increase the interoperability while reducing the configuration process of Things, fostering the determination of air pollution issues in indoor environments.

ACKNOWLEDGEMENTS

This work has been funded by the European Union's Horizon 2020 Research and Innovation Programme under grant agreement No. 768921 and by the University of Oviedo under project 2018/00061/012.

REFERENCES

[1] Lin, J., Yu, W., Zhang, N., Yang, X., Zhang, H. & Zhao, W., A survey on Internet of Things: Architecture, enabling technologies, security and privacy, and applications. *IEEE Internet of Things Journal*, **4**(5), pp. 1125–1142, 2017.

[2] Guinard, D., Trifa, V., Mattern, F. & Wilde, E., From the Internet of Things to the Web of Things: Resource-oriented architecture and best practices. *Architecting the Internet of Things*, eds D. Uckelmann, M. Harrison & F. Michahelles, Springer: Berlin and Heidelberg, pp. 97–129, 2011.

[3] Zeng, D., Guo, S. & Cheng, Z., The Web of Things: A survey (invited paper). *JCM*, **6**(6), pp. 424–438, 2011.

[4] National Emission Ceilings Directive, European Environment Agency. www.eea.europa.eu/themes/air/national-emission-ceilings/national-emission-ceilings-directive. Accessed on: 25 May 2019.

[5] Ibaseta, D., Molleda, J., Díez, F. & Granda, J.C., Indoor air quality monitoring sensor for the Web of Things. *Proceedings*, **2**(23), p. 1466, 2018.

[6] Agranat, I.D., Engineering Web technologies for embedded applications. *IEEE Internet Computing*, **2**(3), pp. 40–45, 1998.

[7] Kindberg, T. et al., People, places, things: Web presence for the real world. *Proceedings of the Third IEEE Workshop on Mobile Computing Systems and Applications*, pp. 19–28, 2000.

[8] Atzori, L., Iera, A. & Morabito, G., The Internet of Things: A survey. *Computer Networks*, **54**(15), pp. 2787–2805, 2010.

[9] Guinard, D., Trifa, V., Pham, T. & Liechti, O., Towards physical mashups in the Web of Things. *2009 Sixth International Conference on Networked Sensing Systems (INSS)*, pp. 1–4, 2009.

[10] Akribopoulos, O., Chatzigiannakis, I., Koninis, C. & Theodoridis, E., A web services-oriented architecture for integrating small programmable objects in the Web of Things. *2010 Developments in E-systems Engineering*, pp. 70–75, 2010.

[11] Ostermaier, B., Kovatsch, M. & Santini, S., Connecting things to the web using programmable low-power Wifi modules. *Proceedings of the Second International Workshop on Web of Things*, pp. 2:1–2:6, 2011.

[12] Hwang, K.I., In, J., Park, N.K. & Eom, D.-S., A design and implementation of wireless sensor gateway for efficient querying and managing through world wide web. *IEEE Transactions on Consumer Electronics*, **49**(4), pp. 1090–1097, 2003.

[13] Jin, W. & Kim, D., Development of virtual resource based IoT proxy for bridging heterogeneous web services in IoT networks. *Sensors*, **18**(6), p. 1721, 2018.

[14] Web of Things (WoT) Architecture. www.w3.org/TR/wot-architecture/. Accessed on: 25 May 2019.

[15] Tran, N.K., Sheng, Q.Z., Babar, M.A. & Yao, L., Searching the Web of Things: State of the art, challenges, and solutions. *ACM Computing Surveys*, **50**(4), pp. 55:1–55:34, 2017.

[16] Web Services Architecture. www.w3.org/TR/ws-arch/. Accessed on: 25 May 2019.

[17] *RESTful Web Services* (Book). www.oreilly.com/library/view/restful-web-services/9780596529260/. Accessed on: 25 May 2019.

[18] IEEE 2413-2019 – IEEE Approved Draft Standard for an Architectural Framework for the Internet of Things (IoT). https://standards.ieee.org/content/ieee-standards/en/standard/2413-2019.html. Accessed on: 25 May 2019.

[19] 14:00-17:00, ISO/IEC 30141:2018, ISO. www.iso.org/cms/render/live/en/sites/isoorg/contents/data/standard/06/56/65695.html. Accessed on: 25 May 2019.

[20] ITU-T Recommendation database, ITU. www.itu.int/ITU-T/recommendations/rec.aspx?rec=12647&lang=en. Accessed on: 25 May 2019.

[21] Open Connectivity Foundation (OCF), https://openconnectivity.org/. Accessed on: 25 May 2019.

[22] Domains that use and develop OGC standards OG'. www.opengeospatial.org/ogc/markets-technologies/swe. Accessed on: 25 May 2019.

[23] oneM2M – Why oneM2M. www.onem2m.org/about-onem2m/why-onem2m. Accessed on: 25 May 2019.

[24] OpenFog Consortium. www.openfogconsortium.org/. Accessed on: 25 May 2019.

[25] Postolache, O.A., Pereira, J.M.D. & Girao, P.M.B.S., Smart sensors network for air quality monitoring applications. *IEEE Transactions on Instrumentation and Measurement*, **58**(9), pp. 3253–3262, 2009.

[26] Bhattacharya, S., Sridevi, S. & Pitchiah, R., Indoor air quality monitoring using wireless sensor network. *2012 Sixth International Conference on Sensing Technology (ICST)*, pp. 422–427, 2012.

[27] Yu T.-C. et al., Wireless sensor networks for indoor air quality monitoring. *Medical Engineering & Physics*, **35**(2), pp. 231–235, 2013.

[28] Yu T.-C. & Lin, C.-C., An intelligent wireless sensing and control system to improve indoor air quality: Monitoring, prediction, and preaction. *International Journal of Distributed. Sensor Networks*, **11**(8), p. 140978, 2015.

[29] Abraham, S. & Li, X., A cost-effective wireless sensor network system for indoor air quality monitoring applications. *Procedia Computer Science*, **34**, pp. 165–171, 2014.

[30] Benammar, M., Abdaoui, A., Ahmad, S., Touati, F. & Kadri, A., A modular IoT platform for real-time indoor air quality monitoring. *Sensors*, **18**(2), pp. 581, 2018.

[31] Marques, G. & Pitarma, R., An Internet of Things-based environmental quality management system to supervise the indoor laboratory conditions. *Applied Sciences*, **9**(3), pp. 438, 2019.

[32] MicroPython – Python for microcontrollers. http://micropython.org/. Accessed on: 25 May 2019.

[33] Sokolovsky, P., Really minimal web application framework for MicroPython. https://github.com/pfalcon/picoweb. Accessed on: 25 May 2019.

[34] Hinch, P., A "resilient" asynchronous MQTT driver. https://github.com/peterhinch/micropython-mqtt. Accessed on: 25 May 2019.

[35] uasyncio, Core Python libraries ported to MicroPython. https://github.com/micropython/micropython-lib. Accessed on: 30 May 2019.

[36] ab – Apache HTTP server benchmarking tool. https://httpd.apache.org/docs/2.4/programs/ab.html. Accessed on: 30 May 2019.

INVESTIGATING OZONE AMBIENT LEVELS: CASE STUDY OF THE FAHAHEEL URBAN AREA, STATE OF KUWAIT

MASUMAH AL-QASSIMI & SULTAN MAJED AL-SALEM
Environment & Life Sciences Research Centre, Kuwait Institute for Scientific Research, Kuwait

ABSTRACT
Continuously monitored ambient concentrations of ozone (O_3), nitrogen monoxide (NO), nitrogen dioxide (NO_2), and nitrogen oxides (NOx) for the years 2004–2005 and 2014–2015 are used to understand the relationships and photochemical rections between ozone (O_3) and nitrogen oxides ($NO_X \approx NO+NO_2$) in one of Kuwait's major urban areas (Fahaheel). The objective of the study is to investigate the fate of O_3 as a secondary pollutant and to determine the chemical coupling and mixing ratio of NO_2 to O_3. The variation of total atmopsheric oxides ($O_X = O_3+NO_2$) concentrations with NO_2 is also assessed to gain an insight into the atmospheric sources of O_X. In addition, the diurnal variations of NO, NO_2, NO_X, O_3, and O_X along with variation of regional and local Ox are examined. The daytime and night-time relationships between O_3 and NO_X and the ozone weekend effect in Fahaheel area are also reported in this communication. This study creates a historical baseline for urban areas subjected to heavy industrial emissions that can significantly improve our understanding of the impacts of future changes in O_3 concentrations.
Keywords: precursors, volatile, pollution, air quality, ozone (O_3).

1 INTRODUCTION

Ozone is a secondary airborne pollutant formed by complex photochemical reactions in air containing nitrogen oxides ($NO_X \approx NO+NO_2$) and a wide range of volatile organic compounds (VOCs) under the influence of sunlight [1]. Complex nonlinear photochemistry is known to drive the relationships between O_3, NO_X and VOCs. With increasing concentrations of VOCs, O_3 formation tends to increase proportionately. On the other hand, the increase in NO_X levels results in variations in O_3 depending on the dominance of VOCs and NO_X [2]. Additionally, the response to NO_X emissions decline is highly nonlinear. Reduction in NO_2 levels is always accompanied by an increase in O_3 levels [3]. VOCs generated from various human activities are also in need of close monitoring namely from industrial related sources.

In some areas, there exists a trend towards higher levels of O_3 concentration on weekends. This is despite low weekend VOCs and NO_X emissions [4]. This phenomenon is known as the ozone weekend effect and was first reported back in the 1970s in the United States of America (USA) [5]. This effect was also assessed in various parts of the USA and the world namely the State of California and North-Eastern Iberian Peninsula [6]–[7]. Past efforts have also been reported in technical literature covering other parts of the world namely the Arabian Gulf region, Indian Peninsula, Southeast Asia and North America [4], [8]–[13]. Literature on the Middle East (ME) region's air quality status is very scarce and outdated. There is also a shortage of such in-depth analysis of O_3 and its precursors in the State of Kuwait, where primary and secondary airborne pollutants are typically associated with the oil and gas industry; unsanitary waste management and unregulated industrial activities. The aim of this study is to investigate the fate of O_3 as a secondary pollutant by evaluating the photochemical reactions in one of Kuwait's major urban areas (Fahaheel).

WIT Transactions on Ecology and the Environment, Vol 236, © 2019 WIT Press
www.witpress.com, ISSN 1743-3541 (on-line)
doi:10.2495/AIR190061

The study also reports on the daytime and night-time relationships between O_3 and NO_X; and investigates the O_3 weekend effect in Fahaheel area.

2 MATERIALS AND METHODOLOGY

2.1 Study area

The Fahaheel urban area (29°05′00″N lat. and 48°07′36″E long.) is one of five major cities in the State of Kuwait under Al-Ahmadi Governorate. The city is associated with downstream and petrochemical industries supporting the oil-based economy of Kuwait. It is characterised by a typical arid climate with harsh summers. During the summer seasons, ambient temperatures typically exceed 55°C in the months June to September. About 100,000 residents are reported to reside in the area which is also adjacent to the state's largest oil refinery (Mina Al-Ahmadi, MAA). All refineries, including MAA comprising Kuwait's three refinery belt alongside Mina Abdullah (MAB) and Shuibah (SHU), are located on the southern side of the main shopping area of Fahaheel (downtown) [14]. On the southern side of the area there are also petrochemical industries and other small private (cottage) industries. Readers are referred to Al-Salem and Khan [14] for a satellite image depicting main routes for commuting and industrial sites around Fahaheel.

2.2 Data sourcing, analysis and processing

Continuously monitored air quality data for the years 2004–2005 and 2014–2015 was acquired from the Fahaheel monitoring station (located 15 m above ground level) by the Air Pollution Monitoring Division, Environment Public Authority, Kuwait (EPA). The ambient air samples were taken from a stationary probe (Group Tek. Model, 3–5 m, stationary photolytic converter, Environment SA and Thermo Models) located at the top of the Fahaheel polyclinic. The samples were examined using various primary pollutant and secondary precursor analysers (Whatman 41, Air sample Grasbey-Anderson Ltd., 1% tolerance, weather station), all linked to EnviDAS software's central online data acquisition system. Continuous hourly averages of pollutant concentrations namely O_3 (ppb), NO (ppb), NO_2 (ppb), and NO_X (ppb) were collected in addition to the metrological parameters of ambient temperature (°C), relative humidity (RH, %), wind speed (WS, ms^{-1}) and wind direction (WD, °). O_3 is often depleted locally by atmospheric titration in high NO_X emission regions. O_X is consequently a more stable entity to study as it shows the potential for renewed production of O_3 [15]. Therefore, work on nitrogen-based pollutants included the Ox filtration procedure where points below its line have been discarded due to photo-oxidation reactions, particulate accumulation, choking or OH^- ion presence malfunctioning of the instruments. The procedure has been established previously by other authors and illustrated as follows [2], [13]–[14], [16]:

$$O_x\,[ppb] = O_3 + NO_2, \tag{1}$$

where, Ox is the total oxide concentration in the ambient atmosphere (ppb). It is well established that the reactions shown below generally dominate the interconversion of O_3, NO and NO_2 under atmospheric conditions [1]:

$$NO_2 + hv\,(\lambda < 398\,mm) \rightarrow NO + O \tag{2}$$

$$O + O_2 + M\,(N_{2,}O_2) \rightarrow O_3 + M \tag{3}$$

$$NO + O_3 \xrightarrow{K3} NO_2 + O_2 \tag{4}$$

$$NO_2 + O_2 \xrightarrow{J4} NO + O_3 \tag{5}$$

$$\frac{[NO][O_3]}{[NO_2]} = \frac{J_4}{k_3} \tag{6}$$

where, hv is the sunlight (solar) intensity; k_3 is the rate coefficient for the reaction of NO with O_3 in eqn (4); and J_4 is the rate of NO_2 photolysis reaction in eqn (5) at equilibrium [17]. The reactions in eqns (4) and (5) characterise a closed system in which components of NO_X ($NO+NO_2$) and oxidant O_X (O_3+NO_2) relate separately but leave unchanged a total mixing ratio of both NO_X and O_X [18]. Over the course of daytime hours, NO, NO_2, and O_3 are typically balanced on a time scale of a few minutes called the photo-stationary state. In this study, determining the J parameter and the NO_2 and O_3 mixing ratio assisted in investigating the photochemical reactions (photolysis) of the pollutants. It also aided in understating the fate of the pollutants in the ambient and their interaction with each other. Diurnal variation of NO, NO_2, NO_X, O_3, and O_X along with variation of seasonal regional and local Ox were also examined in the study.

3 RESULTS AND DISCUSSION

3.1 Ozone and nitrogen oxide(s) diurnal variation and patterns

The O_3, NO, NO_2 and NO_X diurnal variation plots of hourly averages for the years 2004–2005 and 2014–2015 are shown in part (Fig. 1). The daily average maximum level of O_3 reached 69 ppb during the years 2004–2005 compared to 47 ppb in 2014–2015. The diurnal cycle of O_3 concentrations shows high levels with altering morning and late-night concentrations. The lowest concentrations were found at around 5:00 to 8:00 in the early morning hours. The concentration of O_3 started to rise rapidly afterwards, coinciding with the increase in solar radiation until it peaked at around 13:00 to 16:00 during the day. Subsequently, it declined rapidly from daytime to evening (18:00) after reaching its maximum value and continued to decline gradually due to the lack of solar radiation. The same O_3 daily variability was also observed in past research [16], [19]–[20]. This could be attributed to the photochemical reaction of O_3 precursors (e.g. VOCs) with natural ambient air and NO_X long-range transport [21]. It is well established that the inversion layer, solar intensity, wind patterns and sources of emissions have a significant influence on the daily variability of any pollutant [2], [17]. As seen in previous research these factors may also added to the O_3 background levels in the study area [14].

The diurnal variation plots of NO_2 established in Figure 1 show two daily peaks during morning and evening periods. The peak values of NO and NO_2 occurred at the same time during morning hours (e.g. 6:00 to 8:00). This coincided with the school rush hour vehicle emissions. The evening peaks occurred between 19:00 to 21:00 in 2004–2005 and in 2014–2015. The evening peaks corresponded to typical urban activities of Fahaheel area (e.g. traffic, shops and markets opening, restaurants, etc). NO oxidised to NO_2 as morning progresses until it reached maximum values in evening times. NO_2 hourly average concentrations exceeded NO concentrations, showing higher level of environmental oxidation in the area. The rush hours and increased volumes in traffic during morning times and evening periods coincided with the higher NO_X levels. Another factor contributing to

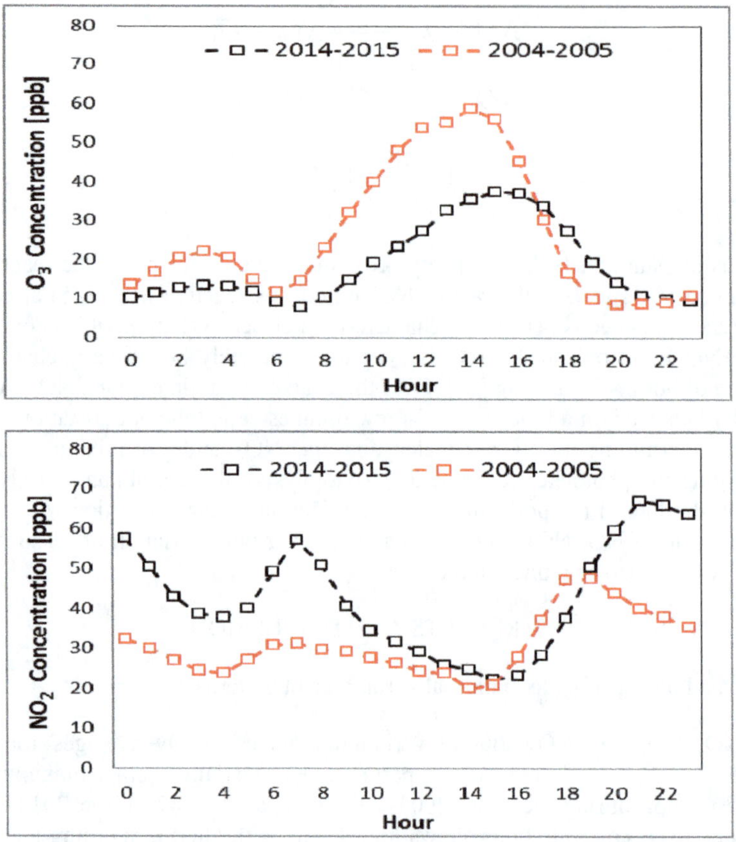

Figure 1: Hourly variations of average concentrations of O_3 (top) and NO_2 (bottom) in the Fahaheel urban area for the years 2004–2005 and 2014–2015.

this phenomenon is night-time inversion. The dilution processes of nitrogen-based chemicals in the atmosphere are hindered by a lower inversion layer [14]. This layer contains emitted pollutants such as NO and NO_2, which may have caused the hourly concentration of NO_X to increase during the night [22]. During the daytime, the lower NO_X levels are mainly due to a higher inversion layer that increases pollutant dilution. In general, the increased rate of NO_2 photolysis at midday causes NO_2 levels to drop and O_3 to rise [23].

The increase in NO and NO_2 levels is associated with a decrease in O_3 concentration [24]. NO is a primary pollutant emitted from various sources around the monitoring receptor point. As stated in eqn (4), NO is converted to NO_2 in the reaction with O_3. As a result of photolysis, NO_2 is reversely produced by stoichiometric balance back to NO during the day leading to O_3 regeneration. Furthermore, the levels of the primary pollutant NO occurred at higher concentrations in 2004–2005 compared to 2014–2015. The same was also detected for O_3 ambient concentration. In general, the lower NO emissions in 2014-2015 are related to the use of catalytic convertors in modern vehicles in addition to lower combustion sources concentration build-up around the vicinity of Fahaheel.

3.2 Chemical coupling of O_3, NO and NO_2

The mixing ratio of NO_2, O_3 and the rate of NO_2 photolysis (J_4) was determined to study the photolysis reaction in the atmosphere. It also established the fate of the pollutants in the ambient and their interaction with each other. The average variations of the value of J_4/k_3 from eqn (6), was obtained for the study duration and is shown in Fig. 2.

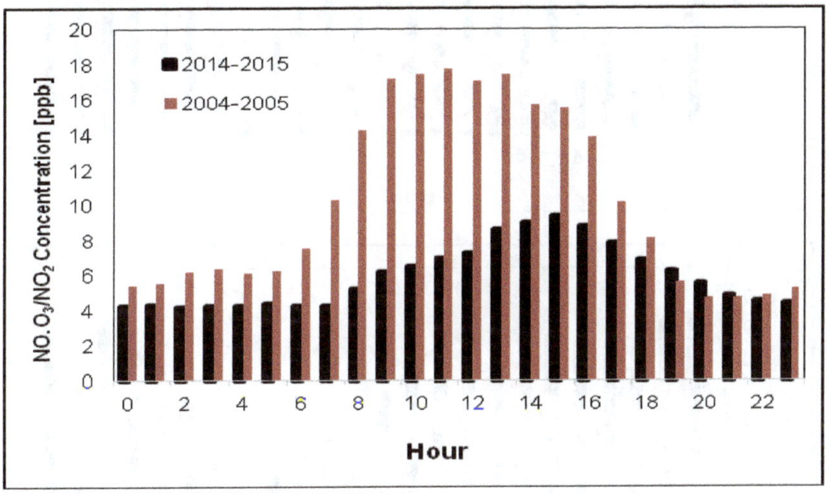

Figure 2: The average variations of J_4/k_3 value in 2004–2005 and 2014–2015.

The mean J_4/k_3 values varied between 4.71 and 17.74 ppb (2004–2005). The maximum value occurred at 15:00. In addition, the mean values of J_4/k_3 varied between 4.18 and 9.38 ppb (2014–2015) with a maximum value also reported at 15:00. Han et al. [17] reported mean J/k values ranging from 0.176 to 5.513 ppb in Tianjin, China, and a maximum value at 10:00. Their work provided an adequate description of diurnal averaged variations of NO, NO_2 and O_3, consistent with the chemical coupling dominated by the previously described reactions in eqns (4) and (5). This indicates that Fahaheel area is a photocatalytic area which is also consistent with high UV radiation sites (e.g. Kuwait).

3.3 Diurnal variation of O_X concentration

O_X levels are influenced by photochemical reactions. The variation in the O_X concentration mean values shows that the concentration of O_X in 2004–2005 was characterised by a midday peak (Fig. 3). The examined O_X diurnal variation attributes the behaviour of O_X to a similar one in the variation of O_3. During daytime, the O_X concentration reached maximum values. Similar results to the O_X pattern observed in 2004–2005 have been previously reported in many locations including Saudi Arabia and India [4], [20]. However, the diurnal variation of O_X in the years 2014–2015 did not fit the same pattern as that of 2004–2005. The NO_2/O_X concentration ratio variation is shown below. NO is converted to NO_2 through an O_3 reaction. This is crucial in establishing levels of NO_2 and O_3. The low ratio of NO_2/O_X could be attributed to the higher O_3 levels during the daytime [13]. This behavioural difference between NO_2 and O_3 may be related to the time they have to react in the atmosphere or the rate of chemical processes in the studied area.

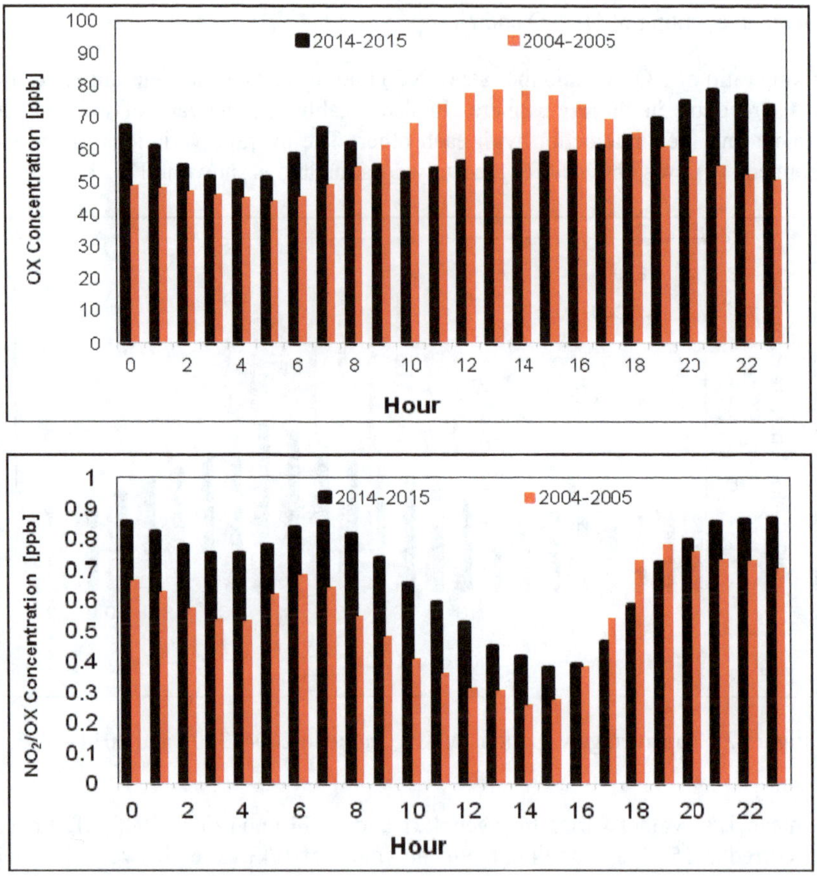

Figure 3: Variations of O_x (top) mean values and NO_2/O_x (bottom) mean value ratio in 2004–2005 and 2014–2015.

3.4 Weekday/weekend differences

The difference between weekends and weekdays of O_3 levels in the Fahaheel area is investigated in this section in order to determine the ozone weekend effect. Differences in ambient air pollutant concentrations such as NO_X and O_3 were extensively studied on weekdays and weekends in urban, suburban and rural sites [25]–[27]. The regional background contributions dominate in sites where concentrations of O_3 on weekdays and weekends are approximately equal [4]. On the other hand, differences on weekdays and weekends appear in sites dominated by O_3 local production. In general, a valuable indicator of whether O_3 originates in local photochemical production or in transportation processes is to study the weekday and weekend differences in O_3 and NO_X levels and NO-titration [4]. The ground concentration of O_3 over Fahaheel urban area depends on the photochemical production of O_3 in relation to its precursors; NO_X and VOCs. The weekly traffic patterns are believed to affect the variations of NO_X and O_3 on weekdays and weekends [4]. Vehicle emissions are the main source of NO_X emissions in the studied urban area; where traffic density on weekends is assumed to be lower than the weekdays due to official days-off of work commuters and school/college students. Weekends were on Thursdays and Fridays in Kuwait between 2004–2005, corresponding to Fridays and Saturdays in 2014–2015. Higher NO, NO_2 and NO_X concentrations and lower O_3 concentration on weekdays were observed during daytime in Fahaheel area in 2004–2005 and 2014–2015 as seen in Table 1. The detected trend towards higher levels of O_3 concentration and lower NO_X emissions on weekends supported the ozone weekend effect in Fahaheel urban area. This corresponded to the change in weekends in Kuwait, and complemented the analysis conducted in this study.

Table 1: Weekdays/weekends daytime diurnal variations of O_3, NO, NO_2, and NO_x (2004–2005) and (2014–2015).

Daytime		O_3 (ppb)	NO (ppb)	NO_2 (ppb)	NO_x (ppb)
2004–2005	Weekdays	42.29	12.64	31.10	43.83
	Weekends	47.67	10.02	27.74	37.88
	All	43.85	11.878	30.13	42.10
2014–2015	Weekdays	28.23	9.76	37.02	47.27
	Weekends	31.35	7.45	30.73	38.71
	All	29.12	9.12	35.28	44.09

4 CONCLUSION

Examining the variations in O_3 and NO_X in the urban area of Fahaheel (Kuwait) over the years (2004–2005) and (2014–2015) resulted in determining their atmospheric behaviour. O_3 levels were clearly more elevated in the summer months. They were seen to gradually increase until reaching the month of July in 2004–2005 and the month of August in 2014–2015. The patterns of hourly diurnal variation of O_3 and NO_X varied between weekdays and weekends. It was observed that O_3 reached higher overall weekdays and weekends concentrations in 2004–2005 compared to 2014–2015. In addition, higher NO, NO_2 and NO_X concentrations and lower O_3 concentration on weekdays were observed during daytime in Fahaheel area in 2004–2005 and 2014–2015. The detected trend towards higher levels of O_3 concentration and lower NO_X emissions on weekends supported the ozone weekend effect Fahaheel urban area. Furthermore, the levels of the primary pollutant NO occurred at higher concentrations in 2004–2005 compared to 2014–2015. This is related to the use of catalytic convertors in modern vehicles as well lower combustion sources concentration build-up around the vicinity of Fahaheel. The present study provided insight into ground level ozone variations in urban Fahaheel, yet the interpretation was limited by meteorological and VOCs data inadequacies. Future studies need to explore more detailed relationships between NO_X, VOCs, and O_3 in the Fahaheel area.

ACKNOWLEDGEMENTS

This study was funded by Kuwait Scientific Research Centre (KISR) in support to Project EM070G. The authors wish to express their gratitude to the Air Pollution Monitoring Division, Environment Public Authority, Kuwait (KEPA) for providing air quality data and to the Kuwait Foundation for the Advancement of Sciences (KFAS) for funding the scientific mission.

REFERENCES

[1] Adame, J.A., Notario, A., Villanueva, F. & Albaladejo, J., Application of cluster analysis to surface ozone, NO_2 and SO_2 daily patterns in an industrial area in Central-Southern pain measured with a DOAS system. *The Science of the Total Environment*, **429**, pp. 281–291, 2012.

[2] Tiwari, S., Dahiya, A. & Kumar, N., Investigation into relationships among NO, NO_2, NOx, O_3, and CO at an urban background site in Delhi, India. *Atmospheric Research*, **157**, 119–126, 2015.

[3] Mazzeo, N.A., Venegas, L.E. & Choren, H., Analysis of NO, NO_2, O_3 and NO_X concentrations measured at a green area of Buenos Aires City during wintertime. *Atmospheric Environment*, **39**(17), pp. 3055–3068, 2005.

[4] Alghamdi, M.A. et al., Temporal variations of O_3 and NOx in the urban background atmosphere of the coastal city Jeddah, Saudi Arabia. *Atmospheric Environment*, **94**, pp. 205–214, 2014.

[5] Cleveland, W.S., Graedel, T.E., Kleiner, B. & Warner, J.L., Sunday and Workday Variations in Photochemical Air Pollutants in New Jersey and New York. *Science*, **186**(4168), pp. 1037–1038, 1974.

[6] Fujita, E.M. et al., Diurnal and weekday variations in the source contributions of ozone precursors in California's South Coast Air Basin. *Journal of the Air & Waste Management Association*, **53**(7), pp. 844–863, 2003.

[7] Jimenez, P., Parra, R., Gasso, S. & Baldasano, J.M., Modeling the ozone weekend effect in very complex terrains: a case study in the North eastern Iberian Peninsula. *Atmospheric Environment*, **39**(3), pp. 429–444, 2005.

[8] Traub., M. & Lelieveld, J., Cross-tropopause transport over the eastern Mediterranean. *Journal of Geophysical Research Atmosphere*, **108**(D23), pp. 4712, 2003.

[9] Lelieveld, J. et al., Severe ozone air pollution in the Persian Gulf region. *Atmospheric Chemistry and Physics*, **9**, 1393–1406, 2009.

[10] Mazzuca, G.M. et al., Ozone Production and Its Sensitivity to NOx and VOCs: Results 1 from the DISCOVER-AQ, *Atmospheric Chemistry and Physics*, **16**, 14463–14474, 2016.

[11] Sharma, A., Sharma, S.K., Rohtash & Mandal T.K., Influence of ozone precursors and particulate matter on the variation of surface ozone at an urban site of Delhi, India. *Sustainable Environment Research*, **26**(2), pp. 76–83, 2016.

[12] Kumar, A., Singh, D., Anandam, K., Kumar, K. & Jain, V.K., Dynamic interaction of trace gases (VOCs, Ozone, and NOx) in the rural atmosphere of sub-tropical India. *Air Quality, Atmosphere & Health*, **10**(7), pp. 885–896, 2017.

[13] Pancholi, P., Kumar, A., Bikundia, D.S. & Chourasiya, S., An observation of seasonal and diurnal behavior of O3-NOx relationships and local/regional oxidant (O_X = O_3 + NO_2) levels at a semi-arid urban site of western India. *Sustainable Environment Research*, **28**(2), pp. 79–89, 2018.

[14] Al-Salem, S.M. & Khan, A.R., Comparative assessment of ambient air quality in two urban areas adjacent to petroleum downstream/upstream facilities in Kuwait. *Brazilian Journal of Chemical Engineering*, **25**(4), pp. 683–696, 2008.

[15] Song, F., Shin, J.Y., Jusino-Atresino, R. & Gao, Y., Relationships among the springtime ground-level NOx, O_3 and NO_3 in the vicinity of highways in the US East Coast. *Atmospheric Pollution Research*, **2**(3), pp. 374–83, 2011.

[16] Al-Salem, S.M. & Khan, A.R., Monitoring and modelling the trends of primary and secondary air pollution precursors: The case of the state of Kuwait, *International Journal of Chemical Engineering*, **2010**, pp. 1–12, 2010.
[17] Han, S. et al., Analysis of the Relationship between O_3, NO and NO_2 in Tianjin, China, *Aerosol and Air Quality Research*. **11**, pp. 128–139, 2011.
[18] Li, K. et al., Smog chamber study on aging of combustion soot in isoprene/SO_2/NOx system: Changes of mass, size, effective density, morphology and mixing state. *Atmospheric Research*, **184**, 139–148, 2017.
[19] Duenas, C., Fernandez, M.C., Canete, S., Carretero, J. & Liger, E., Analyses of ozone in urban and rural sites in Malaga (Spain). *Chemosphere*, **56**(6), pp. 631–639, 2004.
[20] Pires, J.C.M., Alvim-Ferraz, M.C.M. & Martins, F.G., Surface ozone behaviour at rural sites in Portugal. *Atmospheric Research*, **104–105**, pp. 164–171, 2012.
[21] Wang, Y.H., Hu, B., Tang, G.Q., Ji, D.S., Zhang, H.X. & Bai, J.H., Characteristics of ozone and its precursors in Northern China: a comparative study of three sites. *Atmospheric Research*, **132**, pp. 450–459, 2013.
[22] Yadav, R., Sahu, L.K., Jaaffrey, S.N.A. & Beig, G., Distributions of ozone and related trace gases at an urban site in western India. *Journal Atmospheric Chemistry*, **71**(2), pp. 125–144, 2014.
[23] Ismail, M., Abdullah, S., Yuen, F.-S., & Ghazali, N.A., A ten-year investigation on ozone and it precursors at Kemaman, Terengganu, Malaysia. *Environment Asia*, **9**(1), pp. 1–8, 2016
[24] Ramli, N.A., Ghazali, N.A. & Yahaya, A.S., Diurnal fluctuations of ozone concentrations and its precursors and prediction of ozone using multiple linear regressions. *Malaysian Journal of Environmental Management*, **11**(2), pp. 57–69, 2010.
[25] Shutters, S.T. & Balling, R.C., Weekly periodicity of environmental variables in Phoenix. Arizona. *Atmospheric Environment*, **40**(2), pp. 304–310, 2006.
[26] Stephens, S. et al., Weekly patterns of Mexico City's surface concentrations of CO, NOx, PM10 and O3 during 1986–2007. *Atmospheric Chemistry and Physics*, **8**, pp. 5313–5325, 2008.
[27] Semple, D.R., Song, F. & Gao, Y., Seasonal characteristics of ambient nitrogen oxides and ground–level ozone in metropolitan North eastern New Jersey, *Atmospheric Pollution Research*, **3**(2), 247–257, 2012.

INFLUENCE OF SPATIAL RESOLUTION IN MODELING THE DISPERSION OF VOLCANIC ASH IN ECUADOR

RENÉ PARRA
Instituto de Simulación Computacional, Universidad San Francisco de Quito, Ecuador

ABSTRACT

Volcanic ash produces air pollution and other impacts. Regions potentially affected require information about the possible ash dispersion trajectories and affected zones by ash fallout. In the last 19 years, five volcanoes in Ecuador have produced moderate to large explosive eruptions. Information about the volcanic ash dispersion in forecasting time is a priority in Ecuador. Eulerian models can provide results with high spatial and temporal resolutions. However, they need to solve huge amounts of equations, demanding plenty of computational resources when using high spatial resolutions. It is necessary to define a pragmatic spatial resolution, suitable to compute volcanic ash dispersion, both in forecasting time and with enough accuracy. For this purpose, we simulated the meteorology over Ecuador, using the Weather Research and Forecasting (WRF3.7.1) model with spatial resolutions of 36 km, 12 km, 4 km, and 1 km. Meteorological outputs were used into the FALL3DV7.1.4 model to simulate ash dispersion from four eruptions (Tungurahua volcano: 16 December 2012, 14 July 2013 and 1 February 2014; Cotopaxi volcano: 14 August 2015). We compared modeled ash fallout results with records from ash meters around these volcanoes. The coarser resolutions of 36 km and 12 km, provided low modeling performances, with values of the linear correlation coefficient (R^2) between 0.00 to 0.79; and 0.28 to 0.46 respectively. Modeling with 4 km improved the performance, reaching values of R^2 between 0.56 to 0.98. The resolution of 1 km got the best performance, with R^2 between 0.70 to 1.00. Nevertheless, when working with 1 km, it demanded about 20 computational times in comparison with 4 km. These results suggest that for the Ecuadorian case, the resolution of 4 km is a good compromise for generating volcanic ash dispersion in forecasting time, with proper modeling performance.

Keywords: WRF, FALL3D, forecasting time, Cotopaxi, Tungurahua.

1 INTRODUCTION

Volcanic ash produces air pollution and other environmental impacts [1], [2]. Regions potentially affected require information about the possible ash dispersion trajectories and affected zones by ash fallout.

Prevailing winds at different altitudes disperse volcanic ash. So, the meteorological component of Atmospheric Transport Models (ATMs) (e.g. [3], [4]), provides key information when modeling the dispersion of volcanic ash. ATMs also require volcanological inputs, the Eruption Source Parameters (ESP) [5], which include information about particle grain size distribution, and the characterization of the source term (i.e., plume height, eruption duration, mass eruption rate, and vertical distribution of mass along the eruptive column).

In the last 19 years, five volcanoes in Ecuador produced moderate to large explosive eruptions with significant ash plumes (Pichincha 1999–2001, Sangay permanent, Tungurahua 1999 to present, El Reventador 2002 to present, Cotopaxi 2015). Hence, information in forecasting time about the volcanic ash dispersion is a priority in Ecuador.

Eulerian ATMs describe the behavior of the atmosphere into domains composed of three-dimensional arrays of fixed grid cells. For mesoscale studies, ATMs use grid cells with resolutions of few km, providing results with high spatial and temporal resolutions. However, they solve huge amounts of equations, demanding plenty of computational resources. The need for high-resolution simulations (up to 1 km) comes from the better representation of small scale processes of dispersion processes [6]. It is necessary to define a pragmatic spatial

WIT Transactions on Ecology and the Environment, Vol 236, © 2019 WIT Press
www.witpress.com, ISSN 1743-3541 (on-line)
doi:10.2495/AIR190071

resolution, suitable to compute volcanic ash dispersion, both in forecasting time and with proper accuracy.

Tungurahua (lon. 78.446°W, lat. 1.468°S; 5023 m asl), located in the Ecuadorian Andes (Fig. 1), began its activity in October 1999. Since then, ash fallout was the most frequent volcanic hazard [7], [8]. Based on field and modeling studies, ESP were proposed for forecasting ash dispersion due to Vulcanian eruptions at this volcano [4].

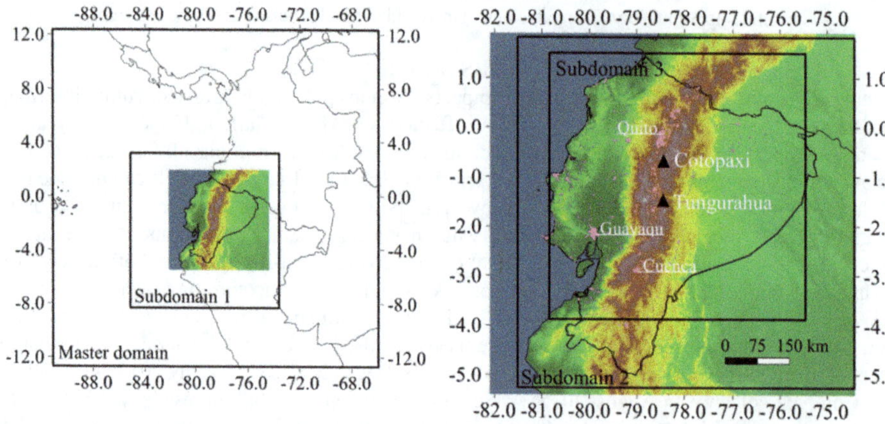

Figure 1: Location of Cotopaxi and Tungurahua volcanoes. Domains for modeling: Master domain (80 × 80 cells, 36 × 36 km), subdomain 1 (109 × 109 cells, 12 × 12 km), subdomain 2 (199 × 199 cells, 4 × 4 km), subdomain 3 (601 x 601 cells, 1 x 1 km).

On 14 August 2015, Cotopaxi (lon. 78.436°W, lat. 0.677°S; 5897 m asl, Fig. 1) awoke after more than 70 years [9], [10]. Cotopaxi volcano is about 50 km south of Quito (capital of Ecuador, Fig. 1). During 2015 its eruption activity continued unevenly until the beginning of December [11]. In the same way, based on field and modeling studies, preliminary ESP were established for modeling ash dispersion at this volcano, for similar eruptions as the one happened on 14 August 2015 [12].

In this study we explore the influence of spatial resolution in modeling the dispersion of volcanic ash in Ecuador, to define a pragmatic one, for computing this information, both in forecasting time and with proper accuracy.

2 METHOD

We modeled the ash dispersion from four historical eruptions, three at Tungurahua (16 December 2012, 14 July 2013 and 1 February 2014) and one at Cotopaxi (14 August 2015). For these days, firstly we simulated the meteorology using the Eulerian Weather Research and Forecasting (WRF3.7.1) model [13], with a master domain of 80 × 80 cells (each of 36 × 36 km) and three nested subdomains (Fig. 1), with spatial resolutions of 12 km (109 x 109 cells), 4 km (199 x 199 cells), and 1 km (601 x 601 km) respectively. In all cases, we used 35 vertical levels (model top pressure at 50 hPa). Initial and boundary conditions came from the GFS forecasts dataset [14].

Meteorological simulations were done using the following physics parameters: WSM5 for microphysics, RRTMG for radiation, Kain–Fritsch for cumulus and YSU planetary boundary layer (PBL) scheme. Previous studies suggested the YSU as one of the PBL schemes with good performance for modeling purposes in Ecuador [12], [15].

The meteorological outputs were used into the Eulerian FALL3DV7.1.4 [16] model to simulate the ash dispersion. Table 1 shows the ESP, emissions, timescales, and options used for modeling these four eruptions [4], [12]. We selected into FALL3DV7.1.4, the model proposed by Mastin et al. [17], to estimate the emissions of volcanic ash.

Modeled ash clouds were qualitative compared with ash clouds detected by the Washington Volcanic Ash Advisory Center (Washington VAAC) at different flight levels (FL) [18]. FL (expressed in 100 feet units) is the height above mean sea level when the pressure at sea level is 1013.2 mb (e.g., FL300 = 30,000 feet, ≈ 9.1 km).

We compared the modeled ash fallout results with records from ash meters around these volcanoes (4 stations for Tungurahua, 14 stations for Cotopaxi), which are operated by the Instituto Geofísico de la Escuela Politécnica Nacional [4], [11] (Fig. 2).

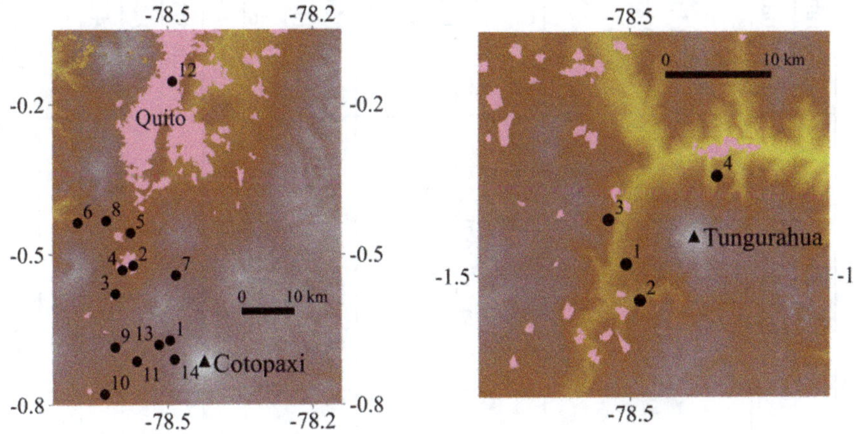

Figure 2: Location of stations and their nomenclature. Cotopaxi volcano (left): 1. Mariscal (Mar), 2. Machachi 1 (Ma1), 3. Jambelí (Jam), 4. Machachi 2 (Ma2), 5. Obelisco (Obe), 6. Aloag (Alo), 7. Santa Ana (San), 8. Gualilagua (Gua), 9. Tiopullo (Tio), 10. Progreso (Pro), 11. Entrada Sur (Ent), 12. Instituto Geofísico (Ins), 13. Agualongo (Agu), 14. BNAS. Tungurahua volcano (right): 1. Choglontus (Cho), 2. Palictahua (Pal), 3. Pillate (Pil), 4. Runtun (Run).

3 RESULTS

Although with differences, for all the spatial resolutions, the direction of modeled ash clouds was consistent with the course of the detected clouds. As an example, Figs 3 and 4 show the detected the corresponding computed ash clouds, for the eruptions at Tungurahua on 1 February 2014 and Cotopaxi on 14 August 2015 respectively.

For the Tungurahua eruption on 1 February 2014, at 23h15 LT (Local time), the detected clouds at FL250, FL400, and FL350 moved SE, S, and SW respectively, while the modeled moved S and SW.

For the Cotopaxi eruption on 14 August 2015, at 18h15 LT, the detected ash clouds at FL460–FL500 and FL360 moved E and N respectively, although the modeled at FL350, FL400, and FL450 moved NE. Also, modeling showed an ash cloud was moving NW at FL200, which was not detected by the Washington VAAC.

Table 1: ESP for modeling volcanic ash dispersion in Ecuador [4], [11].

Volcano	Tungurahua	Tungurahua	Tungurahua	Cotopaxi
Date	16 December 2012	14 July 2013	1 February 2014	14 August 2015
Number of ejections, height above the vent and duration of emissions	1. 2.00 km, 5 min 2. 7.00 km, 5 min	1. 8.80 km, 3 min 2. 4.85 km, 22 min	1. 3.00 km, 9 min 2. 8.00 km, 12 min 3. 8.00 km, 12 min	1. 8.00 km. 0.42 min 2. 8.00 km, 0.42 min 3. 8.00 km, 0.42 min 4. 9.30 km, 0.42 min 5. 9.30 km, 0.42 min 6. 9.30 km, 0.42 min
Time of beginning of each ejection	10h20 UT (05h20 LT) 11h02 UT (06h02 LT)	12h00 UT (07h00 LT) 12h03 UT (07h03 LT)	22h13 UT (17h13 LT) 22h22 UT (17h22 LT) 22h39 UT (17h39 LT)	09h02 UT (04h02 LT) 09h07 UT (04h07 LT) 15h25 UT (10h25 LT) 18h45 UT (13h45 LT) 19h28 UT (14h28 LT) 21h27 UT (16h27 LT)
Estimation of the mass flow rate	Using Mastin et al. model [17]	Using Mastin et al. model [17]	Using Mastin et al. model [17]	Using Mastin et al. model [17]
Estimated volcanic ash emitted	1.3E8 kg	3.5E8 kg	6.1E8kg	1.6E8kg
Source type	Suzuki (A=5, L=10)	Suzuki (A=5, L=10)	Suzuki (A=5, L=10)	Suzuki (A=5, L=10)
Granulometry distribution	Bigaussian	Bigaussian	Bigaussian	Bigaussian
Ø means	0.5, 4.0	0.5, 4.0	0.5, 4.0	1.85, 5.15
Ø range	-1.0 to 4.5	-1.0 to 4.5	-1.0 to 4.5	-0.35 to 7.35
Terminal velocity model	Ganser	Ganser	Ganser	Ganser
Horizontal turbulence model	CMAQ	CMAQ	CMAQ	CMAQ

UT: Universal time. LT: Local time.

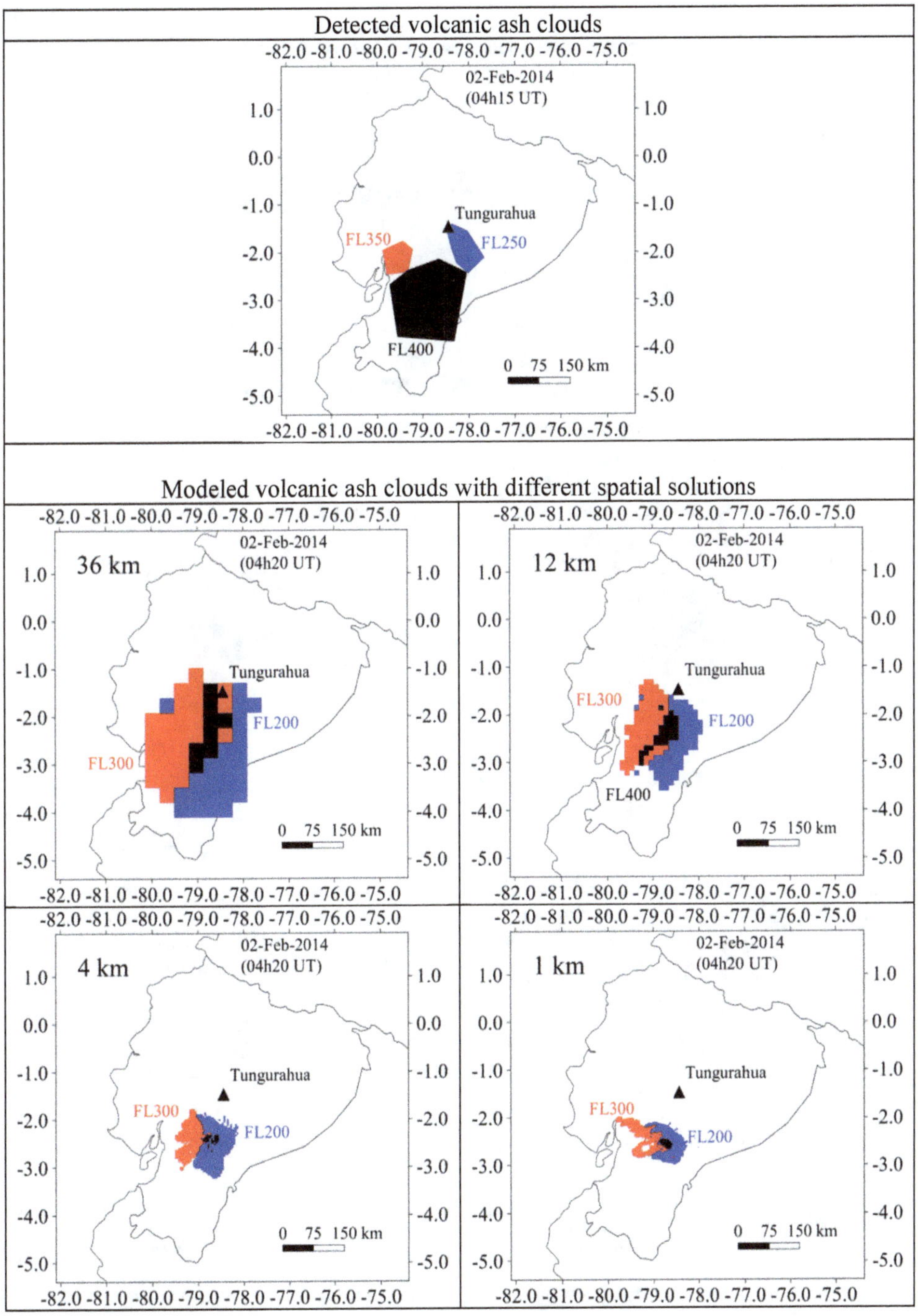

Figure 3: Tungurahua volcano. Eruption on 1 February 2014. Detected [18] versus modeled volcanic ash clouds with different spatial resolutions.

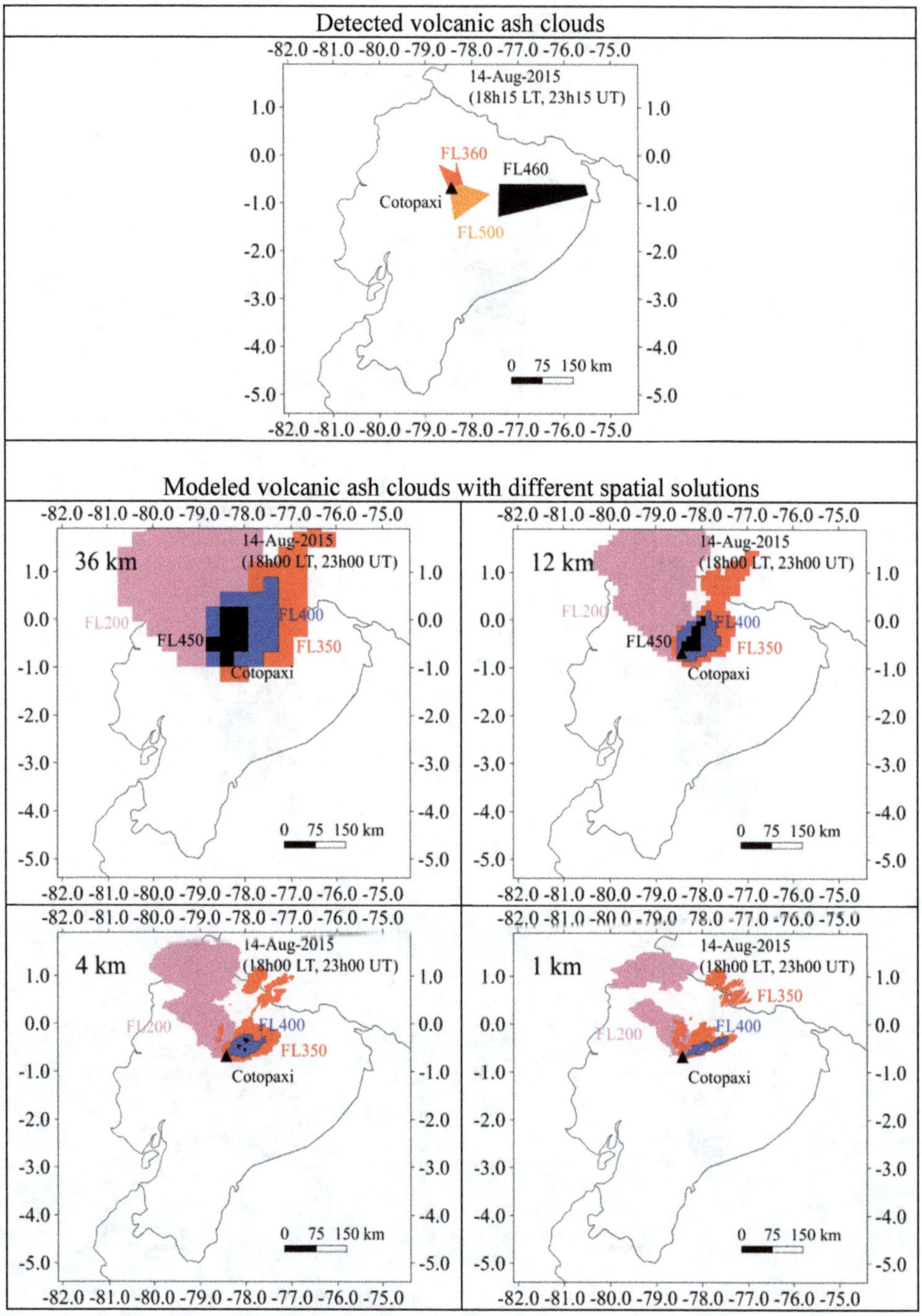

Figure 4: Cotopaxi volcano. Eruption on 14 August 2015. Detected [18] versus modeled volcanic ash clouds with different spatial resolutions.

For all the eruptions, the simulations with coarser resolutions (36 km and 12 km) provided larger areas affected by ash clouds, in comparison with the results when using higher resolutions (4 km and 1 km).

The modeled ash fallout of 16 December 2012, 14 July 2013, and 1 February 2014, affected zones at the N, W, and SW of the crater, respectively (Figs 5–7). The ash fallout of 14 August 2015 took place mainly toward NW (Fig. 8).

When modeling with different spatial resolutions, the main computed direction of ash fallouts was the same. However, the simulations with coarser resolutions (36 km and 12 km), provided larger areas affected by ash fallout in comparison with higher resolutions (4 km and 1 km). Also, coarser resolutions computed areas with ash fallout (red ellipses in Figs 5–8) which were not affected, when modeling with higher resolutions. On the other hand, higher resolutions can identify new affected zones (blue ellipses in Figs 6 and 7) in comparison with coarser resolutions, as show the results of 14 July 2013 and 14 August 2015 eruptions.

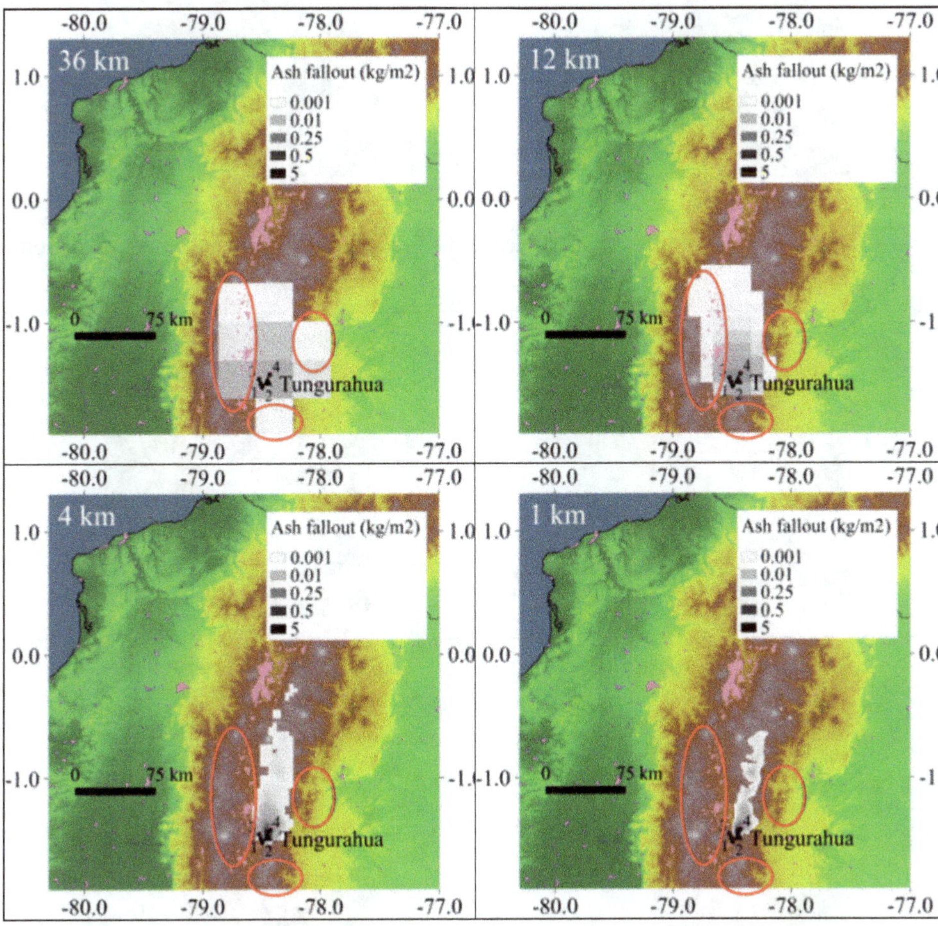

Figure 5: Tungurahua 16 December 2012. Modeled ash fallout using different spatial resolutions.

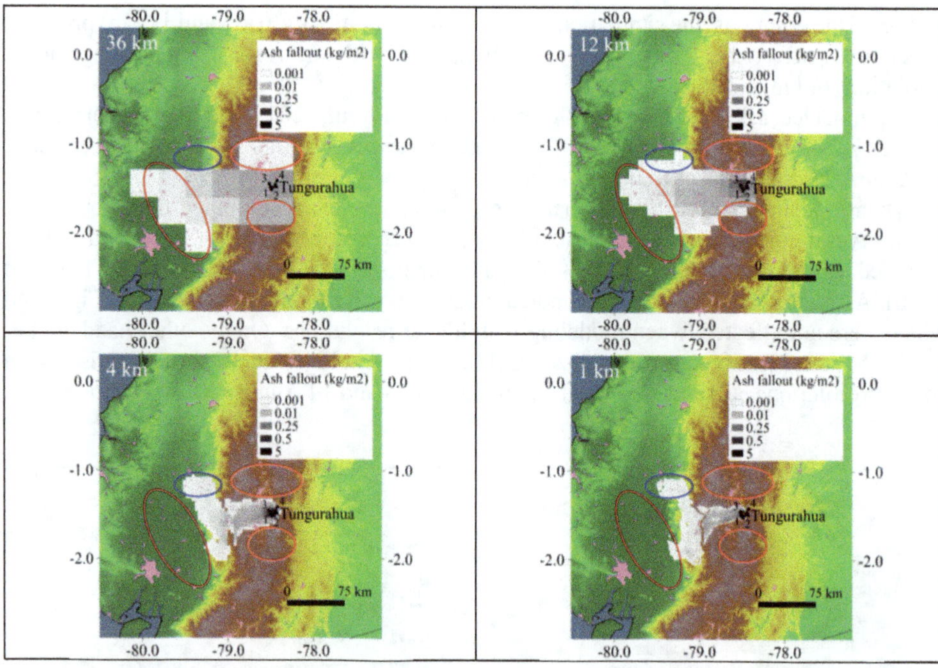

Figure 6: Tungurahua 14 July 2013. Modeled ash fallout using different spatial resolutions.

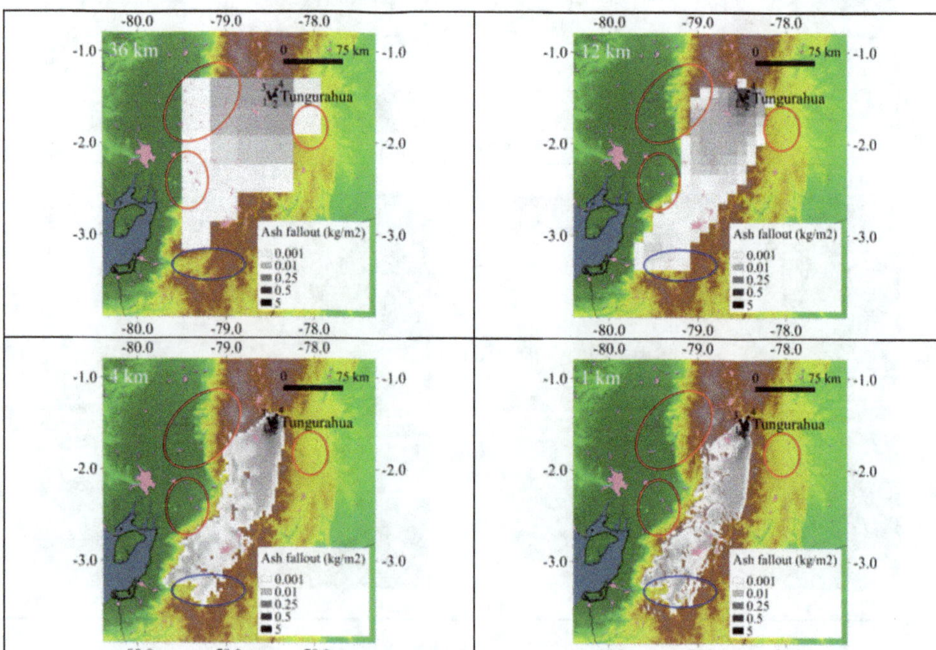

Figure 7: Tungurahua 1 February 2014. Modeled ash fallout using different spatial resolutions.

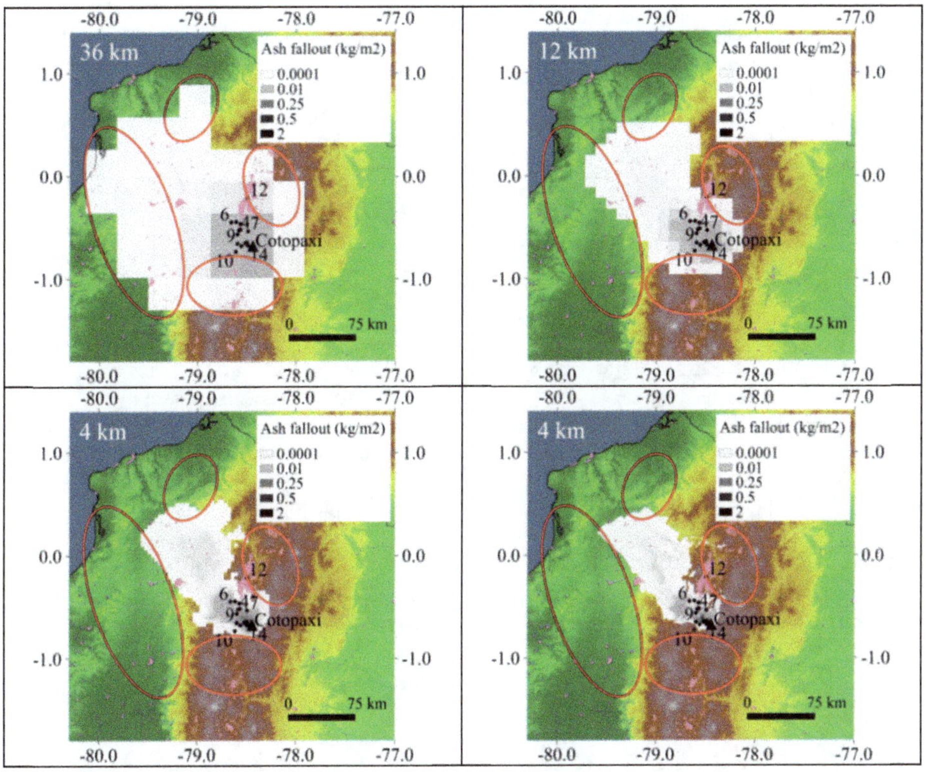

Figure 8: Cotopaxi 14 August 2015. Modeled ash fallout using different spatial resolutions.

The linear correlation coefficient (R^2) between measured and modeled ash fallout improved when using higher spatial resolutions. The coarser resolutions (36 km and 12 km), provided low performances, with values of R^2 between 0.00 to 0.79 (Table 2); and 0.28 to 0.46 respectively. The resolution of 4 km improved, with values of R^2 between 0.56 to 0.98. Modeling with 1 km got the best performance, with R^2 between 0.70 to 1.00.

As an example, Fig. 9 shows the comparison between measured and modeled ash fallout for the eruption of Cotopaxi on 14 August 2015.

Table 2: Values of the linear correlation coefficient (R^2) between measured and modeled ash fallout values using different spatial resolutions.

Volcano	Date of eruption	Spatial resolution			
		36 km	12 km	4 km	1 km
Tungurahua	16 December 2012	0.79	0.46	0.98	1.00
Tungurahua	14 July 2013	0.00	0.28	0.65	0.95
Tungurahua	1 February 2014	0.25	0.33	0.95	1.00
Cotopaxi	14 August 2015	0.27	0.33	0.56	0.70

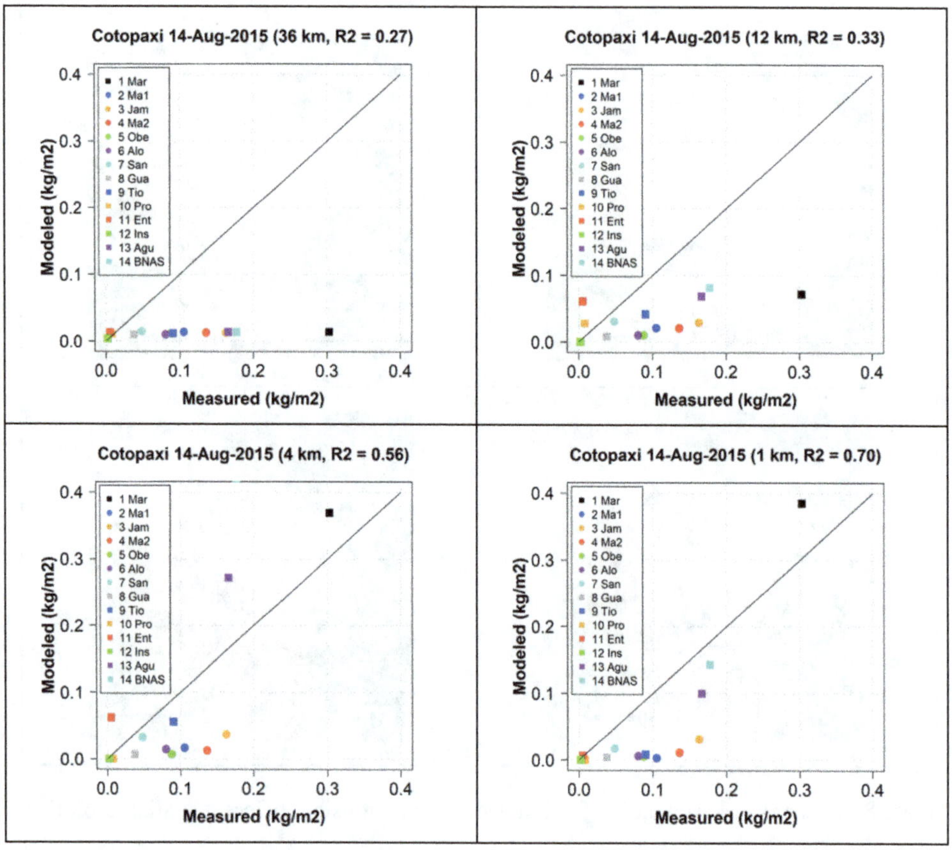

Figure 9: Cotopaxi volcano 14 August 2015. Values of the linear correlation coefficient (R^2) between measured and modeled ash fallout values using different spatial resolutions.

4 DISCUSSION AND CONCLUSIONS

We used spatial resolutions of 36 km, 12 km, 4 km and 1 km, for modeling the dispersion of volcanic ash of four eruptions, which took place in Ecuador in the last seven years.

Although with differences in shape and height, the direction of the modeled ash clouds with these resolutions, were consistent with the course of the detected ash clouds. For all the eruptions, the coarser simulations (36 km and 12 km) provided larger areas affected by ash clouds, in comparison with the results when modeling with higher resolutions (4 km and 1 km). The coarser simulations also provided larger areas affected by ash fallout, in contrast with the results when using higher resolutions.

For all the eruptions, the linear correlation coefficient (R^2) between measured and modeled ash fallout improved when using higher resolutions. The simulations using 4 km and 1 km reached values of R^2 between 0.56 to 0.98, and 0.70 to 1.00 respectively.

Nowadays and based on the WRF3.7.1 and FALL3D7.1.4 models [19], we are using a numerical system for forecasting the potential path of volcanic ash, due to emissions at Tungurahua and Cotopaxi volcanoes. This system is working with 4 km of spatial resolution, under the following schedule: During the first day, WRF generates the meteorology for the next four days. After, FALL3D models the volcanic ash dispersion for the second, third and fourth days. Now, the computational resources dedicated to this task (24 processors, E5, 2.00 GHz and 65.9 GB of RAM [19]) require about 24 h for processing this schedule.

Using the same resources but working with the spatial resolution of 1 km, the computational time increased to about 20 days. So, with this resolution, although with better performance, the computed information will not be available in forecasting time.

These results suggest that for the Ecuadorian case, with the actual computational resources, the resolution of 4 km is a good compromise, which allows the generation of volcanic ash dispersion in forecasting time, with proper modeling performance.

This framework uses an off-line approach. It firstly forecasts the meteorology and after the volcanic ash dispersion.

The advantage of this approach is that a unique database provides the meteorological data for forecasting ash dispersion at different volcanoes. This feature is significant, especially in cases of limited computational resources. Once the meteorology is available, it is possible to forecast in almost real time, the dispersion of volcanic ash, if the beginning time of an actual eruption is known.

Nevertheless, the off-line approach does not consider feedback between volcanic ash and meteorology. The online modeling approach, which works with interactions, could improve the performance. The disadvantage of the online method is the need for larger computational capacities, in part because the ash dispersion of a single eruption demands its meteorology.

ACKNOWLEDGEMENTS

This research is part of the project "Sistema de pronóstico de dispersión de ceniza volcánica en el Ecuador: septiembre 2018–agosto 2019" funded with the USFQ Poli-Grants 2018–2019. Simulations were done at the High Performance Computing system at the USFQ.

REFERENCES

[1] Ayris, P.M. & Delmelle, P., The immediate environmental effects of tephra emission. *Bulletin of Volcanology*, **74**, pp. 1905–1936, 2012. DOI: 10.1007/s00445-012-0654-5.

[2] Horwell, C.J. & Baxter P.J., The respiratory health hazards of volcanic ash: a review for volcanic risk mitigation. *Bulletin of Volcanology,* **69**(1), pp. 1–24, 2006.

[3] Folch, A., Jorba, O. & Viramonte, J., Volcanic ash forecast: Application to the May 2008 Chaitén eruption. *Natural Hazards and Earth System Sciences*, **8**, pp. 927–940, 2008. DOI: 10.5194/nhess-8-927-2008.

[4] Parra, R., Bernard, B., Narváez, D., Le Pennec, J.-L., Hasselle, N. & Folch, A., Eruption source parameters for forecasting ash dispersion and deposition from vulcanian eruptions at Tungurahua volcano: Insights from field data from the July 2013 eruption. *Journal of Volcanology and Geothermal Research*, **309**, pp. 1–13, 2016. DOI: 10.1016/j.jvolgeores.2015.11.001.

[5] Bonadonna, C., Folch, A., Loughlin, S. & Puempel, H., Future developments in modelling and monitoring of volcanic ash clouds: outcomes from the first IAVCEI-WMO workshop on Ash Dispersal Forecast and Civil Aviation. *Bulletin of Volcanology*, **74**, pp. 1–10, 2012. DOI: 10.1007/s00445-011-0508-6.

[6] Falasca, S. & Curci, G., High-resolution air quality modeling: Sensitivity tests to horizontal resolution and urban canopy with WRF-CHIMERE. *Atmospheric Environment*, **187**, pp. 241–254, 2018. DOI: 10.1016/j.atmosenv.2018.05.048.

[7] Le Pennec, J.-L., Ruiz, G.A., Ramón, P., Palacios, E., Mothes, P. & Yepes, H., Impact of tephra falls on the Andean communities: The influences of eruption size and weather

conditions during the 1999–2001 activity of Tungurahua volcano. Ecuador. *Journal of Volcanology and Geothermal Research*, **217–218**, pp. 91–103, 2012. DOI: 10.1016/j.jvolgeores. 2011.06.011.

[8] Bernard, B., Bustillos, J., Wade, B. & Hidalgo, S., Influence of the wind direction variability on the quantification of tephra fallouts: December 2012 and March 2013 Tungurahua eruptions. *Avances en Ciencias e Ingenierías*, **5–1**, pp. A14–A21, 2013.

[9] Hidalgo, S. et al., Cotopaxi volcano's unrest and eruptive activity in 2015: Mild awakening after 73 years of quiescence. *Abstract Volume of the 2016 EGU General Assembly*, p. EGU2016–5043-1, 2016.

[10] Gaunt, H.E. et al., Juvenile magma recognition and eruptive dynamics inferred from the analysis of ash time series: The 2015 reawakening of Cotopaxi volcano. *Journal of Volcanology and Geothermal Research*, **328**, pp. 134–146, 2016.

[11] Bernard, B. et al., Relationship between volcanic ash fallouts and seismic tremor: Quantitative assessment of the 2015 eruptive period at Cotopaxi volcano, Ecuador. *Bulletin of Volcanology*, **78, 80**, 2016. DOI:10.1007/s00445-016-1077-5.

[12] Parra, R., Influence of boundary layer schemes in modeling the dispersion and sedimentation of volcanic ash in Ecuador. *Wit Transactions on Ecology and the Environment*, vol. 230, WIT Press: Southampton and Boston, pp. 83–94, 2018. DOI: 102495/AIR180081.

[13] Weather Research and Forecasting Model, http://wrf-model.org. Accessed on: 10 Dec. 2018.

[14] Global Forecast System (GFS), www.ncdc.noaa.gov/data-access/model-data/model-datasets/global-forcast-system-gfs. Accessed on: 10 Dec. 2018.

[15] Parra, R., Performance studies of planetary boundary layer schemes in WRF-chem for the Andean Region of southern Ecuador. *Atmospheric Pollution Research*, **9**, pp. 411–428, 2018. DOI: 10.1016/j.apr.2017.11.011.

[16] Folch, A., Costa, A. & Macedonio, G., FALL3D: A computational model for transport and deposition of volcanic ash. *Computers and Geosciences*, **35**(6), pp. 1334–1342, 2009.

[17] Mastin, L.G. et al., A multidisciplinary effort to assign realistic source parameters to models of volcanic ash-cloud transport and dispersion during eruptions. *Journal of Volcanology and Geothermal Research*, **186**, pp. 10–21, 2009.

[18] Washington VAAC, www.ssd.noaa.gov/VAAC/messages.html. Accessed on: 10 Dec. 2018.

[19] Parra, R., Numerical system for forecasting volcanic ash dispersion in Ecuador. IEEE Third Ecuador Chapter Meeting (ETCM), pp. 1–5, 2018. DOI: 10.1109/ETCM.2018.8580331.

REAL-TIME ATMOSPHERIC MONITORING OF URBAN AIR POLLUTION USING UNMANNED AERIAL VEHICLES

QINGYUE WANG
Graduate School of Science and Engineering, Saitama University, Japan

ABSTRACT

Unmanned aerial vehicles (UAVs), or Drones, have recently begun to appear frequently in the television, newspapers and social media for their applicability in various fields of science. Although UAVs have been used in a wide range of atmospheric science, there are few reports of using UAVs for lower layer atmospheric observation available. In my laboratory, harmful black carbon (BC) in ambient particulates was determined based on the observation results of harmful chemicals in the lower boundary layer and atmospheric conditions using UAVs. The ambient particulates in the lower layer atmosphere based on heights can be measured by UAVs at the optional location. In this paper, I want to report the lower layer atmospheric observation of ambient particulates with their chemical species for the first time in Japan. The data transfer devices were loaded into UAVs for the transformation of the results measured at different heights during 2018 in order to develop real-time observations of $PM_{2.5}$ and PM_{10} in the lower layer atmosphere of the suburban and urban areas of Saitama city, Japan. Air pollutants transported from the urban areas of metropolitan Tokyo were determined from the real-time results of measurements in the vanity in the morning, noon and afternoon. High concentrations of $PM_{2.5}$ and PM_{10} were observed around noon in the rural area of Yorii-machi, Saitama prefecture on December 5, 2017 which might be influenced by the polluted air mass from the urban atmosphere. It is clear that different air pollution phenomena can be observed easily by using UAVs real-time atmospheric monitoring.

Keywords: unmanned aerial vehicles (UAVs), drone, ambient particulates, $PM_{2.5}$, PM_{10}, black carbon, real-time monitoring, Saitama, Japan.

1 INTRODUCTION

In recent years, increasing attention has been paid to air pollution by fine particles in the urban atmosphere owing to various harmful effects of those particles on human health. Atmospheric pollutants discharged from anthropogenic sources as emissions of the gaseous substance and ambient fine particles of $PM_{2.5}$ (particulate matter with a 50% cut-off aerodynamic diameter < 2.5 μm) [1], the radioactive materials released by the accidents of the nuclear power plants and the chemical plants. Moreover, pollen grains [2], yellow sand dust, volcanic explosion, and a brush fire are also the origins of air pollution. Atmospheric pollutants that emitted from natural and anthropogenic sources can change the composition of ambient air affecting air quality and human health [3].

In Asian countries, severe air pollution phenomena were found with rapid economic development and urbanization during the last several decades [4]–[7]. It is essential to develop methods using unmanned aerial vehicles (UAVs) for the determination of atmospheric pollutants in the lower layer atmosphere or the place where a person cannot go easily. At the same time, UAVs can be used for the observation of the spread mechanism, the source contribution and the generation mechanism of atmospheric pollutants with the vertical distribution.

In this study, my research team has developed several unmanned aerial vehicles (UAVs) and used them for atmospheric monitoring to examine real-time variations and vertical distribution of air pollutants associated with anthropogenic activities or long-range transportation in the lower boundary layer. To investigate the contribution of the chemical components of the traffic emission [6] associated with toxicity of the airborne particulate

WIT Transactions on Ecology and the Environment, Vol 236,
www.witpress.com, ISSN 1743-3541 (on-line)
doi:10.2495/AIR190081

matter, I have especially used those UAVs and determined the vertical distribution of $PM_{2.5}$, ozone (O_3), BC caused by diesel exhaust particles (DEPs) and different carbonaceous particles (OC1~OCx) of ambient particulate matter (as $PM_{2.5}$) at the urban roadside site in Saitama city and background site of Saitama prefecture, Japan.

2 MATERIAL AND METHODS

2.1 Sampling sites

The study was conducted in the Saitama prefecture located in the central Kanto region of Japan having a population of more than one million. Two sites viz. roadside of the city route 463 (Main prefectural road with high traffic volumes of both diesel and gasoline powered around 21,000 vehicles per weekdays) in Saitama city near the metropolitan Tokyo and background site located in the rural area of Yorii-machi in Saitama prefecture, Japan were chosen for observing air pollution.

2.2 Unmanned aerial vehicles (UAVs)

Several UAVs of Model EX1100SA and Model ZION QC730 (Fig. 1) embedded with payload computer were designed and made by Enroute Co., Ltd., Japan. Another Model SA888 was designed and built by Saitama University and Iyobeshoji Co., Ltd., Japan. Although the Ridge Hawk controller has the power to compute and multi-task simultaneous functions, it was built to be energy efficient to minimize battery usage during flights. The computer consists of 4MB of flash memory, a 12-bit A/D converter with 8 input channels, two RS232 channels, a hardware watchdog, a real-time clock and 512 bytes of RAM. During the flights of all UAVs automatically following the GPS system, we can send commands to the computer to control and change the data download feed, each payload instrument and the sensors mentioned in next sections.

Figure 1: A basic unmanned aerial vehicles (UAVs).

The part order containing within the UAVs loading, airborne particulate sensors of $PM_{2.5}$ and PM_{10} including two particle counters (Shibata Scientific Co., Tokyo, Japan) and real-time pollen grains counter equipment were made at the same time. In order to set on all the UAVs of Model EX1100SA, Model ZION QC730 and Model SA888, the connection parts of the all the models were also designed and improved. We had miniaturized each measuring machine module, confirmed for 10% weight saving of all the UAVs and made them of about 10 m/s of survival wind speed to load into UAVs easily.

2.3 Airborne particulate sensors for $PM_{2.5}$ and PM_{10}

Ambient particulate measurement using personal particle sensors of Model LD-6N2 (Shibata Scientific Co., Ltd., Japan) (https://www.sibata.co.jp/products/products17/), is modular in design and can be quickly installed and removed from the UAVs payload bay (Fig. 2) evaluated by Saitama University. The package is powered by the batteries (USB power supply from the same package powered by Li ion battery unit of minipump) for the duration of the flight up to 10 hours [7].

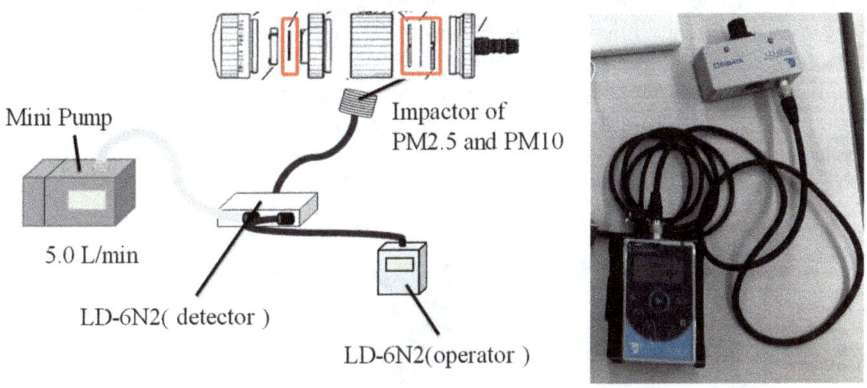

Figure 2: Personal particles sensor (Model LD-6N2) with an impactor of $PM_{2.5}$ and PM_{10}.

The aerosol package includes a passively pumped isokinetic inlet of minipump (Model MP-N∑II, Shibata Scientific Co., Japan) to bring air and particles into the module. An optical particle counter is also under development for the package powered by Li ion battery unit as the field deployment. The operation displays LED indicator for sampling air and features integrated flow measurement functions have been shown in Fig. 2.

The LD-6N2 has a quick response time, grows particles in a butanol-saturated flow and counts particles in different diameter size ranges with the modifications were implemented to address the high-vibration environment of the UAVs. $PM_{2.5}$ and PM_{10} concentrations were also characterized and determined by their impactor and sampling size-selected particles from differential sensors designed by Saitama University [8]–[9] and also developed and hand made by Shibata Scientific Co., Japan.

2.4 Real-time determination of the carbonaceous particles with a $PM_{2.5}$ microCYCLONE

I used the carbonaceous particle monitors (Model AE51 and Model MA 200, AethLabs, Ltd., USA, https://aethlabs.com/microaeth/ma200/overview) for the determination of carbonaceous particles. MA200 monitor is compact, real-time, wearable five-wavelength UV-IR BC monitors with 15 sampling location automatic filter tape advance system allowing up to 2–3 weeks of continuous measurements. The mass concentrations can be detected by light absorbing carbonaceous particles in a sampled aerosol. The instruments have five analytical channels each operating at a different wavelength (880 nm, 625 nm, 528 nm, 470 nm and 375 nm) [10]. Measurement at 880 nm is interpreted as the concentration of BC [11]. Measurement at 375 nm is interpreted as ultraviolet particulate matter (UVPM) indicative of carbonaceous particles like wood smoke, tobacco, biomass burning and DEPs.

In this paper, I have defined those carbonaceous particles as BC, DEPs and different organic carbon (OC1~OCx). The MA series instruments have several important advancements. The instrument automatically controlled the advance of the tape material, moving to a new unused spot when required. This allowed the instrument to run continuously for multiple weeks or months without human intervention. Five-wavelength optical engine enables discrimination between organic and elemental particles which is helpful in source identification when measuring different aerosols. The MA series also features the special loading compensation method, which in real-time measures and adjusts for differing optical properties of particles of varying age and composition. For a particle size cut point of $PM_{2.5}$ BC, the flow rate of the microAeth needs to be set to 50 mL/min. If the flow rate of the microAeth is set to 100 mL/min, the microCYCLONE will provide a particle size cut point of 1.6 microns in diameter shown in Fig. 3. Therefore, BC measurements can be carried out with the corresponding particle size selective cut point of the microCYCLONE at that flow rate.

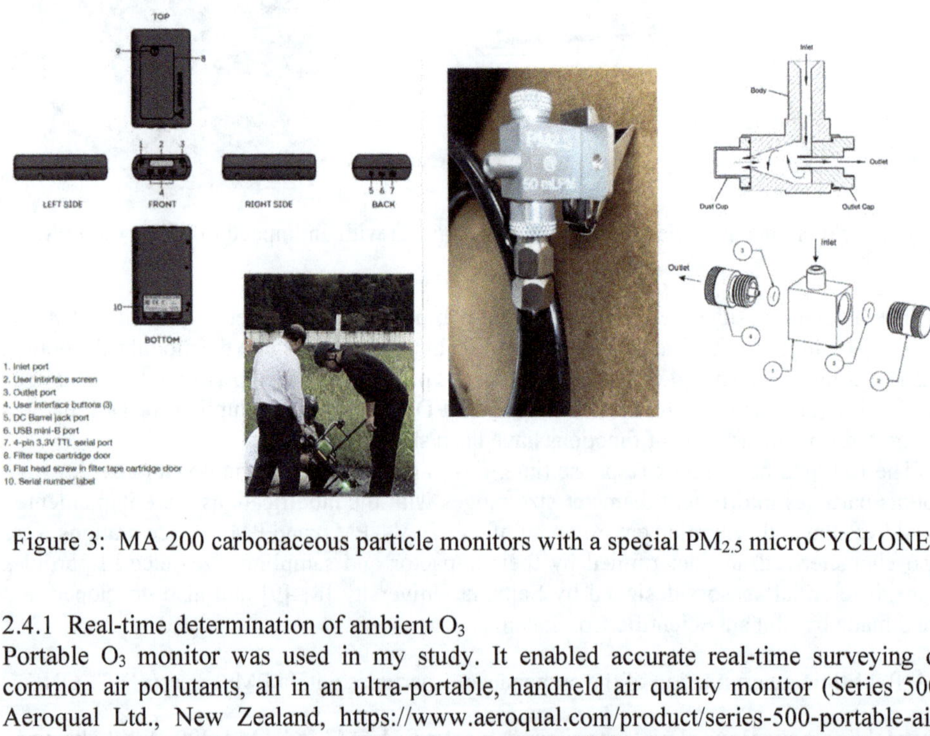

Figure 3: MA 200 carbonaceous particle monitors with a special $PM_{2.5}$ microCYCLONE.

2.4.1 Real-time determination of ambient O_3

Portable O_3 monitor was used in my study. It enabled accurate real-time surveying of common air pollutants, all in an ultra-portable, handheld air quality monitor (Series 500, Aeroqual Ltd., New Zealand, https://www.aeroqual.com/product/series-500-portable-air-pollution-monitor). Data were stored on board the Series 500 with a maximum 8,188 records available. To download the data a USB cable was supplied for connection to PC. Free PC software provided with the Series 500 taken the data and represented that in a chart or table view. Data were downloaded and viewed in excel.

The monitor can also operate in control mode. Upper and lower control limits can be easily set directly on the display. Using the 0-5V output the Series 500 can be used to switch on or off an externally connected device, such as an O_3 generator, or system in the presence of a predetermined level of O_3 gas. By selecting the optional wall-mount bracket and plugging in main power the Series 500 was effectively able to act as a fixed monitor as well as a handheld portable device (Fig. 4).

Preparation

Setting and adjustments

Real time data transmission

Figure 4: Observation using UAVs with the sensors of $PM_{2.5}$, PM_{10}, and O_3 at Yorii-machi rural site, Saitama prefecture, Japan on November 15, 2018.

3 RESULTS AND DISCUSSION

3.1 The real-time variations of $PM_{2.5}$ and PM_{10} concentrations and their patterns

During the last decade, increasing attention has been paid to fine particles such as $PM_{2.5}$ in the urban atmosphere owing to the various harmful effects on human health [6]. $PM_{2.5}$ can penetrate more deeply into the lungs and may reach the alveoli causing serious respiratory diseases. Motor vehicles exhaust is one of the main sources of $PM_{2.5}$ in the urban atmospheric environment. Therefore, we have used UAVs for the first time in Japan to study the spread mechanism of atmospheric pollutants, the source contribution of urban atmospheric pollutants in this article.

In the case of a flight in lower boundary layer up to 140 m, I found out that its inclination was one beyond /100 m clearly 1°C in the temperature profiles in the low-rise sky from the low layer around 70 m. When flight over 100 m of the ground level, one below /100 m went down about 1°C, and it means that the atmosphere was weak stability even in the low layer of 100 m. The stronger atmosphere stability was found over 100 m with a weak inversion layer formed. Therefore, the atmospheric pollutants were concentrated by the pollution accumulation because of the weather condition disadvantageous to the spread of the atmospheric pollutants in the lower boundary layer during winter.

The ambient particulate measurement was carried out using the personal particle sensors in Yorii-machi, Osato-gun, Saitama prefecture on December 5, 2017 is shown in Fig. 5. Since the contaminant air masses were transported from the urban area in the Tokyo metropolitan area, the real-time results of measurement of $PM_{2.5}$ and PM_{10} to a suburb were determined from morning to afternoon of the day. As shown in Fig. 5, it was found out that the influence of short time high concentrated atmospheric pollution occurred. After that the pollution phenomena were also determined again in the afternoon.

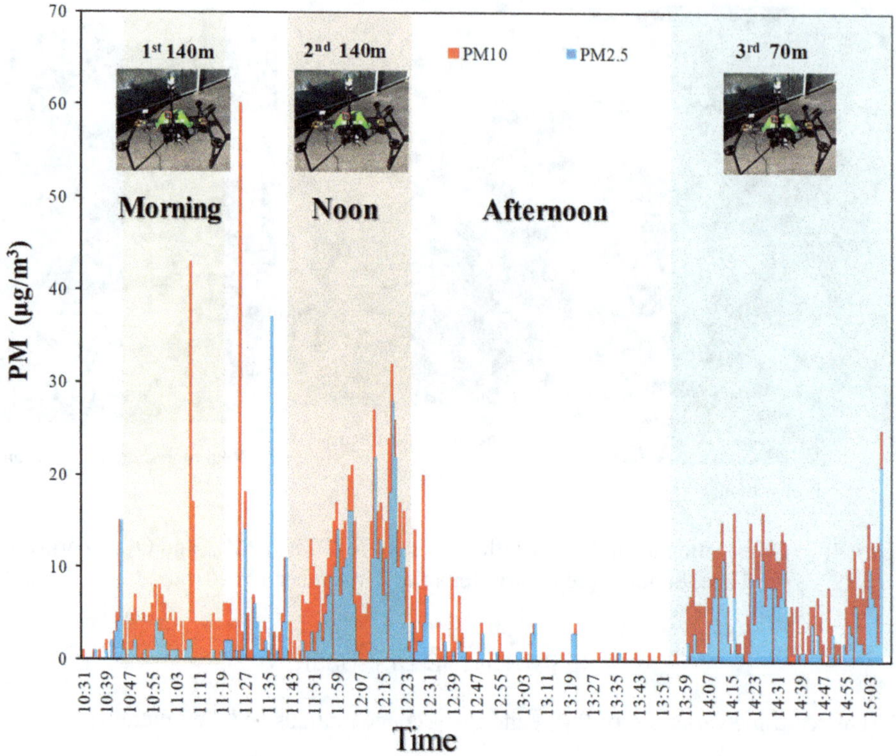

Figure 5: Real-time determination of $PM_{2.5}$ and PM_{10} using UAVs with the personal particle sensor.

3.2 Real-time concentrations of O_3 and $PM_{2.5}$ in 140 m atmospheric layer

As shown in Fig. 6, the concentrations of O_3 and $PM_{2.5}$ were increased in the 140 m atmospheric boundary layer over ground level suggesting that secondary particles contribute considerably as one of the pollutants in Saitama, Japan. NO_3 is usually present as NH_4NO_3 in the atmosphere. During the winter season, the concentration of NO_3 was significantly increased due to diesel exhaust which is two times higher than summer might influence the pollution in the urban atmosphere.

3.3 Real-time variations and vertical distribution of carbonaceous components associated with anthropogenic activities

Diesel exhaust contains substantially more particulate matter than gasoline-fuelled vehicle exhaust and the use of diesel-powered vehicles is leading to the deterioration of air quality in urban areas.

As given in Fig. 7, real-time mass-weighed variations of carbonaceous components were determined during a sampling campaign period in 2018. At the same period, real-time vertical distribution of BC caused by DEPs was mainly determined in $PM_{2.5}$ of the lower boundary layer shown in Fig. 8.

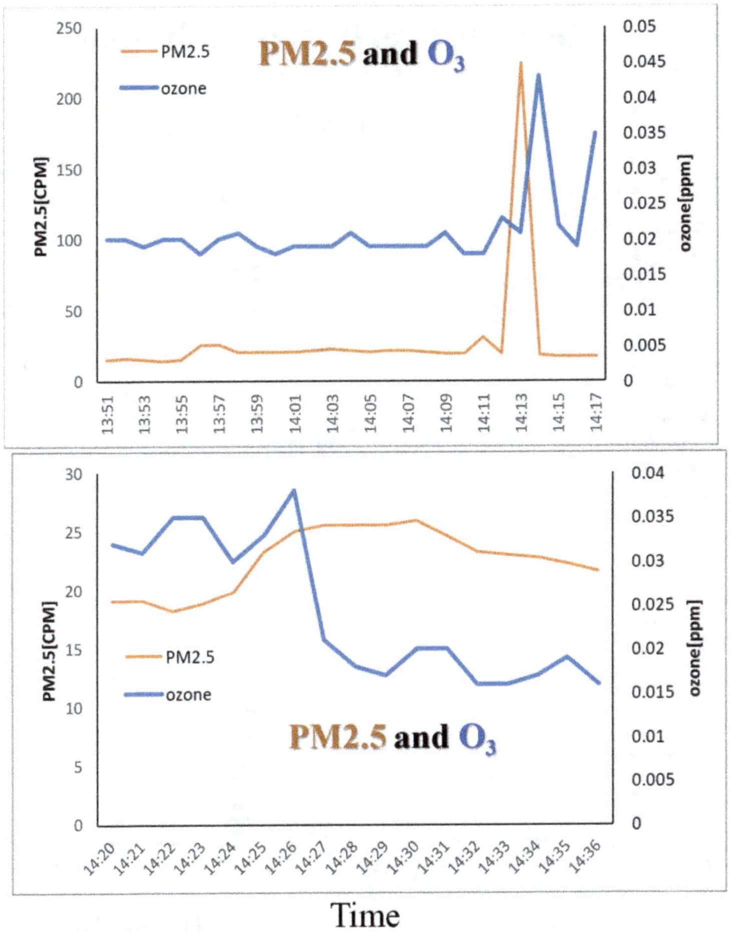

Figure 6: Real-time determination of ambient O_3 in 140 m atmospheric layer over the ground level in a background site.

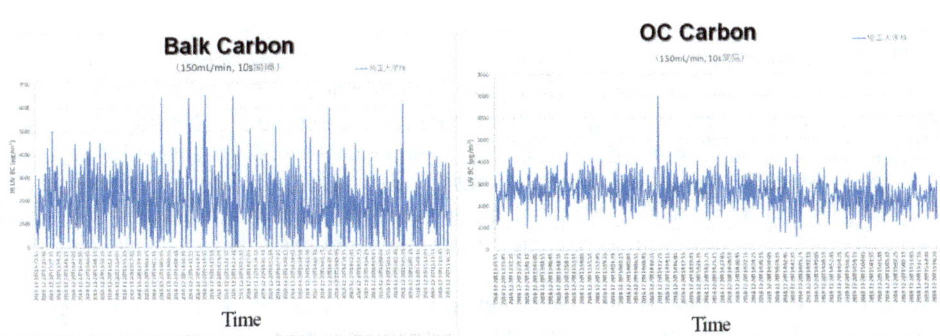

Figure 7: Real-time variations of carbonaceous components during a sampling campaign period in 2018.

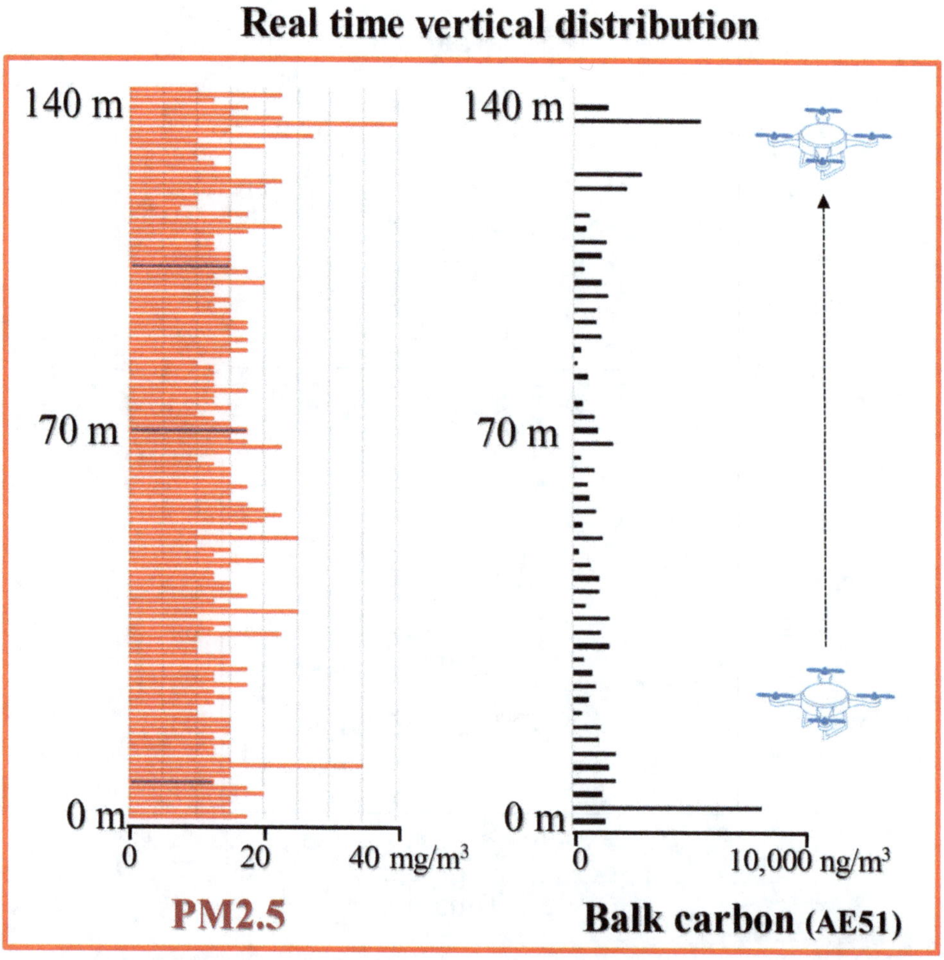

Figure 8: Real-time vertical distribution of BC and $PM_{2.5}$ of the lower boundary layer during a sampling campaign period in 2018.

The high contributions of BC, DEPs and the different organic carbonaceous components (OC1~OCx) were observed in $PM_{2.5}$ during the periods when the wind was coming mainly from the roadside toward the sampling flight site presented in Fig. 9. Therefore, I strongly believe that the contribution of carbonaceous components could be attributed to motor vehicle emissions. Moreover, the concentrations of several chemical components such as BC and OC1~OCx at the roadside were markedly lower on the weekend than during the week. This reduction could be due to the reduction in the number of diesel-powered vehicles on the weekend. The difference was marked in the case of the BC concentrations, though the OC1~OCx concentrations were closer between weekend and weekday. Some species of particulate polycyclic aromatic hydrocarbons (PAHs) were mainly derived from motor-vehicle or petroleum combustion in summer and from coal and biomass combustion in winter [6]. The high molecular weight (HMW) PAHs containing five-ring PAHs and six-ring PAHs were determined in the fine particles of $PM_{2.5}$ as the components of BC.

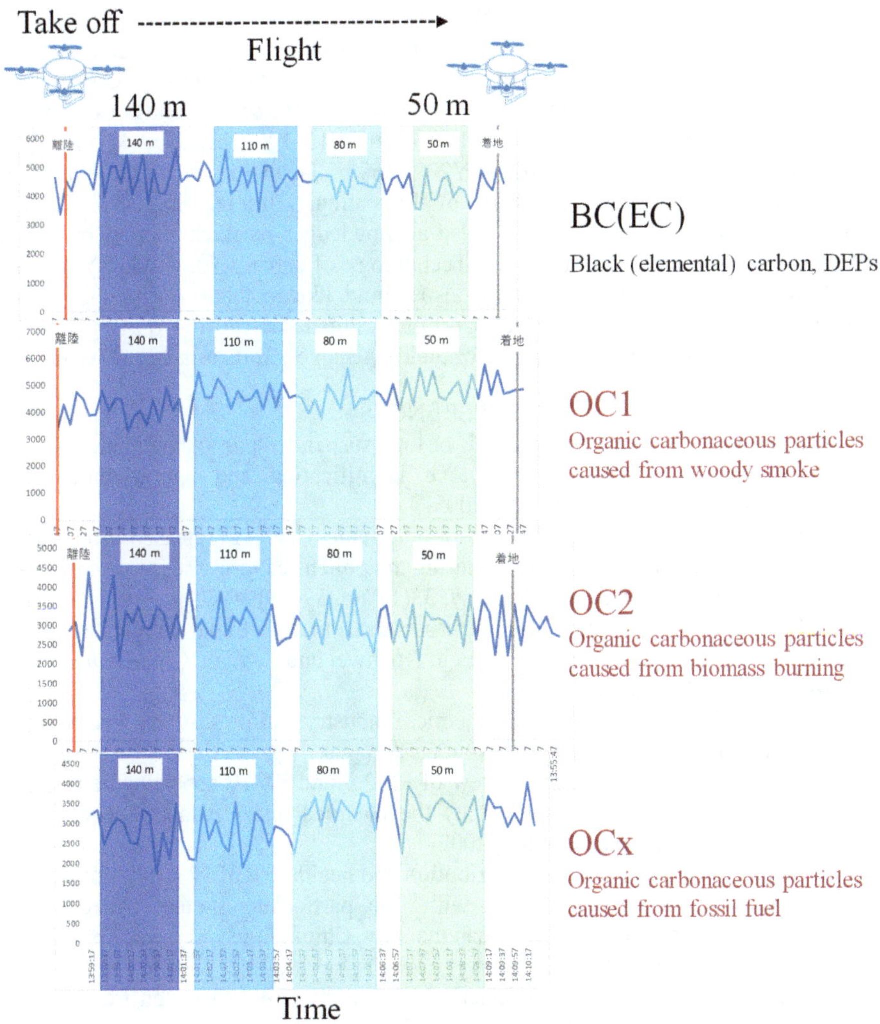

Figure 9: Real-time vertical distribution of BC and OC1~OCx in $PM_{2.5}$ determined by a MA 200 carbonaceous particle monitor during the sampling campaign period in 2018.

4 CONCLUSION

In this article, air quality measurement as $PM_{2.5}$, ozone, black carbon caused by diesel exhaust particles and different carbonaceous particles from personal particles and gas sensors set in several UAVs were carried out. Air quality research is moving towards the more wide-spread use of UAVs to measure those pollutants at different altitudes; compare ground-based data; and autonomously track plumes emitted by combustion sources, revealing the origin of atmospheric pollutants. With the help of our developed UAVs, it is possible to investigate the spread mechanism of atmospheric pollutants, long range transportation, specification of the pollutant sources and the pollution mechanism with the vertical distribution in the lower boundary layer.

ACKNOWLEDGEMENTS

This study was initiated by Saitama University and funded by the Saitama Prefectural Government of Japan accepted as the Saitama leading-edge industry design project (FY 2016~FY2018). Additional support was provided by the Special Funds of Grant-in-Aid for Scientific Research (B) (No. 24310005, FY 2012~FY2014; No. 15H05119, FY 2015~FY2017, No. 18H03384, FY 2018~FY2020) of the Japan Society for the Promotion of Science (JSPS) and Japanese Ministry of Education, Culture, Sports, Science and Technology (MEXT), Japan. The author also acknowledges research support by the Steel Foundation for Environmental Protection Technology of Japan (No. C-33, FY 2015- FY 2017). As this project leader, the author also would like to thank the members of our laboratory, Enroute Co., Ltd., Iyobeshoji Co., Ltd., Shibata Scientific Co., Ltd., and Japan Environmental Research Co., Ltd., Japan for their extensive efforts related to the study.

REFERENCES

[1] Arditsoglou, A. & Samara, C., Levels of total suspended particulate matter and major trace elements in Kosovo: A source identification and apportionment study. *Chemosphere*, **59**(5), pp. 669–678, 2005.

[2] Gong, X. et al., Size distribution of allergenic Cry j 2 released from airborne Cryptomeria japonica pollen grains during the pollen scattering seasons. *Aerobiologia (International Journal of Aerobiology)*, **33**(1), pp. 59–69, 2017.

[3] Lu, S. et al., Single particle aerosol mass spectrometry of coal combustion particles associated with high lung cancer rates in Xuanwei and Fuyuan, China. *Chemosphere*, **186**, pp. 278–286, 2017.

[4] Xie, K., Li, W. & Zhao, W., Coal chemical industry and its sustainable development in China. *Energy*, **35**(11), pp. 4349–4355, 2010.

[5] Chow, J. C. & Watson, J. G., Review of PM2.5 and PM10 apportionment for fossil fuel combustion and other sources by the chemical mass balance receptor model. *Energy Fuels*, **16**(2), pp. 222–260, 2002.

[6] Wang, Q. et al., Studies on size distribution and health risk of 37 species of polycyclic aromatic hydrocarbons associated with fine particulate matters collected in the atmosphere of a suburban area of Shanghai city, China. *Environmental Pollution*, **214**, pp. 149–160, 2016.

[7] Wang, Q., Itoh, S. & Lu, S., Reduction of fine particles exhausted from small-size combustor using agricultural waste residue by controlling burning temperatures. *International Journal of Sustainable Development and Planning*, **9**(5), pp. 717–726, 2014.

[8] Wang, Q., Otsuka, G., Dong, S., Ishihara, K., Lu, S. & Sekiguchi, K., Study on the monitoring method for airborne herbaceous pollen in autumn by a mobile air sampler, *Japanese Journal of Palynology*, **61**(2), pp. 49–55, 2016.

[9] Wang Q., Gong X., Suzuki M., Lu S., Sekiguchi K. & Nakajima D., Size-segregated allergenic particles released from airborne Cryptomeria japonica pollen grains during the Yellow Sand events within the pollen scattering seasons. *Asian Journal of Atmospheric Environment*, 7(4), pp. 191–198, 2013.

[10] Weingartnera, E., Saathoffb, H., Schnaiterb, M., StreitaStreita, N., Bitnarc, B. & Baltensperger, U., Absorption of light by soot particles: determination of the absorption coefficient by means of aethalometers, *Journal of Aerosol Science*, **34**(10), pp. 1445–1463, 2003.

[11] Uematsu, M. et al., Atmospheric transport and deposition of anthropogenic substances from the Asia to the East China Sea. *Marine Chemistry*, **120**(1), pp. 108–115, 2010.

ASSESSMENT OF SOURCE CONTRIBUTIONS TO THE URBAN AIR QUALITY FOR THE BRISTOL CLAIRCITY PILOT CASE

KEVIN OLIVEIRA[1], VERA RODRIGUES[1], SILVIA COELHO[1], ANA PATRÍCIA FERNANDES[1], SANDRA RAFAEL[1], CARLOS FARIA[1], JOANA FERREIRA[1], CARLOS BORREGO[1], TROND HUSBY[2], IASON DIAFAS[2], PER SIEVERTS NIELSEN[3], XIUFENG LIU[3], ANGREINE KEWO[3], CARLO TROZZI[4], ENZO PISCITELLO[4], KRIS VANHERLE[5], SVEIN KNUDSEN[6], EVERT BOUMAN[6], JO BARNES[7], STEPHAN SLINGERLAND[8], ENDA HAYES[7], HANS BOLSCHER[8] & MYRIAM LOPES[1]

[1]CESAM and Department of Environment and Planning, University of Aveiro, Portugal
[2]Planbureau voor de Leefomgeving, The Netherlands
[3]Danmarks Tekniske Universitet, Denmark
[4]TECHNE Consulting SRL, Italy
[5]Transport and Mobility Leuven, Belgium
[6]Norsk Institutt for Luftforskning, Norway
[7]University of the West of England, UK
[8]Trinomics Bv, The Netherlands

ABSTRACT

The world's population has been growing continuously, with most people inhabiting urban settlements. Furthermore, air pollution has become a growing concern, mainly in densely populated cities, where human health is threatened by acute air pollution episodes. The H2020 ClairCity project aims to substantially improve future air quality and carbon policies in European cities by initiating new modes of engaging citizens, stakeholders and policy makers. ClairCity applies an innovative quantification framework developed to assess environmental, health and economic impacts. In this work, the quantification framework was applied and calibrated for the baseline situation in Bristol, the ClairCity pilot city. The second-generation Gaussian model URBAIR was set up to simulate NO_2 and particulate matter (PM) concentrations for the entire year of 2015. An analysis of source contribution was performed providing information on the contributions of different source sectors (e.g. road transport, industrial, residential and commercial) to NO_2 and PM concentrations. The results point to a predominant contribution of road transport sector of 53% to NO_2 concentrations in Bristol, while the residential sector is the main contributor (with a contribution of 82%) to particulate matter concentrations, mainly linked with a high use of solid biomass combustion in this sector. These results can be powerful to support the design of air quality management plans and strategies and to forecast potential benefits of reducing emissions from a particular source category.

Keywords: H2020 programme, ClairCity, air pollution reduction, citizens engagement, European cities, urban areas.

1 INTRODUCTION

As the population continues to grow exponentially and moving from rural to urban areas this results in a large population density in urban settlements, which contributes to a high pressure over the environment. Furthermore, air pollution represents a global threat that leads to major impacts on human health and ecosystems. It represents a public health concern mainly due to population's long-term exposure to high levels and acute air pollution episodes. According to the World Health Organization (WHO), in 2015, in the European region exposure to air pollution contributed in 391 000 premature deaths attributed to $PM_{2.5}$ and to 76,000 premature deaths linked to NO_2 [1]. The most common reasons for premature death attributable to air pollution are due to noncommunicable diseases – notably cardiovascular diseases, stroke, chronic obstructive pulmonary disease and lung cancer. Air pollution also increases the risks for acute respiratory infections. In terms of exposure, both short and long-

WIT Transactions on Ecology and the Environment, Vol 236, © 2019 WIT Press
www.witpress.com, ISSN 1743-3541 (on-line)
doi:10.2495/AIR190091

term exposure to air pollution can lead to reduced lung function, respiratory infections and aggravated asthma, where children and older people, as well as, people with previous health conditions, are more vulnerable and therefore more likely to be affected by air pollution. In Europe, currently the average loss in life expectancy due to fine particles is about 5 months but can be more than 12 months in some urban areas [1]–[4].

According to the European Environment Agency's estimates of the urban population exposure to air pollution between 2014 and 2016, a substantial percentage of the urban population in the EU-28 was exposed to concentrations of certain air pollutants above EU limit and this number is even higher when the WHO air quality guideline values were applied. In terms of fine particulate matter ($PM_{2.5}$), 6–8% of the EU-28 urban population was exposed to concentrations above the EU limit value, while 74–85% was exposed to concentrations above the WHO guideline value. For PM_{10} the exposure estimates were 13–19% above the EU limit value and 42–52% above the WHO guideline value and for Nitrogen dioxide (NO_2) the estimates were around 7–8% of the EU-28 urban population were exposed to levels above the limit value, both from the EU and the WHO guidelines values [1], [5].

ClairCity project aims to improve future air quality and carbon policies in six pilot European cities (Amsterdam, Bristol, Sosnowiec, Genoa, Ljubljana and the Aveiro Region) by introducing new approaches to engage citizens, stakeholders and policymakers. Placing the citizen and their behaviours at the centre of both the problems and the solutions of the decision-making will allow citizens and other stakeholders to discuss the role of air quality and carbon policies for citizens' health, general well-being and future quality of life.

This work applies and calibrates for the baseline situation of Bristol city a quantification framework to assess the impacts on air quality and population exposure focus on NO_2, PM_{10} and $PM_{2.5}$ concentrations. The second-generation Gaussian model URBAIR was set-up to simulate NO_2 and particulate matter concentrations for the entire year of 2015. For these pollutants, an analysis of source contribution was performed providing information on the concentration contributions of different source sectors (e.g. road transport, industrial, residential and commercial sectors).

2 METHODOLOGY

Bristol, the pilot case study of ClairCity project, is a city located in South West England with an estimated population of roughly 454,000 inhabitants.

The assessment of source contributions from distinct emission sectors to the simulated air quality is performed for the pilot case study of the ClairCity project, based on the quantification framework established under the project. This framework consists of an assessment of the impact of a set of policies on emissions, air quality, human health and related costs. The air quality assessment is performed covering distinct spatial scales: the WRF modelling system is applied to an European domain using a horizontal grid resolution of 0.25 degrees, and then to a regional domain covering the urban area of Bristol using as horizontal resolution 0.05 degrees, both in an hourly basis. The air quality at urban scale is simulated using the second-generation Gaussian model URBAIR for the computational domain over the urban area of Bristol of 20 km x 20 km with a horizontal grid resolution of 200 m x 200 m. The URBAIR model has been implemented in previous works for a set of urban applications [6], [7] and tested against measured data.

The baseline simulations were performed using as input data the meteorological vertical profiles provided by the WRF model. The air quality simulations were performed for the full-year in an hourly basis. The concentrations of NO_2, PM_{10} and $PM_{2.5}$ were simulated using the emission rates available on the ClairCity emission database. The ClairCity emission database includes point sources with the emission rates of the large industry sources, the line sources with the road traffic emissions and the area sources covering the residential, commercial and industrial emissions. The database is built from distinct emission sectors, in line with statistics by sector, by time of day, establishing the link with citizen's behaviour. The air quality is assessed for the current situation through modelling tools, evolving towards the comparison of the simulation outputs with observations.

2.1 Background concentrations

The quantification framework covered the simulation of the transport, commercial and residential, and the industrial emission sectors. As a result of the absence of other relevant emission sectors within the domain and also due to important transboundary pollution, it lead to a clear underestimation of the simulated concentrations. Therefore, to overcome this issue, other remaining sectors were accounted as background. For the background quantification, a methodology was applied based on the background concentration maps published by the UK's Department for Environment Food and Rural Affairs (Defra). The air quality maps produced by Defra present concentration values by different sectors and corresponded to an annual mean in a spatial resolution of 1 km x 1 km. For this case study, to the simulated NO_2 concentrations were added the concentrations available on the Defra database linked with the contributions from the categories of aircraft, rail, other and rural, whereas for PM_{10} and $PM_{2.5}$ the added categories where: rail, other, secondary PM, residual and salt.

2.2 Model adjustment procedure

The calibration of the simulations results along with the added background concentrations were made through an adjustment method by comparing the obtained values with measured data from 2015. For NO_2 concentrations the measured data available include 107 diffuse tubes distributed mainly over Bristol's urban area of which 96 classify as roadside, four as kerbside and seven as urban background tubes. In addition, four continuous measurement points (two roadside, one kerbside, and one urban background sites) together with St Paul's urban background station. The values from the diffusion tubes indicate an average concentration of 42.1 $\mu g.m^{-3}$, while the maximum concentration stands around 91.2 $\mu g.m^{-3}$. At St Paul's station, the NO_2 annual average concentration was equal to 22.5 $\mu g.m^{-3}$. The continuous measuring devices measured an average value of 36.0 $\mu g.m^{-3}$ with a maximum concentration of 44.2 $\mu g.m^{-3}$.

The adjustment corresponds to a linear regression between the simulated values and the corresponding matching points for all the measured data. The result of the linear regression was a slope of 1.615, which was used as a correction factor to be applied over the whole domain.

In case of particulate matter, there is limited measured data, and consequently the adjustment performed only takes into account measurements from St Paul's station. In 2015, the annual average concentration for PM_{10} was 14.9 $\mu g.m^{-3}$ and for $PM_{2.5}$ was 10.6 $\mu g.m^{-3}$.

3 RESULTS

ClairCity focuses particularly on the transport and energy related behaviour of Bristol citizens and its contribution to air pollution and carbon emissions. The application of the quantification framework comprises the simulation of the transport, industrial, residential and commercial emission sectors. The assessment of impacts was performed in terms of air quality and population exposure focus on NO_2, PM_{10} and $PM_{2.5}$ concentrations.

3.1 Air quality maps and exposure

Fig. 1 shows the annual average concentration fields for NO_2, PM_{10} and $PM_{2.5}$ for Bristol with the adjustment factor applied and the background added over the domain. For each contour map, the scale limits are set by the EU limit for each pollutant.

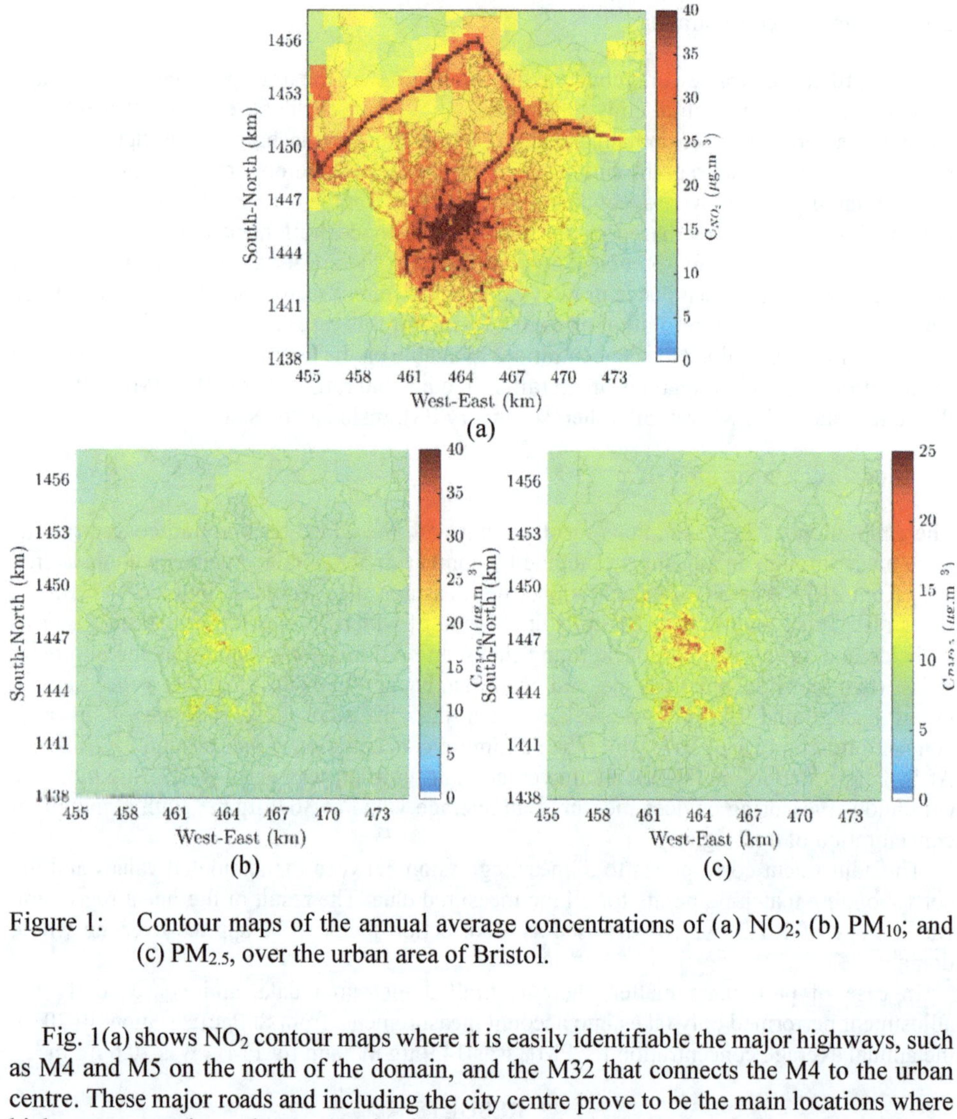

Figure 1: Contour maps of the annual average concentrations of (a) NO_2; (b) PM_{10}; and (c) $PM_{2.5}$, over the urban area of Bristol.

Fig. 1(a) shows NO_2 contour maps where it is easily identifiable the major highways, such as M4 and M5 on the north of the domain, and the M32 that connects the M4 to the urban centre. These major roads and including the city centre prove to be the main locations where higher concentration values are attained. Fig. 1(b) and (c) shows contour maps for PM_{10} and $PM_{2.5}$, respectively. For both maps, the higher values are achieved in the centre of the urban area, mainly caused by residential and commercial combustion but also increased by the existing heavy traffic. However, for particulate matter no exceedances of the EU legal limits are registered. Air quality results indicate a maximum value of 91.2 $\mu g.m^{-3}$ for NO_2, of 25.1 $\mu g.m^{-3}$ for PM_{10} and for $PM_{2.5}$ of 22.3 $\mu g.m^{-3}$.

As previously mentioned there was a total of 111 measurement points with available data for NO_2. The applied methodology is the same for all the selected pollutants, therefore only NO_2 is presented. Fig. 2 shows the simulated values of NO_2 for each computational cell corresponding to the location of the measurement point, the simulation results with the added background concentrations and the final concentrations, with the background value added to the simulated concentrations as well as the adjustment factor. For the computational cells corresponding to the location of a measurement equipment the final concentration (simulation + background + adjustment) is equal to the real measured concentration at the same location.

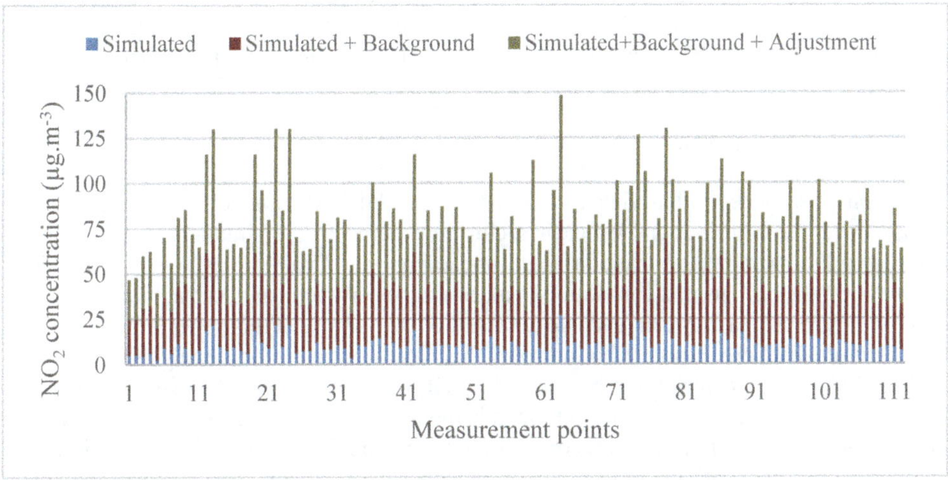

Figure 2: Comparison of NO_2 concentrations for each computational cell corresponding to the location of a measurement point, including the simulated values, the simulation values with added background and the simulation values with the background concentrations together with the adjusted values.

Fig. 2 denotes the underestimation between the simulation results and the measurements, already identified before. Although the simulated maximum values are achieved in the same points as the maximum measured concentrations. The underestimation reduces significantly when the background concentrations are added.

Table 1 presents the population of Bristol that is potentially exposed to concentrations above the EU and WHO Guidelines limits. For NO_2 the limits established by the EU and the WHO are equivalent, being 40 $\mu g.m^{-3}$ for the annual mean. As for particulate matter, the limits diverge between both standards, with WHO showing much firmer limits. PM_{10} values under the EU annual mean limits are 40 $\mu g.m^{-3}$ and under WHO guidelines are 20 $\mu g.m^{-3}$, for $PM_{2.5}$, the EU established for the annual mean limit value of 25 $\mu g.m^{-3}$ and for the WHO limits it's established at 10 $\mu g.m^{-3}$.

As presented in Table 1, 5% of Bristol population are exposed to concentration values of NO_2 above the EU limits and, although the particulate matter are below the EU limits, when stricter limits (WHO guidelines) are applied the number of inhabitants exposed becomes significant. When looking at the domain grid cells with values over the WHO limits, for $PM_{2.5}$ the area covered is 7% and for PM_{10} the area covered is less than 1% of the whole

Table 1: Population potentially exposed to levels above EU limits and WHO guidelines to NO_2, PM_{10} and $PM_{2.5}$.

Pollutant	Population potentially exposed (%)	
	EU limits	WHO guidelines
NO_2	5%	5%
PM_{10}	0%	1%
$PM_{2.5}$	0%	25%

domain, while for NO_2 the total area is around 2%. It is important to note that comparing the area covered to the whole domain, it may seem low but the cells with higher values are mainly located in the centre of the Bristol urban area, which combines with the higher population density.

3.2 Source apportionment

Based on the concentration fields obtained by the URBAIR model, it is important to comprehend the contributions by each sector considered in the simulations. Overall, the background concentrations have a high impact on the final values. On the obtained simulations, the background contribution values for the peripheral area can achieve over 80% and for the urban area they can reach up to 60%. For the purpose of this paper it is not relevant to present the background contribution so by this, it was left out of this analysis.

Fig. 3 shows the contribution by sector for NO_2 concentrations and since the sector that contributes the most for NO_2 concentrations is the transport, it is also presented the average contribution by each transport category. The fleet composition was divided into six major categories: car, bus, van, motor (includes mopeds and motorcycles) and medium and heavy truck.

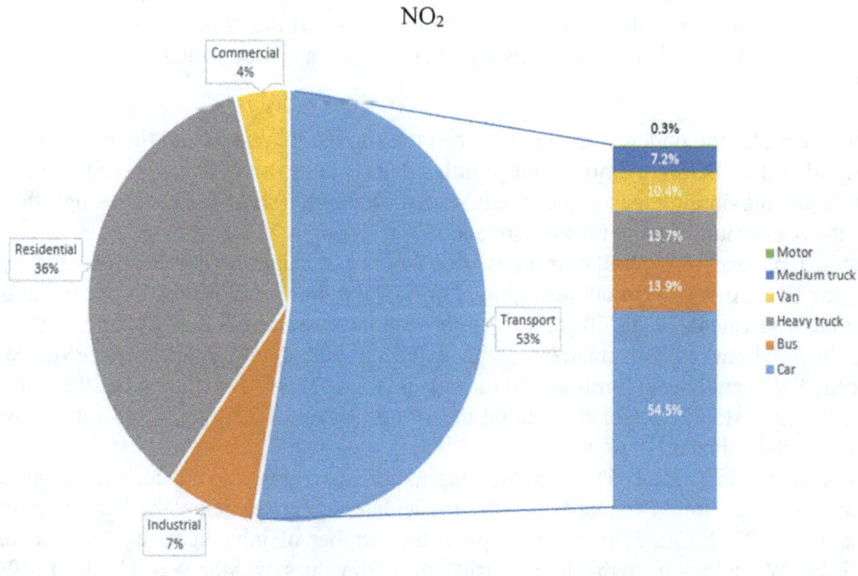

Figure 3: Contribution by sector for NO_2 concentrations and contribution by category for transport sector.

As revealed by Fig. 3, the major contributor is the transport sector. Therefore, to understand behavioural aspects the total concentration for this sector was decomposed into different fleet categories, although for the simulations the fleet composition was divided into weekdays and weekends, only the average composition is presented. Analysing the contribution by category for an average day, as expected the biggest contribution comes from the cars category (54.5%), followed by bus (13.9%) and heavy truck (13.7%).

The total transport concentration for the whole domain indicates a decrease of 25.4% between a typical weekday and a typical weekend day. In terms of category contribution is noticeable a difference between weekends and weekdays for each category. On weekdays, medium and heavy trucks represent twice the contribution of a typical weekend, where occurs a decrease of the contribution by car, bus and van, while for motor the differences are residual.

Fig. 4 shows the contribution by sector for PM_{10} concentrations. Residential accounts for 82% of the total concentration being the major simulated source of particulate matter. The contribution for $PM_{2.5}$ is assumed to be the same as PM_{10}, therefore only PM_{10} analysis is presented. The figure below also presents the disaggregation of the total residential concentrations by category. Residential sector was divided into 8 main categories: energy efficient fireplaces/stoves using solid biomass, combustion plants using solid biomass, advanced stoves using solid biomass, combustion plants using hard coal, conventional stoves using solid biomass, fireplaces using solid biomass; and a category denominated "other" which aggregates combustion plants using LPG (liquefied petroleum gas), combustion plants using gas/diesel oil, combustion plants using natural gas and advanced fireplaces using solid biomass.

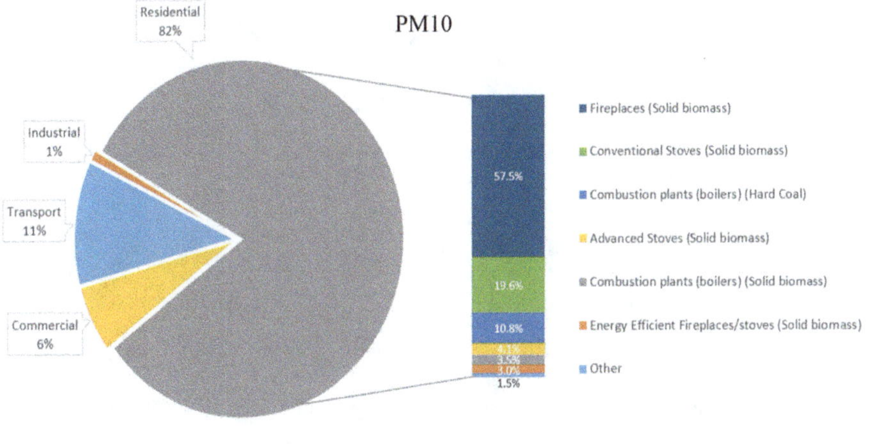

Figure 4: Contribution by sector for PM_{10} concentrations and contribution by category for the residential sector.

For the residential sector, the category with the highest contribution for particulate matter concentration is fireplaces burning solid biomass (57.8%), followed by conventional stoves burning solid biomass (19.6%) and combustion plants burning hard coal (10.8%).

To analyze the behavioral tendencies for the residential sector it was important to examine the daily concentration variation during the year. Fig. 5 represents the concentration on a daily average for particulate matter for the year of 2015.

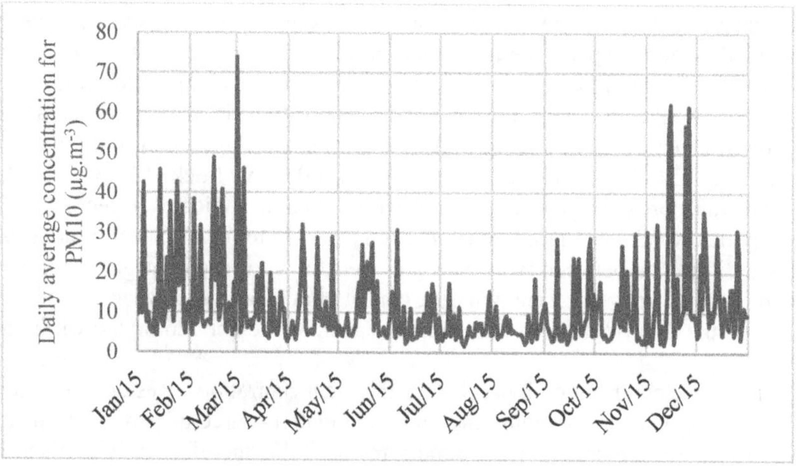

Figure 5: Daily average of PM_{10} concentrations for 2015. The average of PM concentrations result from a spatial average over all the domain.

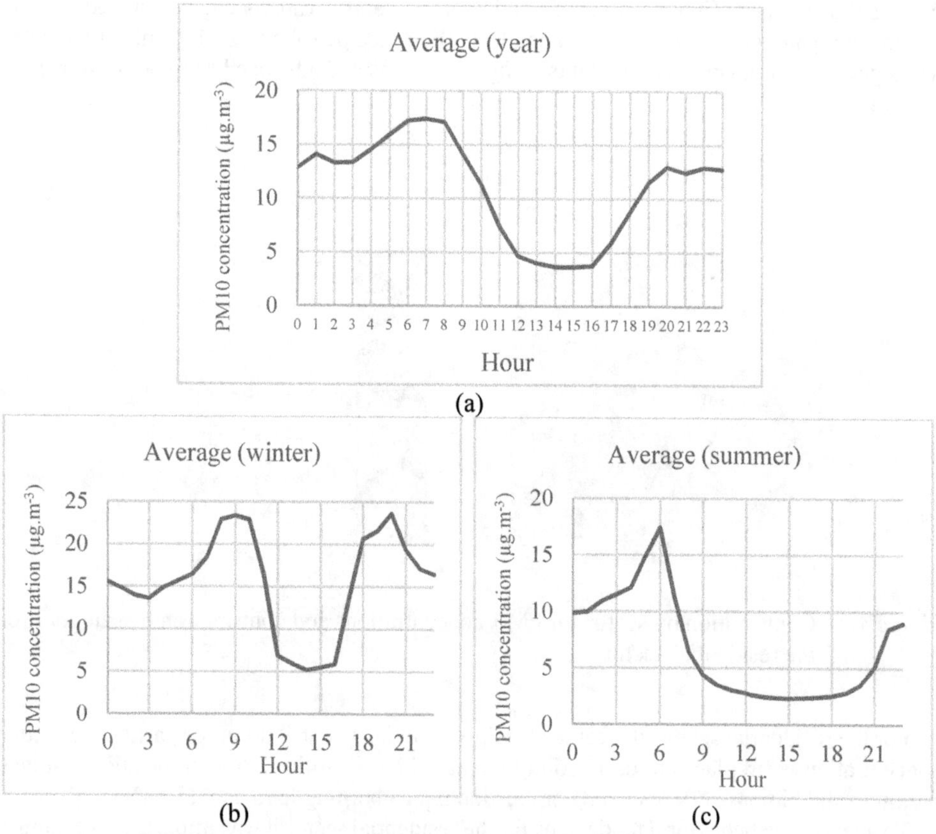

Figure 6: Average particulate matter concentration for the residential sector by hour for (a) All year; (b) Winter; and (c) Summer.

The daily average distribution of PM_{10} concentrations over the year denotes significant differences between winter and summer months. During the colder periods maximum concentrations are achieved and the values are commonly superior to warmer months. This outcome can be easily explained due to the high demand for solid biomass burning for heating purpose during colder periods. This proves to be a good indicator of the behavior of citizens related to energy consumption but also recurring air pollution episodes during winter.

Fig. 6(a) shows the concentration average for the whole year for each hour. Fig. 6(b) and (c) show the concentration average for winter months and for summer months, respectively.

The hourly annual average on Fig. 6 easily distinguishes the peak hours when maximum values are achieved, but also the lowest values periods. The maximums are reached mainly during early mornings between 6 h and 8 h, and the minimums are roughly at the afternoon around 13 h and 16 h. When comparing plot (b) and (c), the tendencies are diverse but having a common peak early in the morning. For winter months the values are overall much higher than summer months, showing two maximum periods during the day, being those in the morning between 7 h and 10 h and at the end of the day around 18 h and 21 h. For summer months, the trend is significantly different comparing to the winter. A maximum is reached early in the morning and it is the only peak during the day, having a larger minimum period during the day. These trends can be explained by being majorly incited for heating purposes and related to the population behavior, for example, working and school time schedule.

4 CONCLUSIONS

This work focused on the behaviour of the Bristol citizens and its impacts on air quality. The assessment of impacts on air quality was based on NO_2, PM_{10} and $PM_{2.5}$ concentrations, taking into account the emissions from residential, commercial, industrial and transport sectors.

Looking at air quality results it is visible for NO_2 concentrations the existence of hotspots centred mainly on the urban area. Furthermore, high values are equally observed near the major highways. In terms of exceedances of the EU concentration limits, the only pollutant in disagreement with the law is NO_2, but when stricter limits are applied, as established by the World Health Organization, the particulate matter becomes a significant problem for the population exposed where 25% of the Bristol population is potentially exposed to harmful concentrations of $PM_{2.5}$.

Focusing on the contribution by sector, the major contributor for NO_2 is the transport sector and for particulate matter is the residential sector. Analysing the category contribution for NO_2 for the transport sector it was possible to conclude that the major contribution comes from cars (54.5%), followed by bus (13.9%) and heavy truck (13.7%). The distinction between weekday and weekend showed a decrease of concentration of 25.4% from a typical weekday to weekend. The contribution by category also changed where the biggest difference comes from medium and heavy trucks, where the weekday contribution is twice the values of the weekend.

When considering the contribution of the residential sector and disaggregating into categories, for PM_{10}, the major contribution comes from fireplaces burning solid biomass (57.8%) followed by conventional stoves burning solid biomass (19.6%) and combustion plants burning hard coal (10.8%). The analysis of the concentration profile by day showed bigger values for colder months comparing to warmer periods, proving that heating is the main purpose. As a result, during winter the concentration values are significantly bigger than during the summer. When observing the hourly variation, the results demonstrate that, as concluded before, the differences are caused by outside temperature but the work/school schedule define the profile tendencies. Therefore, during winter the peak values are during morning and at the afternoon/night, achieving low values between 13 h and 16 h. For summer periods, the results show only one brief peak during morning and followed by a large period of low values.

Future developments under ClairCity project include the continuation of this work by evaluating the impact of citizen's behaviour on air pollution; i.e., considering the disaggregation by citizen's behaviour patterns for example, transport to school, to work, to shopping, etc. Therefore, the main objective is to influence people's knowledge, to encourage change in citizen's behaviour by promoting a more participative society on solving air quality problems and reducing carbon footprint.

ACKNOWLEDGEMENTS

This work was supported by the ClairCity project. ClairCity has received funding from the European Union's Horizon 2020 research and innovation programme under grant agreement 689289. J. Ferreira is funded by national funds (OE), through FCT – Foundation for Science and Technology, I.P., in the scope of the framework contract foreseen in the numbers 4, 5 and 6 of the article 23, of the Decree-Law 57/2016, of 29 August, changed by Law 57/2017, of 19 July. Thanks are also due for the financial support to the PhD grant of S. Coelho (SFRH/BD/137999/2018), and to CESAM (UID/AMB/50017/2019), to FCT/MCTES through national funds, and the co-funding by the FEDER, within the PT2020 Partnership Agreement and Compete 2020.

REFERENCES

[1] European Environmental Agency (EEA), Air quality in Europe: 2018 Report, EEA Report No 12/2018, 2018.

[2] Maas, R. & Grennfelt, P. (eds), Towards cleaner air. Scientific Assessment Report 2016. EMEP Steering Body and Working Group on Effects of the Convention on Long-Range Transboundary Air Pollution, Oslo, 2016.

[3] European Environmental Agency (EEA), Air quality in Europe: 2014 Report, EEA Report No 5/2014, 2014.

[4] World Health Organization (WHO), *Ambient Air Pollution: A Global Assessment of Exposure and Burden of Disease*, WHO, 2016.

[5] European Environmental Agency (EEA), Europe's urban air quality: Re-assessing implementation challenges in cities, EEA Report No 24/2018, 2018.

[6] Dias, D. et al., Assessing the importance of transportation activity data for urban emission inventories. *Transportation Research Part D-Transport And Environment*, **62**, pp. 27–35, 2018.

[7] Borrego, C. et al., Urban scale air quality modelling using detailed traffic emissions estimates. *Atmospheric Environment*, **131**, pp. 341–351, 2016.

ASSESSMENT OF PRIMARY AIR POLLUTANTS IN A TROPICAL METROPOLITAN REGION BY COMBINING LOCAL AND GLOBAL EMISSIONS INVENTORIES

YASMIN KAORE LAGO KITAGAWA[1], ERICK GIOVANI SPERANDIO NASCIMENTO[2], LÍLIAN LEFOL NANI GUARIEIRO[2], TACIANA TOLEDO DE ALMEIDA ALBUQUERQUE[1,3] & DAVIDSON MARTINS MOREIRA[1,2]
[1]Federal University of Espírito Santo (UFES), Brazil
[2]Manufacturing and Technology Integrated Campus (SENAI CIMATEC), Brazil
[3]Federal University of Minas Gerais (UFMG), Brazil

ABSTRACT

The community multiscale air quality (CMAQ) model was evaluated in the metropolitan region of Salvador (MRS), which is the capital of the state of Bahia, Brazil. Once located at the tropics, two episodes were selected (one in the dry season and another in the rainy season), in order to perform the assessment. The meteorological information required for air quality modeling was driven by the weather research and forecasting (WRF) model: its performance was evaluated against observational data collected by monitoring stations located in Salvador city, Brazil. For the emissions inventory, we applied the Emissions Database for Global Atmospheric Research (EDGAR), in conjunction with an estimated emissions inventory of local point sources located in the major industrial complexes, in addition with the biogenic emissions estimated by the model of emissions of gases and aerosols from nature (MEGAN). To compute these different emissions sources, we used the sparse matrix operator kernel emission (SMOKE). Following that, the CMAQ model was applied, in order to simulate the chemical transport and formation of air pollutants based on the meteorological and emissions input previously described. The required boundary conditions were determined by also applying the CMAQ model, which used data from the MEGAN and EDGAR inventories, to a larger domain that incorporates the domain in which the MRS is represented. The application of these modeling tools was shown to be a good practice in the assessment of air quality in urban areas; thus, this work aims to be the first great effort to study, simulate and validate the chemical transport of pollutants over the MRS, based on a hybrid emissions inventory and using the state of the art in the computational atmospheric modeling field.

Keywords: Brazil, WRF, SMOKE, CMAQ, EDGAR, MEGAN, comparative emissions, emissions inventory, air quality.

1 INTRODUCTION

Despite the efforts put into practice in seeking for strategies and actions that contribute to the improvement of air quality in Brazil, there is a scarcity of information that is still a troublesome, which counts with only 1.7% of coverage of Brazilian cities by air monitoring systems [1], and most of these cities count with a single station for monitoring a whole urban region. Beyond that, several regions of South America face serious deficiencies in local emission data, measurement campaigns, and the application of air quality models [2] that are critical for generating information; and consequently, to have a better understanding of the air quality status of Brazil.

The Brazilian legislation was updated through resolution number 491/2018 [3], which sets new air quality standards with intermediate and progressive goals, until reaching the final standard suggested by the World Health Organization (WHO).

We can say that a first step was taken towards stricter public policies, that can probably lead us to an improvement in the air quality in Brazil. As we know, air quality management is based on a proper site diagnosis that involves a reliable description of the initial atmosphere

WIT Transactions on Ecology and the Environment, Vol 236, © 2019 WIT Press
www.witpress.com, ISSN 1743-3541 (on-line)
doi:10.2495/AIR190101

and boundary conditions, the representation of land use, the coverage of changes, and the development of an emissions inventory; however, even with a great lack of information of air quality from meteorological stations, radiosondes, radars, an emissions inventory, particulate matter speciation, etc. The endeavour to understand the atmospheric chemical conditions over industrialized Brazilian cities has been done by employing different approaches and methodologies to estimate and generate these pieces of information. Thus, different chemical transport models (CTMs) and also input data are being used in air quality assessment. In addition, CTMs can cover larger regions, when compared with the performance of air monitoring stations that have specific time–space scales and can be economically inviable to be implemented over vast areas like Brazil.

Most studies focus on urban–industrial areas located in the Brazilian south–southeast regions [4]–[8]; while almost none have addressed the other parts of Brazil, such as the northeast region, where there had been a boom in industrial activities and growth in population, over the last decades. Therefore, one of the objectives of the present work was to employ the best practices seen in others' work that employ CTMs to comprehend the local air pollution over the Metropolitan Region of Salvador (MRS). For that, the WRF-SMOKE-CMAQ modeling system was used to assess the chemical transport of air pollutants (like SO_2, CO, NO_2) over the MRS during two different seasons, dry and rainy, based on a hybrid emissions inventory applying a local inventory in conjunction with the global inventories called EDGAR and MEGAN, and to compare these modeling results with the observed air quality data from six monitoring stations located within the region.

2 METHODOLOGY

2.1 Study area description

In our study area, there have been no quantitative studies yet that evaluate the use of CTMs through observational data over the MRS. In addition, there was also no application of inventories that could account for different emission sources, as well as the chemical transport of air pollutants from other regions. MRS is an urban–industrial and costal area, located in the state of Bahia in the north east of Brazil, formed by 13 cities with a total population of over 4 million inhabitants. Beyond the activities related to tourism and commerce, which play a very important role in the region's economy, this area also counts with several industries, such as: petrochemical, automotive, metallurgical and construction, whose activities and major companies are displayed in Fig. 1.

The Cristal factory manufactures chemical products, including titanium dioxide and sulfuric acid, that are used in the paint, paper, plastic bottle, vinyl siding and packaging materials industries. The Camaçari Industrial Complex (CIC) includes producers of petrochemical products that are used to make thermoplastic resins, fertilizers and copper metallurgy, among others.

The Aratu Industrial Center (CIA), with its operation covering the areas of Salvador, Simões Filho and Candeias counties, is a multisectoral industrial complex that has several enterprises: chemical segments, footwear, food, metallurgy, furniture, plastics, fertilizers, electronics, beverages and textiles; and the Landulpho Alves Refinery (RLAM), in São Francisco do Conde and Madre de Deus counties, which belongs to Petrobrás S.A. and is also a petrochemical complex.

To distinguish seasonal differences in the MRS, we analyzed the observed data for precipitation rainfall. Then, we selected two periods: a dry one (from 16 January–15 February) and a rainy one (23 June–23 July) within the year 2015, totaling 30 days for each

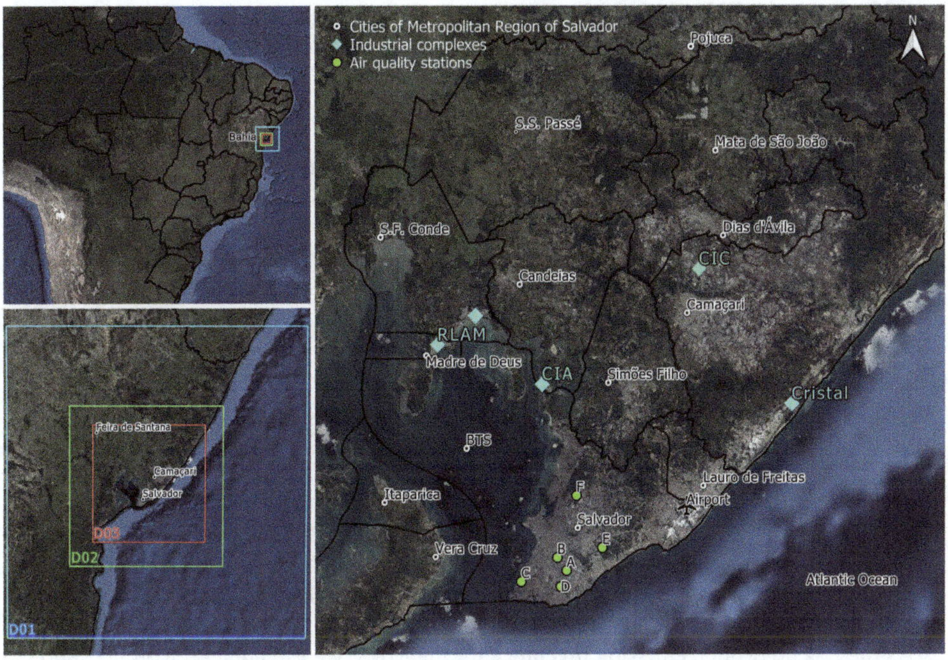

Figure 1: The location of Bahia state in Brazil (upper left), and the domains used in the configuration of the meteorological and photochemical models (lower left). The major industrial complexes (cyan diamonds), and the air quality stations (green dots) are shown within the MRS.

period of analysis. Both periods comprise part of a vacation period in which vehicle emissions may differ substantially and influence air pollutant concentrations; however, it is worth mentioning that vehicular emissions were not accounted for in this work.

2.2 Setup of the WRF model

The meteorological fields were generated by the WRF model [9], version 3.9.1. Its outputs were used by the MCIP model, in order to create input data that are needed by SMOKE and by CMAQ. A brief description of the spatial and physical configuration of the WRF model is given below.

The initial and boundary conditions came from the GDAS/FNL operational model global tropospheric analyses data set [10], available at a resolution of 0.25 degrees every 6 hours. The land use data were provided by the US Geological Survey at 5 min, 2 min and 30 s resolutions. We used three nested domains with a 9 km (39×39 cells), 3 km (60×60 cells), and 1 km grid resolution (132×132 cells), as shown in Fig. 1, with 23 vertical levels with the model's top one set at 50 hPa. The physical configurations adopted were: Lin microphysics, Eta similarity surface layer scheme, Mellor–Yamada–Janjić planetary boundary layer scheme, Grell–Devenyi cumulus parametrization, Dudhia shortwave radiation, Rapid Radiative Transfer Model longwave radiation and the Noah land-surface model. Even though WRF performance in this study had reasonably good agreement between the observed and simulated data for the dry and rainy periods, it was decided not to show the hourly variation plots and statistical metrics of this WRF performance, since detailed evaluation of the WRF model results to the MRS can be found in previous studies [11].

2.3 Hybrid emissions inventory details

To build the hybrid emissions inventory, we used information from EDGAR, MEGAN, and a local estimated emission inventory based on point sources. SMOKE, which is an emissions processor used by CMAQ, was prepared to run for the coarse domain (D01) and the domain of interest (D03) for Fig. 1. The EDGAR and MEGAN data was applied in the D01, in order to generate boundary conditions for the D03.

The EDGAR is a global inventory [12] that contains annual emissions of greenhouse gases and air pollutants from 1970 to 2012. We used the newest version, 4.3.2 (https://cidportal.jrc.ec.europa.eu/ftp/jrc-opendata/EDGAR/datasets/v432_AP/), with a spatial resolution of 0.1 degree. Datasets for CO, NO_x, NMVOC, NH_3, SO_2, BC, OC, PM_{10}, $PM_{2.5}$ with base year of 2012 were used (the closest year to 2015).

The MEGAN is a global model used to estimate fluxes of biogenic compounds, between terrestrial ecosystems and the atmosphere, that can be used for regional air quality modeling and global earth system modeling studies [13]. In this study, MEGAN version 2.04 was run in an offline mode, which means that the MEGAN driving variables used meteorology and land cover data from WRF to provide emissions in a format that is suitable for the SMOKE processor.

As there is no official emission inventory for the MRS, we used other's work to build the local inventory [14], with estimated emission rates for SO_2, NO_x and CO based on data of consumption, fuel composition, type of boiler/furnace, hours of operation per year and oil/gas density. The emission factors were taken from US EPA AP-42 guidelines. Again, because of the absence of full information about the other source types in his work, we decided to count only the point sources category, which information about latitude–longitude coordinates and stack parameters (e.g. stack height and diameter, exit gas velocity, exit gas temperature. Thus, we used information from 42 companies located at CIC and RLAM (Fig. 1), numerating 203 point sources, which were only computed in the D03.

2.4 Setup of CMAQ model

The chemical transport model CMAQ [15] version 5.2.1 was applied in this study. To promote the assessment of air quality, information about meteorology, emissions, and initial (IC) and boundary conditions (BC) were necessary. The BC used by the D01 were a default setup of the BCON processor and had static concentrations. The BC used by the D03 were generated by the simulations performed with D01, and had time-dependent concentrations. The IC for both domains had static concentrations and were also estimated by the CB05 gas mechanism with sixth-generation to aerosol, also an option in ICON processor. The temporal profiles used to convert the annual emissions estimated [14] to hourly emissions were defined uniformly over time. The spatial resolution comes from the WRF settings. The chemical speciation of NO_x was of 95% NO and 5% NO_2, as previously suggested [16]. The CMAQ configurations are summarized in Table 1.

The performance of the CMAQ model was done by comparing the hourly observed data from six different stations located in Salvador city (to be found at (a) Avenue ACM; (b) Barros Reis; (c) Campo Grande; (d) Itaigara; (e) Paralela; and (f) Pirajá in Fig. 1). Despite that the air quality monitoring network of the MRS has been operating since 2013, it is still limited in its spatial coverage and lacks detailed information. We calculated the statistical

Table 1: Spatial and physical configuration of the CMAQ model.

Domain	**Dry period**	**Rainy period**
Initial date	15 Jan. 2015 (00 UTC)	22 Jun. 2015 (00 UTC)
Final date	23 Feb. 2015 (06 UTC)	23 Jul. 2015 (06 UTC)
Grid resolution	1 km	
Column and row numbers	130 × 130	
Chemical mechanism of the IC	cb05e51_ae6_aq (used in D01 and D03 domains)	
Chemical mechanism, BC-D01	cb05e51_ae6_aq (default setup)	
Chemical mechanism, BC-D03	Outputs of simulations run with D01	
Chemical mechanism, CCTM	cb6r3_ae6_aq	

metrics of the observed mean (OBS), modeled mean (MOD), root mean square error (RMSE), normalized mean bias (NMB) and correlation coefficient (r). Graphical plots of the mean hourly variation comparing the observed and modeled SO_2, CO and NO_2 concentrations in both the dry and rainy period at each station were also presented, in order to aid in the models' performance evaluation. All plots are presented given the local time and were generated by Python packages found within the Google Earth program.

3 RESULTS AND DISCUSSION

3.1 Sulfur dioxide (SO_2) concentrations

The mean observed SO_2 concentrations presented values approximately up to1 ppb, with seasonal differences between dry and rainy periods being hardly seen; however, the dry period showed values a bit higher than the wet period in hourly variation. We noted that there were slight changes in the observed SO_2 concentrations at 6 a.m., in the dry period at Itaigara station (Fig. 2(d)); and also at 11–12 h at the Barros Reis station (Fig. 2(b)). The lowest difference of mean observed SO_2 concentrations between the dry and rainy periods was seen at Pirajá station (Table 2 (ID f)). Among the others, the dry period presented average values that were higher than twice those of the wet period, except for the Barros Reis station (Table 2 (ID b)). By comparing with the Brazilian air quality standards, we found that the observed SO_2 was far from the limit established by the environmental council (which is 7 ppb for a 24 h mean); the limits were up to the value of 2 ppb, i.e. they did not exceed critical levels. It was expected that there would be larger observed SO_2 concentrations, due to the presence of considerable vehicular fleet activity and also because of many sources from petrochemical activities.

The CMAQ model has overestimated all SO_2 concentrations, except in the first hours of the morning, between midnight and 3–4 a.m., in some sites. The largest discrepancies occurred especially at 6 and 7 a.m. within the rainy and dry period, respectively.

The increasing SO_2 concentrations were shown by all stations for the wet period (around 3 a.m.), and in the dry period (with a time lag of 1 h). These peaks observed in the mean modeled SO_2 data are maybe most probably caused by industrial chimneys represented in the local inventory, since the maximum values produced by local SO_2 were about seven times greater than the maximum values estimated by the EDGAR inventory (MEGAN does not estimate SO_2 concentrations as input data). Because local inventory is the major contributor

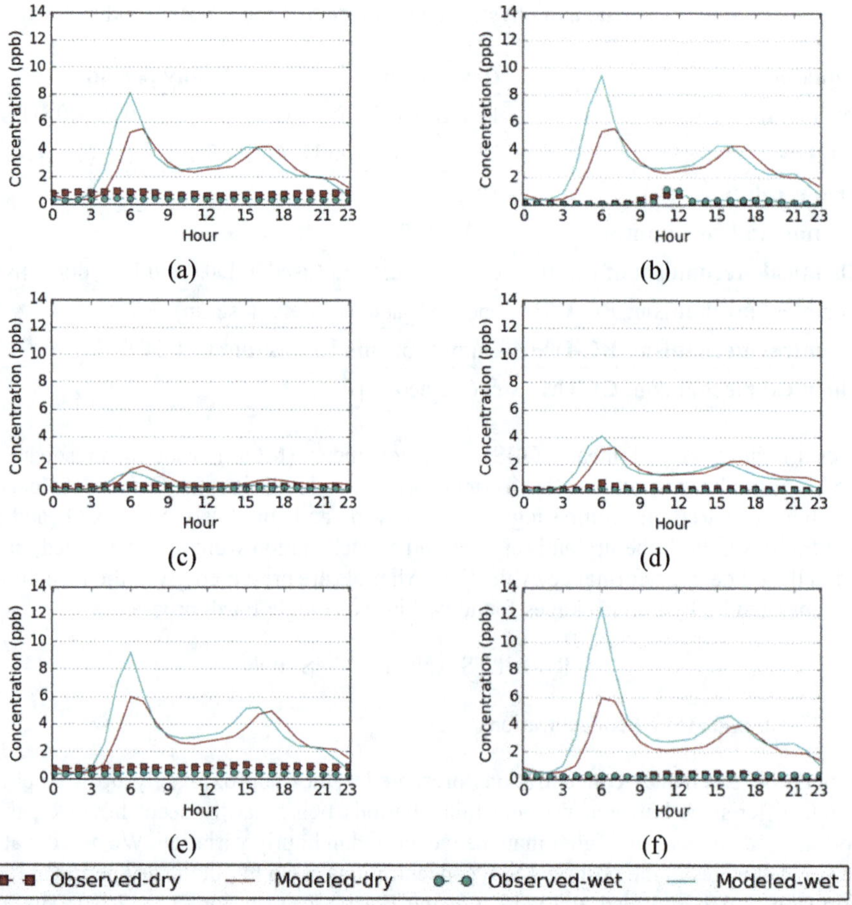

Figure 2: Comparison of mean hourly variation of the modeled (solid lines) and observed (dashed lines) SO_2 concentrations at the following stations. (a) ACM; (b) Barros Reis; (c) Campo Grande; (d) Itaigara; (e) Paralela; and (f) Pirajá.

Table 2: Statistics comparing modeled and observed SO_2 for each air quality station, in ppb units.

ID	OBS		MOD		RMSE		NMB		r	
	Dry	Wet	Dry	Wet	Dry	Wet	Dry	Wet	Dry	Wet
a	0.79	0.35	2.51	2.78	2.39	3.33	2.19	6.86	0.03	0.02
b	0.18	0.23	2.58	3.00	2.88	3.91	13.10	12.28	0.08	0.01
c	0.37	0.13	0.64	0.39	0.73	0.91	0.68	1.98	0.11	-0.01
d	0.24	0.10	1.40	1.40	1.44	1.85	4.77	12.82	0.35	0.25
e	0.83	0.37	2.80	3.21	2.64	3.93	2.36	7.76	0.07	0.05
f	0.24	0.26	2.47	3.55	2.74	4.82	9.26	12.54	0.04	0.01

ppb: parts per billion; SO_2: Sulfur dioxide.

to SO_2 emissions, we can say that the emissions factors of AP-42 were assumed [14], inducing the overestimation of SO_2 emissions rates. Beyond that, the temporal profiles were assumed to be uniform over time.

The mean modeled SO_2 concentrations were higher during the rainy period than the dry period, except for the Campo Grande station (Table 2 (ID c)); however, overall, the hourly variation of modeled results presented almost the same behavior in the different sites of Salvador city, with the peaks occurring during rush hours, even though mobile sources were not accounted for in the local inventory. The highest errors were most associated with the rainy season, showing the great dependence of the CTMs on the meteorological model results.

All sites had very poor values of correlation coefficient: the highest one was shown by Itaigara ($r_{dry} = 0.35$; $r_{wet} = 0.25$). Campo Grande had the lowest errors in statistical indexes, while the others presented almost the same error values.

3.2 Carbon dioxide (CO_2)

The seasonal differences of mean observed CO_2 concentrations were noted at the Barros Reis and Paralela stations (Fig. 3(b) and (e)); however, they had the inverse behavior of the highest values. Seasonal variation was slightly different among the others stations. The observed CO_2 concentrations' limits were up to 1 ppm in the dry period and up to 2 ppm in the rainy season, which means it did not reach the value of 9 ppm, established by the Brazilian air quality council. The mean diurnal variation in both periods was more prominent than the results presented for SO_2. The peaks of observed CO_2 occurred in the beginning of the morning (6–7 a.m.) and in the late afternoon (at 18 h–19 h).

The CMAQ model has clearly underestimated two sites (Fig. 3(b) and (c)); however, the behavioral changes of the diurnal variations were captured by the model, meaning that the CMAQ model showed a better agreement for CO_2 results. The peaks of mean modeled CO_2 also may be or are probably caused by the industrial chimneys that are represented in the local inventory, as the maximum values produced by local CO_2 were about three times greater than the maximum values estimated by the EDGAR inventory. In the MEGAN estimate, we found almost no CO_2 concentrations (with values around 0.05 to 0.1 moles/s of CO_2). By the way, all species data provided by MEGAN had concentrations values close to zero, with the dry season yielding twice as much an amount as the rainy season. Inversely, EDGAR provided higher concentrations during the rainy season.

Table 3 presents the statistical metrics calculated for CO_2 results. The mean observed CO_2 concentrations between both periods were similar, except for Barros Reis, where CO_2 concentrations were 60% higher during the rainy season; and at Paralela, where CO_2 concentrations were double during the dry period. Mean modeled CO_2 concentrations were virtually identical for both periods, as well as the statistical errors, but with slight variations. The correlation coefficient values for CO_2 were higher than SO_2 values, indicating a better agreement for this pollutant.

3.3 Nitrogen oxides (NO_x)

The analyses of NO_x results were made though the assessment of NO_2 concentrations, since only the air monitoring stations provide this kind of data. Despite that, the daily variation of observed NO_2 presented a similar behavior to CO_2, with the maximum occurring in the early morning (from 6–7 a.m.), decreasing during the day, and increasing again in the late afternoon. Another finding was that, unlike the SO_2 and CO_2 results, the NO_2 concentrations

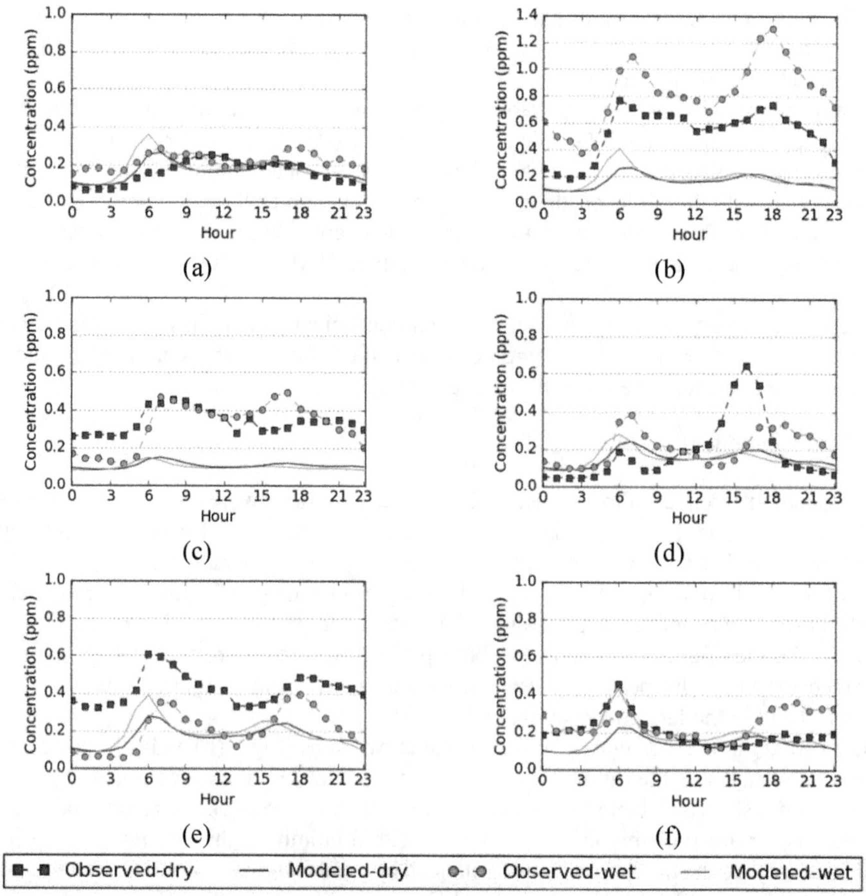

Figure 3: Comparison of mean hourly variations of modeled (solid lines) and observed (dashed lines) CO_2 concentrations at the monitoring stations in: (a) ACM; (b) Barros Reis; (c) Campo Grande; (d) Itaigara; (e) Paralela; and (f) Pirajá.

Table 3: Statistics comparing modeled and observed CO_2 in ppm for each air quality station.

ID	OBS		MOD		RMSE		NMB		r	
	Dry	Wet	Dry	Wet	Dry	Wet	Dry	Wet	Dry	Wet
a	0.16	0.22	0.16	0.17	0.14	0.18	0.05	-0.21	0.19	0.19
b	0.52	0.82	0.16	0.18	0.41	0.76	-0.69	-0.78	0.68	0.32
c	0.34	0.32	0.11	0.10	0.32	0.32	-0.69	-0.70	0.23	0.23
d	0.18	0.21	0.15	0.15	0.24	0.24	-0.18	-0.29	0.31	0.16
e	0.42	0.20	0.17	0.18	0.27	0.20	-0.60	-0.09	0.48	0.23
f	0.20	0.24	0.14	0.18	0.18	0.25	-0.30	-0.27	0.14	0.08

ppm units: parts per million.

were significantly different between seasons: the lowest difference was seen in Itaigara station (Obs_{dry} = 8.03 ppb; Obs_{wet} = 10.31 ppb). The rainy season exhibited higher NO_2 concentrations than the dry season, and that was probably caused by the increase of emissions. Paralela station (Fig. 4(e) and Table 4 (ID e)) was the only station that did not present this behavior, most likely because it is located on an important route of vehicular traffic for Salvador city; hence, the concentrations measured at this station are directly influenced by traffic congestion.

The observed time series showed NO_2 concentrations varying up to 30 ppb in the dry season (with some peaks between 40–50 ppb); and reaching up to 40 ppb in rainy period (with some peaks between 50–70 ppb). In any case, the air quality standard (which is set at 100 ppb for 1 h mean) was not exceeded in either of both periods.

The CMAQ model captured the diurnal variation of NO_2; though the increase in the late afternoon was underestimated, especially during the wet period. The peaks of modeled NO_2 values again were provided by the local inventory, which presented the maximum NO_2 values about fifteen times greater than the maximum values estimated by EDGAR during the dry season (and thirteen times the values in the rainy period). Although knowing that biogenic sources can contributor to NO_x emissions, the MEGAN inventory estimated no quantities of NO_2. This is due to the fact that most of the NO_x emissions to the atmosphere in it was set up in the form of NO. Even so, the NO concentrations generated by MEGAN had low values, when compared to local and EDGAR inventories. Overall, the NO_2 results provided by the CMAQ model seemed to agree a little better with the monitoring stations, when qualitatively compared to the other pollutants (CO_2 and SO_2).

Table 4 shows that the mean modeled NO_2 concentrations were virtually similar for both periods; however, the statistical errors were found to be greater in the rainy season and the correlation coefficient values were lower. These discrepancies are associated with the high concentrations of NO_2 that were recorded during the rainy period.

4 CONCLUSIONS

To the best of our knowledge, there have not yet been any studies evaluating air pollutants (SO_2, CO_2, and NO_2) measured at surface of the MRS, using the WRF-SMOKE-CMAQ modeling system, nor combining different approaches of emission inventories.

Regarding the inventories, the data that came from MEGAN had values close to zero, meaning that it did not significantly contribute to the emissions rates read by the CMAQ model. Despite this, the EDGAR inventory provided slight concentrations, which were much lower than the local inventory, and it seemed not to have correctly represented the spatial distribution of the emissions (e.g. where one of the biggest petrochemical complexes exists in Latin America, the CIC, there is no representation of one of the most important primary air pollutants. The local inventory only has counted with industrial point sources, missing the representation of other types of sources. Therefore, we found nothing more appropriate than the development of the fully local emissions inventory for MRS.

Seasonal differences of the observed data between dry and rainy periods were quite difficult to note, this may be associated with the fact that the seasons in the tropics are not very well defined, such as they are in the middle latitudes/temperature climate zones. Important meteorological variables, such as the temperature and water vapor content that influence chemical reactions in the atmosphere do not present significant changes between both periods, so consequently, to the environmental conditions over the MRS. Beyond the meteorological conditions, we also can say that the emission rates of these air pollutants did not noticeably change between both periods. Regardless, the air quality standards established by the Brazilian environment council were not overtaken by them.

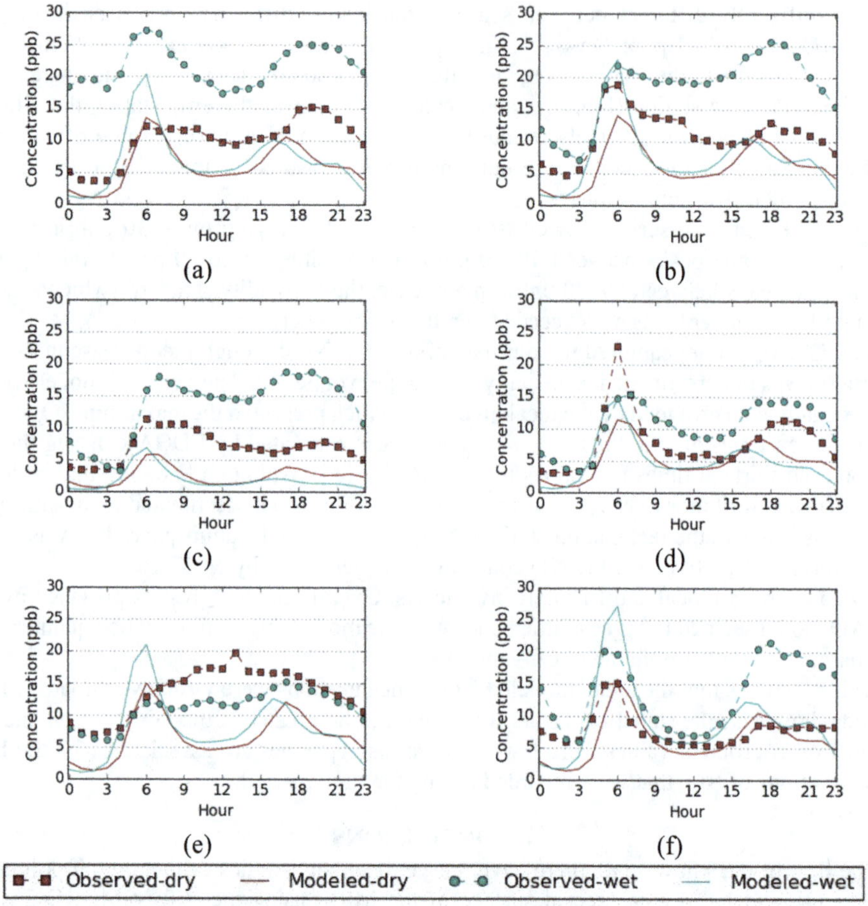

Figure 4: Comparison of hourly variation of modeled (solid lines) and observed (dashed lines) NO_2 concentrations at the stations in: (a) ACM; (b) Barros Reis; (c) Campo Grande; (d) Itaigara; (e) Paralela; and (f) Pirajá.

Table 4: Statistics comparing modeled and observed NO_2 for each air quality station, in ppb units.

ID	OBS		MOD		RMSE		NMB		r	
	Dry	**Wet**	**Dry**	**Wet**	**Dry**	**Wet**	**Dry**	**Wet**	**Dry**	**Wet**
a	10.04	21.65	6.00	6.94	5.47	17.80	-0.40	-0.67	0.56	0.23
b	11.05	18.31	6.20	7.53	6.18	13.45	-0.44	-0.59	0.64	0.32
c	6.98	13.39	2.37	1.77	6.10	13.32	-0.67	-0.87	0.45	0.13
d	8.03	10.31	5.07	4.94	5.43	7.89	-0.37	-0.52	0.67	0.35
e	13.50	11.06	6.62	7.70	8.44	6.89	-0.51	-0.30	0.36	0.29
f	7.80	13.55	6.29	9.04	4.12	9.77	-0.19	-0.33	0.47	0.36

NO_2: nitrogen dioxide; ppb units: parts per billion units.

As meteorology and topography play an important role over the MRS, and since the region experiences the influence of sea breezes that possibly aid to disperse the air pollutants concentrations, it is worth mentioning that the evaluation of this work was also done using monitoring stations which represented single points within a whole region, but do not exclude the possibility of having areas where the concentrations exceed the air quality standards.

The discrepancies found between the observed and modeled data can be attributed to the poor representation of the emissions inventories; and the deficiencies in periodic maintenance and bad location of the monitoring stations, the latter by reason of growing industrialization and urbanization over the MRS. Furthermore, it would be necessary to include monthly, daily and weekly variability of the emissions; and to improve the chemical speciation profiles. It is also fundamental to perform experimental campaigns, in order to evaluate the vertical profiles, especially of secondary pollutants. Nevertheless, we found that the CMAQ model was able to represent the hourly and daily variability of the air pollutants at the surface, despite having some large variances.

ACKNOWLEDGEMENTS

The authors thank the Supercomputing Center for Industrial Innovation (Yemoja-CIMATEC), the FAPESB, the NQualiAr of UFES, and also the Coordenação de Aperfeiçoamento de Pessoal de Nível Superior (CAPES) for granting the PhD scholarship to the first author.

REFERENCES

[1] Vormittag, E.M.P.A., Costa, R.R., Braga, A.A., Miranda, M., Nascimento, N.C. & Saldiva, P.H.N., Monitoramento da Qualidade do Ar no Brasil (Monitoring of Air Quality in Brazil). Instituto Saúde e Sustentabilidade (Institute of Health and Sustainability), 2014.

[2] Alonso, M.F. et al., An urban emissions inventory for South America and its application in numerical modeling of atmospheric chemical composition at local and regional scales. *Atmos. Environ.*, **44**, pp. 5072–5083, 2010. DOI: 10.1016/j.atmosenv.2010.09.013.

[3] CONAMA, Environmental National Committee, New national air quality standards, Resolution number 491/2018. www2.mma.gov.br/port/conama/legiabre.cfm?codlegi=740. Accessed on: 5 Feb. 2019.

[4] Borrego, C. et al., Modelling the photochemical pollution over the metropolitan area of Porto Alegre, Brazil. *Atmos. Environ.*, **44**, pp. 370–380, 2010. DOI: 10.1016/j.atmosenv.2009.10.027

[5] Gidhagen, L. et al., Experimental and model assessment of PM2.5 and BC emissions and concentrations in a Brazilian city: The Curitiba case study. *Atm. Chem. Phys. Dis.*, 2018. DOI: 10.5194/acp-2018-1094.

[6] Albuquerque, T.T.A. et al., WRF-SMOKE-CMAQ modeling system for air quality evaluation in São Paulo megacity with a 2008 experimental campaign data. *Env. Sci. Pollut. Res.*, **25**, pp. 36555–36569, 2018. DOI: 10.1007/s11356-018-3583-9.

[7] Gavidia-Calderón, M., Vara-Vela, A., Crespo, N.M. & Andrade, M.F., Impact of time-dependent chemical boundary conditions on tropospheric ozone simulation with WRF-Chem: An experiment over the Metropolitan Area of São Paulo. *Atmos. Env.*, **195**, pp. 112–124, 2018. DOI: 10.1016/j.atmosenv.2018.09.026.

[8] Pedruzzi, R. et al., Performance evaluation of a photochemical model using different boundary conditions over the urban and industrialized metropolitan area of Vitória, Brazil. *Env. Sci. Pollut. Res.*, 2019. DOI: 10.1007/s11356-019-04953-1.

[9] Skamarock, W.C. et al., *A Description of the Advanced Research WRF Version 3*. National Center for Atmospheric Research, 2008.

[10] National Centers for Environmental Prediction (NCEP)/National Weather Service/ National Oceanic and Atmospheric Administration (NOAA)/U.S. Department of Commerce. Updated daily. GDAS/FNL 0.25 degree global tropospheric analyses and forecast grids. Research data archive at the National Center for Atmospheric Research, Computational and Information Systems Laboratory. DOI: 10.5065/D65Q4T4Z.

[11] Kitagawa, Y.K.L., Nascimento, E.G.S., De Souza, N.B.P., Moreira, S., Albuquerque, T.T.A. & Moreira, D.M., Evaluation of the chemical transport of air pollutants in the metropolitan region of Salvador, Brazil. *WIT Transactions on Ecology and the Environment*, vol. 230, WIT Press: Southampton and Boston, pp. 519–530, 2018. DOI: 10.2495/AIR180481.

[12] Crippa, M. et al., Gridded emissions of air pollutants for the period 1970–2012 within EDGAR v4.3.2. *Earth Syst. Sci. Data*, **10**, pp. 1987–2013, 2018. DOI: 10.5194/essd-10-1987-2018.

[13] Guenther, A.B. et al., The model of emissions of gases and aerosols from nature version 2.1 (MEGAN2.1): An extended and updated framework for modeling biogenic emissions. *Geosci. Model Dev.*, **5**, pp. 1471–1492, 2012. DOI: 10.5194/gmd-5-1471-2012.

[14] Lyra, D.G.P., Modelo Integrado de Gestão de Qualidade do Ar da região Metropolitana de Salvador (Integrated Model of Air Quality for the Metropolitan Region of Salvador). PhD thesis, State University of Campinas, São Paulo, Brazil, 2008.

[15] Byun, D. & Schere, K.L., Review of the governing equations, computational algorithms, and other components of the Models-3 community multiscale air quality (CMAQ) modeling system. *Appl. Mech. Rev.*, **59**(2), pp. 51–77, 2006. DOI: 10.1115/1.2128636.

[16] Carpenter, M.L. & Nokleby, J., In-stack measurement of NO_2 partitions from various combustion sources. Presented at *Annual Meeting for the Environmental Professionals of Arizona*, Scottsdale, AZ, USA, 2012.

COMPARISON OF TWO $PM_{2.5}$ FORECASTING MODELS IN OSORNO, CHILE

PATRICIO PEREZ & BYRON NUÑEZ
Departamento de Fisica, Universidad de Santiago de Chile, Chile

ABSTRACT

According to a recent study of the World Health Organization (WHO), Osorno, a medium sized city in the south of Chile is among the most polluted city in South America. With 150,000 habitants, the city has unfavorable conditions for pollutant dispersion. It is located in a valley between the Andes mountains and a coastal range. Thermal inversions that trap particles emitted mainly by wood combustion used for heating are frequent during fall and winter. Air pollution forecasting models will be useful for authorities to implement a policy of restrictions to emissions when necessary. These models are also useful for the habitants so that they have the possibility to avoid places where air quality is critical, and also so they can choose to restrict physical exercise. The specific meteorological and pollution variables (mostly associated with wood combustion) that can be used as input for statistical $PM_{2.5}$ forecasting models are identified. Results obtained with two models based on artificial neural network techniques: a multilayer perceptron (MLP) and a radial basis function model (RBF) are presented. Both models show comparable accuracy. With them it is possible to anticipate, 24 h in advance, more than 80% of the high concentration episodes during 2018.
Keywords: air quality forecasting, neural networks, air pollution episodes.

1 INTRODUCTION

Osorno is a medium size city located in the south of Chile at 40°57'S and 73°08'W and it is 900 km south of the capital of the country, Santiago. Average altitude is 19 m over sea level and lies in a valley between the Andes mountains and a coastal range. At present, the city has a population approaching 150,000 habitants. Fig. 1 shows a satellite picture of the city in which we can observe the Andes mountains to the right and the coastal range and the Pacific Ocean to the left. Annual average temperature is 11°C, average wind speed is 1.6 m/s and average precipitation is 1,300 mm. Most of this rain is concentrated in the cold season that can be defined by the period between April and September. It is during this cold season that episodes of $PM_{2.5}$ pollution are observed (annual average is 37 $\mu g/m^3$). Most homes use wood stoves for heating and this represents of the order of 90% of $PM_{2.5}$ emissions. Fine particulate matter emitted by stoves is not easily dispersed on dry days because of unfavourable topographic and atmospheric conditions. Low temperatures facilitate the occurrence of thermal inversions due to surface cooling. Under these conditions pollutants become trapped bellow an altitude of no more than 100 m [1]. Daily concentrations often exceed 170 $\mu g/m^3$ which is defined as an Emergency condition. In order to take actions in order to preview these high pollution episodes it seems convenient to have in operation a forecasting model. A statistical forecasting model uses historical data of meteorological and pollution data in order to fix a number of adjustable parameters and rests on the assumption that future behaviour will obey similar functional relation [2]. Most used statistical models for particulate matter forecasting are multi linear regressions (MLR), multilayer perceptrons (MLP) and radial basis function networks (RBF). Stadlober et al. [3] have shown that a MLR model is efficient for the forecasting of daily PM_{10} in three cities in the Alps region. MLP has proven to be a useful tool for NO_2 and PM_{10} forecasting in a populated area in China [4]. Lu et al. [5] report a calculation using a RBF model, and they claim that this method is faster and more effective than more traditional neural network

WIT Transactions on Ecology and the Environment, Vol 236, © 2019 WIT Press
www.witpress.com, ISSN 1743-3541 (on-line)
doi:10.2495/AIR190111

models for the forecasting of particulate matter and nitrogen oxides. In this paper the performance of two $PM_{2.5}$ forecasting models based on artificial neural networks, a multilayer network (MLP) and a radial basis function network (RBF), adapted to Osorno conditions is presented. Emphasis is given to forecast high concentrations episodes.

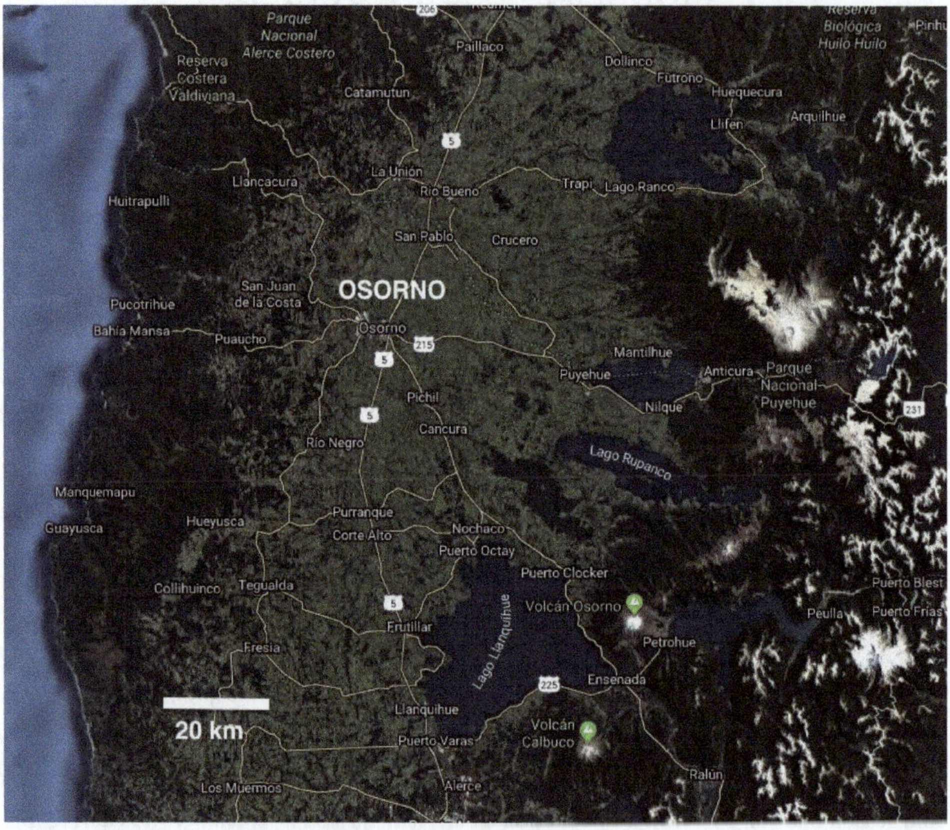

Figure 1: Satellite picture of the city of Osorno.

2 DATA

Data analyzed corresponds to years between 2015 and 2018, with emphasis on the cold period (April–September). Pollution and meteorological information is obtained from an official monitoring station located in the downtown area. According to Chilean pollution legislation, $PM_{2.5}$ 24 h average concentrations are classified into five ranges, which from low to high concern may be labelled in the following manner: range A, corresponding to concentrations bellow 50 μg/m^3, range B, concentrations between 50 μg/m^3 and 80 μg/m^3, range C, concentrations between 80 μg/m^3 and 110 μg/m^3, range D, concentrations between 80 μg/m^3 and 110 μg/m^3 and range E concentrations greater than 170 μg/m^3. The Chilean standard for this quantity is 50 μg/m^3, so only days in range A are considered "safe" days. The Ministry of the Environment has established a plan for the management of pollution in the city which applies for the period between April and September (cold season). During cold season, wood stoves do not properly certified are permanently prohibited. The same applies for bush burning. For days in ranges D and E, additional restrictions to emissions are enforced. During range D days, industrial and residential boilers identified as responsible of high emissions are not allowed to operate. Also, outdoor physical exercise in schools is prohibited. In the case of range E days, in addition to restrictions applied on D days, the emission of visible smoke from residential heating is banned.

3 MODELING

The goal of this study is to forecast the maximum of the 24 h moving average for the next day (which defines the range of the day) based on pollution data and meteorology available at 19 h of the present day. In this way it is possible to generate an air quality report at 20 h. In order to identify the predictor variables to feed the statistical forecasting model, we did a correlation analysis of candidate variables with the target variable. The selected predictor variables are:

1) Average $PM_{2.5}$ concentration between 1 am and 7 pm of present day.
2) Hourly $PM_{2.5}$ measured at 7 pm of the present day.
3) Hourly PM_{10} measured at 7 pm of the present day.
4) Average PM_{10} concentration between 1 am and 7 pm of present day.
5) Minimum temperature for the next day.
6) Temperature at 8 am of the present day
7) Wind speed at 7 pm of present day.
8) Maximum wind speed for the next day.
9) Maximum relative humidity for the next day.
10) Amplitude of wind direction of present day.

From this list, variables 5, 8 and 9 are forecasted by an independent meteorological model. Variables 1–4 are associated with the tendency of concentrations and are important in the evaluation of 24 h averages on the next day. They are indirectly related to the intensity of wood combustion. Temperature variables are correlated with wood stove usage. Wind related variables are associated with probability of dispersion of particulate matter contained in smoke emitted by stoves. Relative humidity is an indicator of the presence of rain.

Two models based on artificial neural networks are developed. One of them is the traditional Multilayer Perceptron trained with the Backpropagation algorithm [6] (see Fig. 2).

Here, the 10 input variables are connected to units or neurons in the hidden layer by means of weights $w^{(1)}_{ij}$ and activation of these units is calculated by a sigmoid function of a linear combination. Every neuron of the hidden layer is connected to the output by means of weights $w^{(2)}_{k}$ and activation of this unit is calculated again by a sigmoid of a linear combination. Weights calculated during a training phase by an optimization algorithm based on a sample of the available historical data. In our case, 2015, 2016 and 2017 data are used for training. Best results are obtained with a hidden layer with 40 neurons. With weights fixed, an independent test is performed with 2018 data.

The second model is the Basis Radial Function Network (RBF).

The main difference between RBF and MLP is that in RBF the inputs are not equally connected to the hidden units, but only significantly to those that are within a given distance (calculated on the basis of a vector of inputs) from position vectors (that have the dimension of the number of inputs) of these hidden neurons [7], [8]. Output value is calculated through a linear combination of Gaussian functions centered on the position

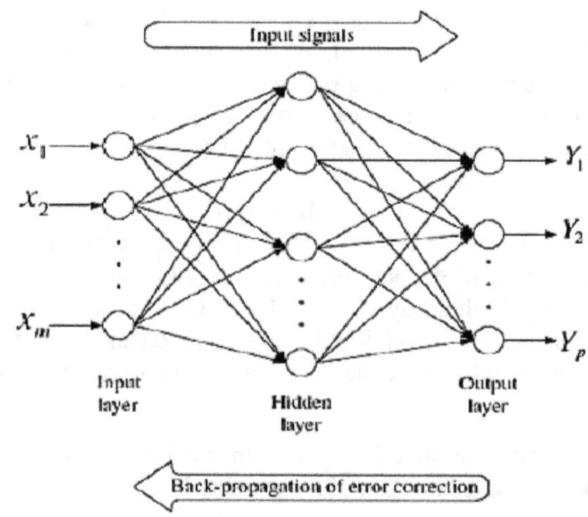

Figure 2: The MLP network.

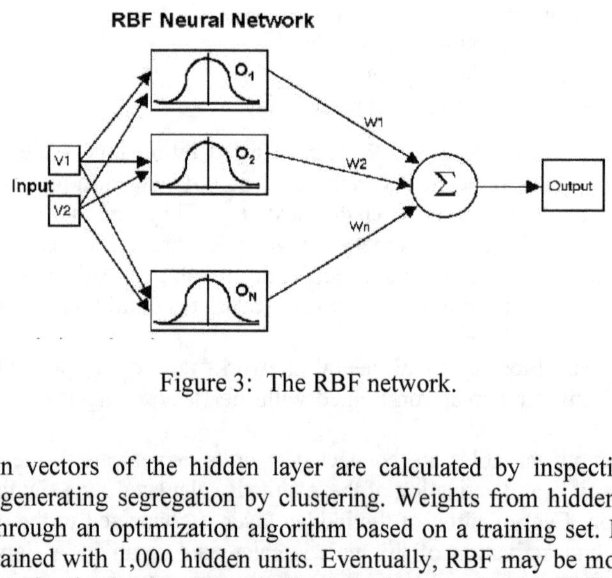

Figure 3: The RBF network.

vectors. Position vectors of the hidden layer are calculated by inspection of the input variable space, generating segregation by clustering. Weights from hidden layer to output are calculated through an optimization algorithm based on a training set. In our case, best results were obtained with 1,000 hidden units. Eventually, RBF may be more accurate than MLP if this clustering is clearly present in the data. In general, with RBF we have a faster training phase and with MLP a faster test phase. MLP has proved to be an efficient algorithm for air pollution forecasting in two Chilean cities, for both daily and hourly averages and for particulate matter and gas pollutants [9]–[12], but RBF has shown to be more accurate than MLP for nitrogen oxide forecasting in Spain [13].

4 RESULTS

The performance of the models is evaluated by testing 2018 data. Indicators of this performance are the mean absolute percentage error (MAPE), root mean squared error (RMSE) [14] and the global model quality observed in contingency tables, which is important to visualize range forecasting.

Table 1 shows the results of the best forecasting MLP and RBF models developed with 2015, 2016 and 2017 data and tested with 2018 cold season Osorno data. It is observed that both models are of comparable accuracy.

Table 1: MAPE and RMSE for 2018 Osorno data using MLP and RBF models.

	MLP	RBF
MAPE	20%	21%
RMSE	39.6	39.5

Although from Figs 4 and 5, an apparently similar performance for both models may be concluded by plotting observed and forecasted values, a more detailed analysis provided by contingency tables (Tables 2 and 3) shows that the MLP model is more accurate on range forecasting. This is more evident for range D, 80% agreement using MLP against 71% with RBF model, and for range C, 48% agreement with MLP against 39% with RBF. Other useful information can be extracted from the contingency tables. As an example, for the MLP model, from the 28 observed range E days during 2018, 23 were correctly forecasted, 3 of them were forecasted as range D and 2 as range C days. It is worth to mention that to get the displayed results with the MLP network, a long time for training and adjustment with different initial conditions was needed. In the case of the RBF network, the fact the best results were obtained with a hidden layer with a number of units of the order of magnitude the size of the training set implies that not significant clustering of the input data was found.

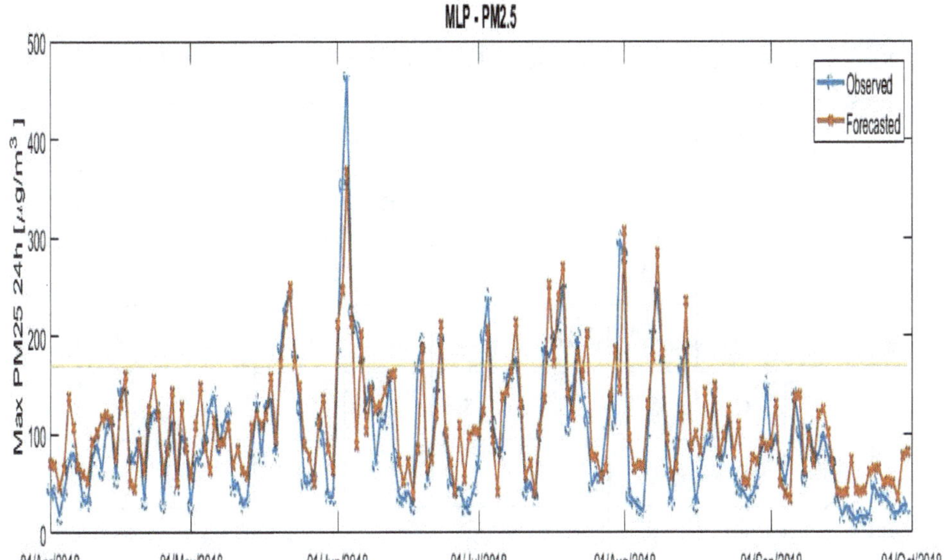

Figure 4: Observed and forecasted maximum of next day $PM_{2.5}$ 24 h average using MLP model. 2018 Osorno test data. Horizontal line indicates level of range E.

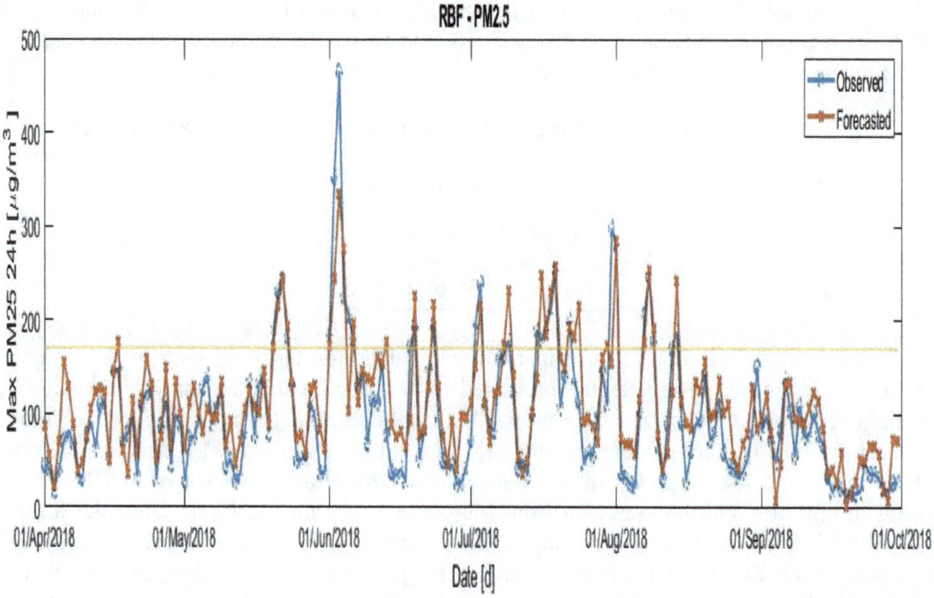

Figure 5: Observed and forecasted maximum of next day $PM_{2.5}$ 24 h average using RBF model. 2018 Osorno test data.

Table 2: Contingency table for observed and forecasted ranges using MLP model.

		Forecast MLP model						
		A	B	C	D	E	TOT	%O
OBS	A	**13**	33	8	0	0	54	24
	B	1	**16**	12	6	0	35	46
	C	2	3	**15**	10	1	31	48
	D	0	1	**5**	**28**	1	35	80
	E	0	0	2	3	**23**	28	82
	TOT	26	53	42	47	25	*183*	*47*
	% F	81	30	35	60	92		

Table 3: Contingency table for observed and forecasted ranges using RBF model.

		Forecast RBF model						
		A	B	C	D	E	TOT	%O
OBS	A	**20**	24	9	1	0	54	37
	B	4	**6**	17	7	1	35	17
	C	1	4	**12**	13	1	31	39
	D	0	1	5	**25**	4	35	71
	E	0	0	2	3	**23**	28	82
	TOT	25	35	45	49	29	*183*	*47*
	% F	80	17	27	51	70		

5 CONCLUSIONS

This study shows that with an appropriate choice of input variables it is possible to implement operational $PM_{2.5}$ statistical forecasting models based on artificial neural network algorithms for the city of Osorno, Chile. The best predictors that apply to this city differ from those used for $PM_{2.5}$ forecasting models in other cities, which may be associated with particularities of this locality. This may help the authorities to take actions in order to protect the population from high levels of pollution. All indicates that shift to heating systems not based on wood combustion would improve air quality significantly.

ACKNOWLEDGEMENT

We would like to thank the support of the Research Department of Universidad de Santiago de Chile (DICYT) through project 091931PJ.

REFERENCES

[1] Molina, C., Toro, A., Morales, R., Manzano, C. & Leiva-Guzman, M., Particulate matter in urban areas of south-central Chile exceeds air quality standards. *Air Qual. Atmos. Health*, **10**, pp. 653–667, 2017.

[2] Shahraiyni, H.T. & Sodoudi, S., Statistical modelling approaches for PM_{10} prediction in urban areas: A review of 21st studies. *Atmosphere*, 7(2), p. 15, 2016.

[3] Stadlober, E., Hörmann, S. & Pfeiler, B., Quality and performance of a PM_{10} daily forecasting model. *Atmos. Environ.*, **42**, pp. 1098–1109, 2008.

[4] Liu, W. et al., Land use regression models coupled with meteorology to model spatial and temporal variability of NO_2 and PM_{10} in Changsha, China. *Atmos. Environ.*, **116**, pp. 272–280, 2015.

[5] Lu, W.Z., Wang, W.J., Wang, X.K., Yan, S.H. & Lam, J.C., Potential assessment of a neural network model with PCA/RBF approach for forecasting pollutant trends in Mong Kok urban air, Hong Kong. *Environ. Res.*, **96**, pp. 79–87, 2004.

[6] Rumelhart, D.E., Hinton, G.E. & Williams, R.J., Learning internal representations by error propagation. *Parallel Distributed Processing: Explorations in the Microstructure of Cognition*, eds D.E. Rumelhart & J.L. McClelland, vol. 1, pp. 318–362, MIT Press: Cambridge, MA, 1986.

[7] Powell, M.J.D., Radial basis functions for multivariable interpolation: A review. *Algorithms for Approximation*, eds J.C. Mason & M.G. Cox, vol. 10, Institute of Mathematics and its Applications Conference Series, pp. 143–167, Oxford University Press: Oxford, 1987.

[8] Moody, J. & Darken, C., Fast learning in networks of locally tuned processing units. *Neural Computation*, **4**, pp. 740–747, 1989.

[9] Perez, P. & Reyes, J., An integrated neural network for PM_{10} forecasting. *Atmos. Environ.*, **40**, pp. 2845–2851, 2006.

[10] Perez, P. & Trier, A., Prediction of NO and NO_2 concentrations near a street with heavy traffic in Santiago, Chile. *Atmos. Environ.*, **35**, pp. 1783–1789, 2001.

[11] Perez, P., Menares, C. & Ramirez, C., $PM_{2.5}$ forecasting in the most polluted city in South America. *WIT Transactions on Ecology and the Environment,* vol. 230, WIT Press: Southampton and Boston, pp. 199–204, 2018.

[12] Perez, P. & Gramsch, E., Forecasting hourly $PM_{2.5}$ in Santiago de Chile with emphasis on night episodes. *Atmos. Environ.*, **124**, pp. 22–27, 2016.

[13] Capilla, C., Application of radial basis functions compared to neural networks to predict air pollution. *WIT Transactions on Ecology and the Environment,* vol. 198, WIT Press: Southampton and Boston, pp. 41–50, 2015.

[14] Botchkarev, A., Performance metrics (error measures) in machine learning regression, forecasting and prognosis: properties and typology, 2018. https://arxiv.org/pdf/1809.03006.

ESTIMATING CONCENTRATIONS OF SUSPENDED PARTICULATE MATTER OVER THE METROPOLITAN AREA OF MEXICO CITY USING SATELLITE AND GEOSPATIAL IMAGERY: PRELIMINARY RESULTS

RODRIGO T. SEPÚLVEDA-HIROSE, ANA B. CARRERA-AGUILAR,
MAGNOLIA G. MARTINEZ-RIVERA, PABLO DE J. ANGELES-SALTO & CARLOS HERRERA-VENTOSA
Faculty of Engineering, Universidad Nacional Autónoma de México, Mexico

ABSTRACT

In order to diminish health risks, it is of paramount importance to monitor air quality; however, this process is accompanied by high costs of physical and human resources. This research is carried out in this context with the main objective of developing a predictive model for concentrations of inhalable particles (PM_{10} and $PM_{2.5}$) using remote sensing. To develop the model, satellite images, mainly from Landsat 8, of the Mexico City's metropolitan area, were used. Using historical PM_{10} and $PM_{2.5}$ measurements of the RAMA (Environmental Monitoring Network of Mexico City) and processing the available satellite images, a preliminary model was generated in which it was possible to observe critical opportunity areas that would allow for the generation of a robust model. After applying this preliminary model to the scenes of Mexico City, three areas of great interest were identified due to their presumed high concentration of PM. The zones of interest all presented high plant density, bodies of water and bare soil without buildings or construction, or vegetation. To date, work continues along these lines to improve the proposed preliminary model. In addition, a brief analysis of six distinct models was made and presented in articles developed in different parts of the world in order to visualize the optimal bands for the generation of a suitable model for Mexico City. It was found that infrared bands have helped modelling in other cities, but the effectiveness of these bands for the specific geographic and climatic conditions of Mexico City is still in need of evaluation.

Keywords: air quality, modelling pollution, particulate matter, remote sensing.

1 INTRODUCTION

Particulate matter (PM) has a variety of sources and compositions. It appears naturally in the atmosphere due to volcanic activity and fires; however, human activity also contributes to high levels of PM pollution. Particles form in the atmosphere as a result of complex reactions of chemicals such as sulphur dioxide and nitrogen oxides. Particle pollution is composed of: PM_{10}, particles with a diameter of 10 micrometers and less, and $PM_{2.5}$ with a diameter of 2.5 micrometers and less. Due to their small size, particles can be easily inhaled causing serious health problems [1]. PM_{10} particles, after inhaled, can travel to the lungs and can even reach the bloodstream, while $PM_{2.5}$, being even smaller, pose a greater health risk.

Access to air pollution monitoring equipment is restricted due to its elevated cost; this explains the use of remote sensing techniques as an alternative method for measurement. Remote sensing provides advantages such as global and exhaustive coverage of land surface, multi-scale and non-destructive observations and repetitive coverage.

Several investigations that have taken place, mainly in Asia, show the feasibility of using satellite images for the estimations of particulate material concentrations [1]–[10]. According to Chu et al. [4], the four most used models for the prediction of particulate material are: Lineal Multiple Regression, Mixed Effects Model (MEM), Model of Chemical Transport and Pondered Geographic Regression.

The project's main objective is to develop air quality models for the estimation of both PM_{10} and $PM_{2.5}$ from both locally measured data and remote sensing.

WIT Transactions on Ecology and the Environment, Vol 236, © 2019 WIT Press
www.witpress.com, ISSN 1743-3541 (on-line)
doi:10.2495/AIR190121

2 STUDY AREA AND DATA SETS

2.1 Study area

Mexico City's metropolitan area comprises an area of 2,370 km^2 and has a population of 20,565,000 inhabitants which generates a density of 8,700 people per square kilometer [11]. The city, the capital of Mexico, is located at 19 29 52 N and 99 7 37 O in the central region of Mexico. Dominant wind flow in Mexico City moves from north to south with a confluence line in the southeastern corner of the city. This wind behavior indicates a flow of atmospheric pollution from the north and east to the south and southwest part of the metropolitan area. As a consequence of these orographic conditions in the metropolitan area, it is subject to air stagnation and high concentrations of pollutants, especially in the southern region of the city. Average wind speeds oscillate between 0.5 and 1.5 m/s, with maximum speeds reaching to 2.5 m/s [12].

Mexico City has an automatic air quality monitoring system comprised of 29 stations with a capability of measuring PM_{10} and $PM_{2.5}$ hourly year-round. The coverage of these stations will define the area of study for the project. According to Mexican regulation, the allowed limits for PM pollution are shown in Table 1.

Table 1: Mexican regulations allowed limits for particulate matter pollution.

Pollutant	**Period of measurement**	**Limit (μg/m^3)**
$PM_{2.5}$	Annual average	12
$PM_{2.5}$	24 hour average	45
PM_{10}	Annual average	40
PM_{10}	24 hour average	75

2.2 Data sets

A statistical analysis was carried out to determine the spatial and temporal distribution of the concentrations of particulate matter ($PM_{2.5}$ and PM_{10} particles) present in Mexico City's metropolitan area. Precipitation was also considered as a variable to determine a possible relationship between the behavior of both variables (precipitation and particulate matter concentrations) in a certain time span.

A selection of dates when the satellite Landsat 8 passed over the ZMCM was made. Said satellite made contact with the ZMCM for the first time on May 20, 2013 at 5:02 pm (GMT+0), 11:00 am local time. From that date on, all later dates were considered in this study up until the month of May 2018, accounting the satellite's temporal resolution of 16 days for a total of 115 dates during the 5 years in which the analysis took place.

Regarding the pollutants PM_{10} and $PM_{2.5}$, the data was extracted from the database generated by the measurements of all Automatic Network of Environmental Monitoring stations (Fig. 1) under the Ministry of the Environment of Mexico City.

Both sets of data (PM_{10} and $PM_{2.5}$) were processed. Fig. 2 shows the boxplot graphic for PM_{10}, at 11:00 am, on the dates compatible with the passing of the Landsat 8 platform. As can be observed, the concentrations have an annual cyclical behavior related to the rainy season (May to November) and dry months (November to May) in Mexico City.

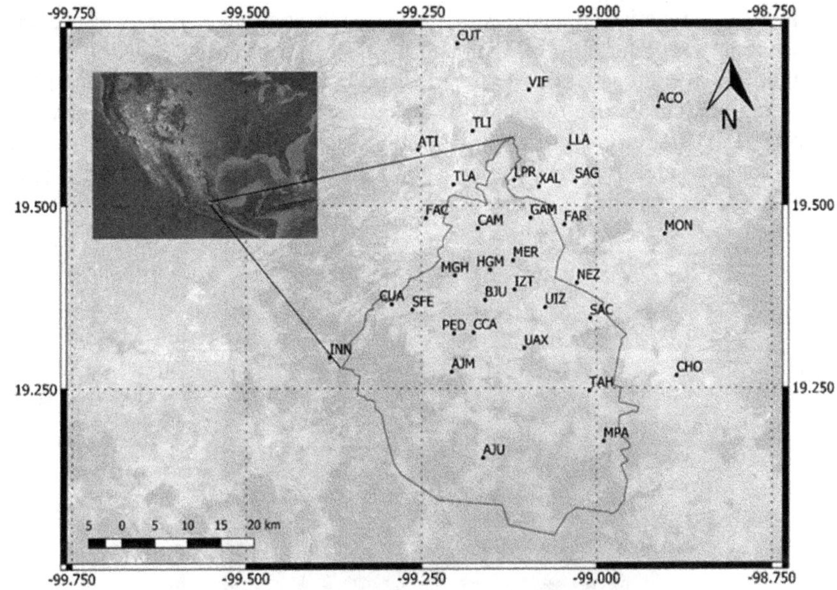

Figure 1: Location of the stations of the Automatic Network of Environmental Monitoring. *(Source: Ministry of the Environment of Mexico City.)*

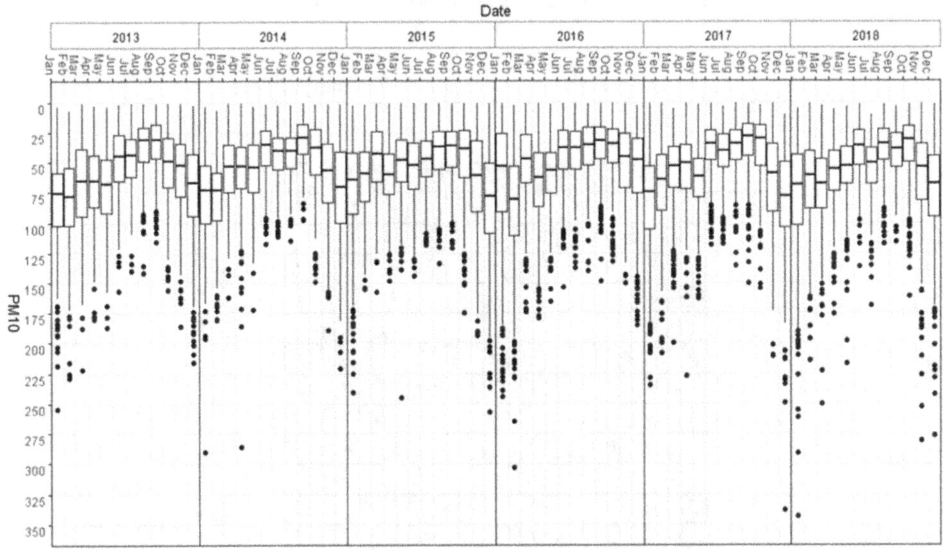

Figure 2: Boxplot graph showing PM_{10} concentrations (µg/m3) in Mexico City on the dates in which satellite Landsat 8 passed over the city in the period between May 2013 and May 2018. The median of measured data is represented by the line in the box, the interquartile range box represents the middle 50% of the data, and the whiskers represent the ranges for the top and bottom 25% of the data. The dots represent outliers.

In addition to the statistical analysis per season, a spatial analysis was also elaborated upon, first based on zoning by quadrants and then, later, spatial interpolation of the data, using the monthly average values as a reference and thus obtaining the pollutants $PM_{2.5}$ and PM_{10} and interpolating the data gathered by the entirety of the stations located within the study area. The evaluation period lasted from the year 2013 until 2018, as it tied in with the information provided by the Landsat 8 platform. Figs 3 and 4 show the mentioned interpolations for $PM_{2.5}$ and PM_{10} respectively.

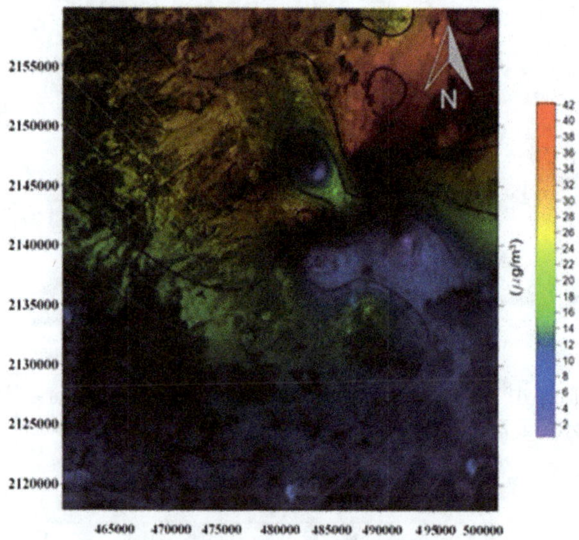

Figure 3: $PM_{2.5}$ concentration, February 2018.

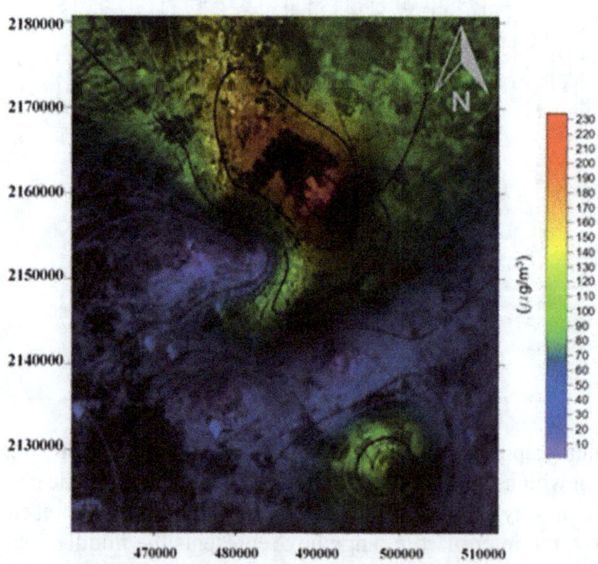

Figure 4: PM_{10} concentration, January 2018.

3 METHODOLOGY

The project methodology begins with the collection of historical data from the SEDEMA databases regarding PM_{10} and $PM_{2.5}$. Once the historical data is collected, a statistical analysis is carried out. The analysis is comprised of one study on the temporal relation and another on the spatial relation to the system. For the temporal study, the monthly averages of particulate matter PM and precipitation are plotted against time and thus the relationship between both variables analyzed. With respect to the spatial study, a spatial representation of PM pollution is made in each of the 4 zones of the Mexico City's metropolitan area and the results are analyzed in conjunction.

Together with the collection of historical data, the satellite image platform is selected. Once a platform is selected, the satellite image bank is made. These images have to be subjected to pre-processing which allows for the retrieval of the corrected reflectance data. The next step is to analyze and classify viable images for the study. In order to do so, the values of cloud percentage in the images were obtained and a visual inspection was made by previewing the images through the digital platform Earth Explorer.

Based on this analysis it was found that, of the 119 scenes available to date, only 44 can be used. Of these 44 scenes, the products of digital numbers (Level 1) are not corrected for atmospheric conditions, however reflectance TOA (Level 2) and surface reflectance (Level 2) were downloaded, as shown in Table 2.

Table 2: Summary Landsat 8 scenes acquired.

Type of images	Number	Percentage
Useful	44	35.77%
Not useful	74	60.16%
Not available	5	4.07%
Total	123	100.00%

3.1 Processing and digital analysis of images

In order to validate the products of Level 2, pre-processing was performed on the May 20, 2013 scene with the help of the software QGIS (Quantum GIS). The following eqn was applied for the calculation of reflectance on the top of the atmosphere (TOA)

$$\rho'_\lambda = M_p Q_{cal} + A_p, \tag{1}$$

where ρ'_λ is top of atmosphere (TOA) reflectance without solar angle correction, M_p is specific rescaling factor by band obtained in metadata, A_p is specific resizing factor by band obtained in metadata, and Q_{cal} is pixel values of standard products quantified and calibrated (DN).

In order to obtain the TOA reflectance corrected for solar angle in the center of the scene, it must be divided by the sine of that angle, information provided in the metadata of the scene. Once this pre-processing was concluded, the reflectance values were compared with the respective Landsat 8 product, and differences of the order of 3% were found. This is due to the Landsat 8 algorithm which performs the correction of the solar angle per pixel in the scene, offering a more precise result. Atmospheric corrections were made via different methods, to verify the surface reflectance. First, the subtraction of dark objects (DOS) correction was made, which identifies the objects that "should be dark" in some bands, having as a consequence the supposition that the reflectance measured by the sensor is only that caused by the atmosphere. Another correction also took place with the help of the module FLAASH (fast line-of-sight atmospheric analysis of hypercubes) of ENVI; the problem with these methods is that their correction is constant and when applied to the whole image, they do not give precise information of the atmospheric reflectance.

As a conclusion of these analyses, the use of the Level 2 product of Landsat 8 was chosen. This decision was made based on the fact that it offers a more precise atmospheric correction due to its better calibration. Due to the location of monitoring stations, a vector layer is generated and associated to each image in order to begin processing data. The atmospheric reflectance data is extracted for each location and associated with the historical field data obtained from the monitoring stations.

In order to determine the optimal bands to use in the generation of the model for Mexico City, an analysis of reported models was made. The scene was analyzed using 6 reported models (Table 3). Table 4 shows the mean squared error obtained from each of them.

Table 3: Proposed equations to model PM_{10} concentrations.

Reference	Proposed equation	Study area
Torres and Vivanco [5]	$PM_{10} = -31.56 + 111.4NDVI - .03RA.B1 + 0.089RA.B3 - .019RA.B4 - 5.42Season$	Quito, Ecuador
Abad and Mejía-Coronel [6]	$PM_{10} = -126.9 + 0.005ND.B11 + 582.7TOA.B2 - 207.1TOA.B5$	Cuenca, Ecuador
Saraswat et al. [7]	$PM_{10} = -0.8689 - 94.22RA.B1 + 166.48RA.B2 + 21.01RA.B3 - 78.98RA.B4$	Delhi, India
Shaheen et al. [8]	$PM_{10} = 10008RA.B2 - 21356RA.B3 + 10965RA.B4$	Gaza Strip, Palestine
Nadzri et al. [9]	$PM_{10} = 396RA.B2 + 253RA.B3 - 194RA.B4$	Mecca, Saudi Arabia
Ozelkan et al. [10]	$PM_{10} = 232.66\left(\frac{RA.B6}{RA.B7}\right) - 78.673$	Izmir, Turkey

Table 4: Mean squared errors reported after modelling PM_{10} concentrations.

Reference	Reported mean squared error	
	Authors	**Mexico City's Metropolitan Area**
Torres and Vivanco [5]	NA	54.71
Abad and Mejía-Coronel [6]	77.70	45.90
Saraswat et al. [7]	18.98	75.5
Shaheen et al. [8]	9.71	74.33
Nadzri et al. [9]	NA	55.02
Ozelkan et al. [10]	NA	49.59

After processing, classifying and analyzing field measurements as well as the extracted data from remote sensing, best fitted models for air quality parameters in Mexico City can be propounded. Significant variations in air quality across space and time are observed.

4 RESULTS

A multiple linear regression was made in order to model the dependency of PM_{10} on reflectance values obtained from remote sensing. The model that was generated is preliminary and used to observe the performance by means of the parameters R^2, with an adjusted R and root mean squared error.

After applying the preliminary model to the scene for the 44 available dates, it was observed that the model presents temporal variations in the concentration of PM_{10}, which was an expected result.

Model for June 5, 2013:

$$PM10 = 0.553RA.B1 - 1.764RA.B2 + 0.524RA.B3 + 0.432RA.B4 + 3.7RA.B5 - 0.757RA.B6 + 0.3017RA.B7 + 31.646TOA.B10 - 41.306TOA.B11 + 3310.923$$

$$R^2 = 0.93.\ R^2 - adjusted = 0.73.\ RSME = 4.84\ \mu g/m^3. \tag{2}$$

Model for December 1, 2014:

$$PM10 = -1.835RA.B1 + 29358RA.B2 - 0.279RA.B3 + 0.319RA.B4 - 1.034RA.B5 + 1.796RA.B6 - 0.587RA.B7 - 197.454TOA.B10 + 247.547TOA.B11 - 1537.326$$

$$R^2 = 0.86.\ R^2 - adjusted = 0.45.\ RSME = 10.10\ \mu g/m^3. \tag{3}$$

Figs 5 and 6 show the PM_{10} concentrations for June 5, 2013 and December 1, 2014. On these dates, low and high concentration values, respectively, were observed. The areas circled in red reflect the zones in which the model did not work properly.

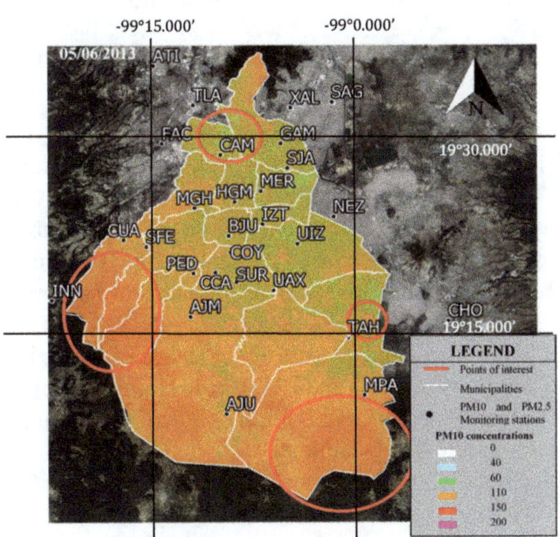

Figure 5: PM_{10} modelled concentrations in Mexico City on June 5, 2013.

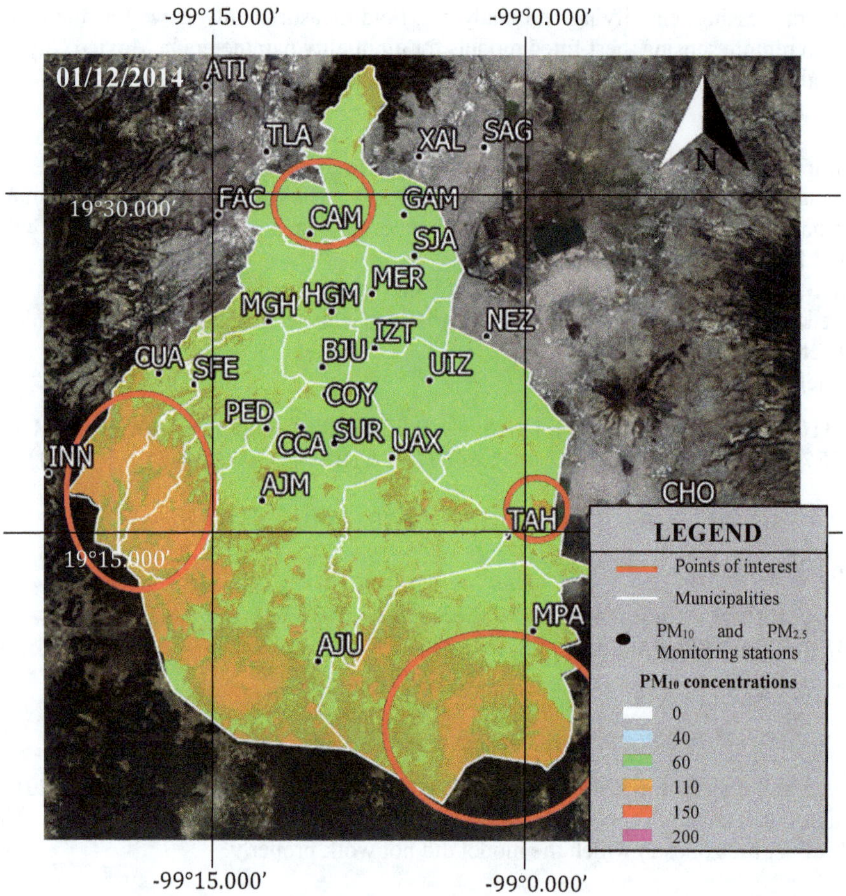

Figure 6: PM_{10} modelled concentrations in Mexico City, December 1, 2014.

5 DISCUSSION

The model for June 5, 2013 corresponds to the rainy season (May to November), whereas the model for December 1, 2014 corresponds to the dry season (November to May). The determined relationship between correlation coefficients were quite strong suggesting a high correlation. Figs 7 and 8 show the observed vs. predicted PM_{10} concentrations for June 5, 2013 and December 1, 2014, respectively. The performance of both models was very good, with a mean R^2 for June of 0.93 and for December of 0.86. As discussed above, the concentration of PM varies considerably across space and time. Thus, it is essential to match the spatial–temporal resolutions of AOD and PM data as closely as possible. Fortunately, for Mexico City, the monitoring system provides hourly air quality data and the acquisition time of Landsat 8 satellite is at 11:00 AM local time.

Other alternatives for remote sensing atmospheric modelling have been tested by several authors, including AOT retrieved from the operational MODIS algorithm. Although kilometric resolution works well in regional scales, the detailed aerosol spatial distribution at city scale may be missing. The develop of the predicting models for PM_{10} concentrations at a 30 m resolution across Mexico City will be a very useful tool for more accurate long-term estimations of PM exposure in this urban area. Although the determined correlation values are good respective to linear regressions, our future research will be geared towards improving algorithms for modelling PM including artificial neural networks.

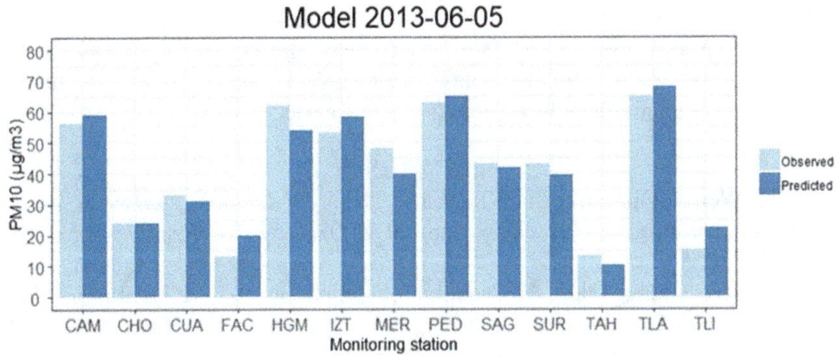

Figure 7: Comparison between observed and predicted values, June 5, 2013.

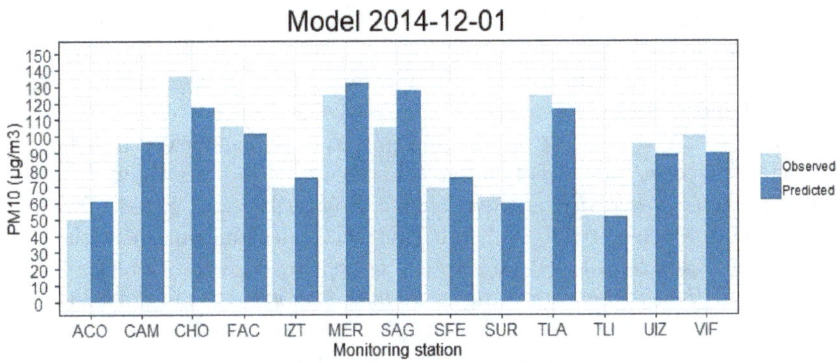

Figure 8: Comparison between observed and predicted values, December 1, 2014.

6 CONCLUSION

One of the main aspects in air pollution modelling is the quality and quantity of data used. An enormous amount of existing information regarding the parameters of this study has been made available thanks to the constant monitoring of SEDEMA during the past decades. Transversal and active cooperation of different governmental instances in the development of this work allowed for the access to high quality data. On the other hand, satellite-based measures of aerosol optical depth (AOD) as compared to fixed ground monitors allows for greater coverage of PM_{10} estimations. In this paper, a method for retrieving PM_{10} from Landsat 8 Operational Land Imager (OLI) images over urban areas is proposed. Important variations in the air quality within space and time can be observed, which will be incorporated into the second phase of this work.

The models presented in this work can be used in various disciplines for estimating PM and may be particularly useful for environmental epidemiology studies. The obtained PM_{10} distribution maps can be integrated into a geographical information system and analyzed with relational data to obtain more reliable interpretations that act against air pollution effects.

ACKNOWLEDGEMENTS

The authors thank the Ministry of Education, Science, Technology and Innovation of Mexico City for the funds, advice and support given during the development of the Project SECITI/089/2017 "Development of Models for the Assessment of Suspended Particulate Material in the Metropolitan Area of Mexico City, with the Utilization of Satellite Images and Geospatial Information." We also thank all the students and professors on our team for their enthusiasm which motivated the success of this project.

REFERENCES

[1] Sun, L., Wei, J., Bilal, M., Tian, X., Jia, C., Guo, Y. & Mi, X., Aerosol optical depth retrieval over bright areas using Landsat 8 OLI images. *Remote Sensing*, **8**(23), pp. 1–14, 2015.

[2] Hamzelo, M., Gharagozlou, A., Sadeghian, S., Baikpour, S.H. & Rajabi, A., Modelling of carbon monoxide air pollution in large cities by evaluation of spectral LANDSAT 8 images. *International Archives of the Photogrammetry, Remote Sensing and Spatial Information Sciences – ISPRS Archives*, **40**(1W5), pp. 281–285, 2015.

[3] Nguyen, N.H. & Tran, V.A., Estimation of PM_{10} from AOT of satellite Landsat 8 image over Hanoi City. Presented at *International Symposium on Geoinformation for Spatial Infrastructure Development in Earth and Allied Sciences*, Danang, Vietnam, 2014.

[4] Chu, Y. et al., A review on predicting ground $PM_{2.5}$ concentration using satellite aerosol optical depth. *Atmosphere*, **7**(10), p. 129, 2016.

[5] Torres, N. & Vivanco, V. Comparación en la estimación de material particulado PM_{10} usando imágenes satelitales Landsat 7, Landsat 8 y Modis en Quito. (Undergraduate thesis). *Atmospheric Measurement Techniques Discussions*, pp. 1–47, 2018.

[6] Abad, L. & Mejía-Coronel, D., Estimación de la concentración de material particulado menor a 10 micras a través de sensores remotos en el área urbana de la Ciudad de Cuenca. *XVI Conferencia de Sistemas de Información Geográfica*, pp. 381–390, 2017.

[7] Saraswat, I., Mishra, R.K. & Kumar, A., Estimation of PM_{10} concentration from Landsat 8 OLI satellite imagery over Delhi, India. *Remote Sensing Applications: Society and Environment*, **8**(April), pp. 251–257, 2017.

[8] Shaheen, A., Kidwai, A.A., Ain, N.U., Aldabash, M. & Zeeshan, A., Estimating air particulate matter 10 using Landsat multi-temporal data and analyzing its annual temporal pattern over Gaza Strip, Palestine. *Journal of Asian Scientific Research*, **7**(2), pp. 22–37, 2017.

[9] Nadzri, O., Zubir Mat Jafri, M., San Lim, H., Othman, N. & Hwee San, L., Estimating particulate matter concentration over arid region using satellite remote sensing: A case study in Makkah, Saudi Arabia. *Modern Applied Science*, **4**(11), 2010.

[10] Ozelkan, E., Karaman, M., Mostamandy, S., Avci, Z.D.U. & Toros, H., Derivation of PM_{10} levels using obra on Landsat 5 TM images: A case study in Izmir, Turkey. *Fresenius Environmental Bulletin*, **24**(4B), pp. 1585–1596, 2015.

[11] Demographia World Urban Areas, Built up urban areas or world agglometarions, 14th annual ed. http://demographia.com/db-worldua.pdf. Accessed on: 14 Mar. 2019.

[12] Secretaría de Medio Ambiente, Informe climatológico ambiental del Valle de México. Environmental climatological report on the Valley of Mexico, 2006. www.aire.cdmx.gob.mx/descargas/publicaciones/flippingbook/informe_anual_climatologico_2006/#p=1. Accessed on: 12 Mar. 2019.

HOURLY TROPOSPHERIC OZONE CONCENTRATION FORECASTING USING DEEP LEARNING

LUCAS ALVES, ERICK GIOVANI SPERANDIO NASCIMENTO & DAVIDSON MARTINS MOREIRA
Senai CIMATEC, Brazil

ABSTRACT

The purpose of this work is to build, train and evaluate a deep learning-based model to forecast tropospheric ozone levels hourly, up to twenty-four hours ahead, using data gathered from the automatic air quality monitoring system in the metropolitan region of Vitória city, Espírito Santo (ES), Brazil. Observational data of air pollutant concentrations and meteorological parameters were used as the input variables of the model once they represented the state of the atmospheric fluid in terms of its properties and chemical composition throughout the time. Several topologies of multilayer perceptron neural networks were tried and evaluated using statistics of the predictions over unseen data. The best architecture was compared with reference models and the results showed that deep learning models can be successfully applied to hourly forecasting of ozone concentrations for urban areas. Once such models are fitted to the data, the forecasting procedure has a very low computational cost, meaning that it can be used as an alternative approach in comparison with numerical modelling systems, which require much more computational power.

Keywords: air quality forecasting, ozone, neural networks, deep learning.

1 INTRODUCTION

Ozone (O_3) is a secondary pollutant in the troposphere and one of the photochemical oxidants causing air quality problems. It is formed from chemical reactions between gases emitted by natural and anthropogenic sources, such as nitrogen oxides and volatile organic compounds in the presence of solar radiation. O_3 can irritate the respiratory system, reduce lung capacity, and aggravate asthma problems [1]. Moreover, it can damage plants and affect agricultural production [2]. The World Health Organization (WHO) air quality guidelines provide thresholds for health-harmful pollution levels and the 2005 publication sets the recommended value for ozone concentration at 100 $\mu g/m^3$ for a daily maximum 8-hour average [3]. Therefore, it is important to develop a powerful forecasting model that could help authorities and the population to take preventive measures and avoid imminent health risks, even before the recommended limits are reached.

2 DATA

The data used in the experiments is publicly available and was gathered from the Automatic Air Quality Monitoring Network (*Rede Automática de Monitoramento da Qualidade do Ar – RAMQAr*) owned by the State Institute of Environment and Water Resources of Espírito Santo (*Instituto Estadual de Meio Ambiente e Recursos Hídricos do Espírito Santo – IEMA-ES*). The monitoring station chosen for this study is located in Cariacica, a city in the metropolitan region of Vitória, ES, Brazil. This station measures hourly averages of the twelve atmospheric pollutant concentrations and meteorological parameters displayed on Table 1.

Data from the years 2001 to 2015 were collected and treated to eliminate records with invalid or missing measurements in one or more sensors. Furthermore, only valid data and hourly sequences with at least twenty-four consecutive samples were kept in order to make possible the generation of ozone concentration ground truth targets, required on supervised machine learning algorithms for the model fitting. These steps discarded most part of the

WIT Transactions on Ecology and the Environment, Vol 236, © 2019 WIT Press
www.witpress.com, ISSN 1743-3541 (on-line)
doi:10.2495/AIR190131

data, including all the measurements from the years 2007, 2009, 2010 and 2013, resulting in a data set with 30,492 samples and their corresponding targets. Fig. 1 shows the amount of useful data by year and month.

Table 1: Parameters measured in Cariacica's air quality monitoring station.

Parameter	Parameter characteristics	
	Type	Unit
Particulate matter below 10μm (PM_{10})	Atmospheric pollutant	(μg/m³)
Total suspended particulate matter	Atmospheric pollutant	(μg/m³)
Sulphur dioxide (SO_2)	Atmospheric pollutant	(μg/m³)
Nitrogen monoxide (NO)	Atmospheric pollutant	(μg/m³)
Nitrogen dioxide (NO_2)	Atmospheric pollutant	(μg/m³)
Nitrogen oxides (NO_x)	Atmospheric pollutant	(μg/m³)
Carbon monoxide (CO)	Atmospheric pollutant	(μg/m³)
Ozone (O_3)	Atmospheric pollutant	(μg/m³)
Temperature	Meteorological	(ºC)
Humidity	Meteorological	(%)
Scalar wind direction	Meteorological	(º)
Scalar wind speed	Meteorological	(m/s)

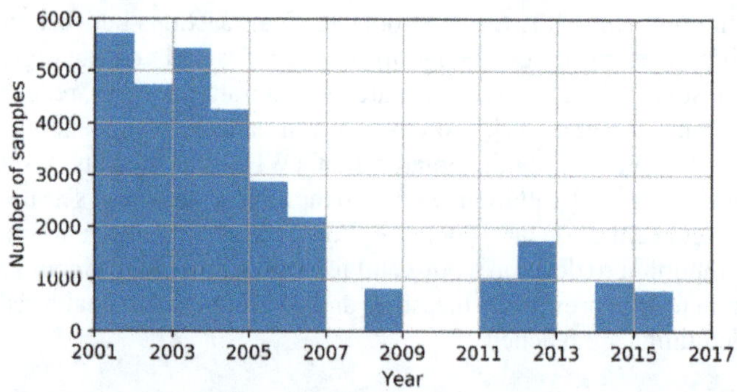

(a)

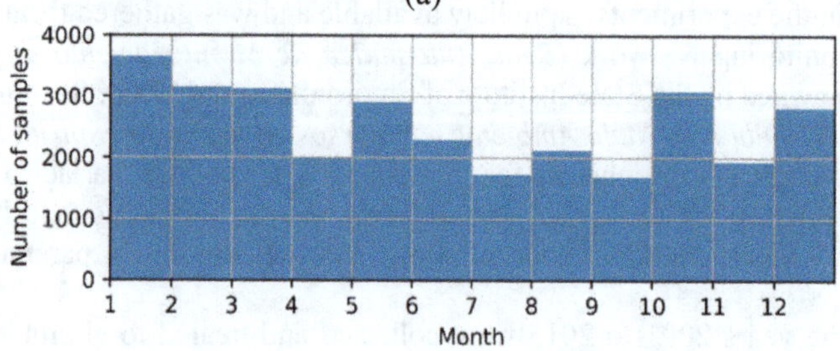

(b)

Figure 1: Histograms showing the distribution of useful data by time period. (a) Number of samples by year; and (b) Number of samples by month.

In the machine learning domain, one of the main objectives is to create computational models with the ability to generalize well the extracted attributes to new data. Poor generalization is often characterized by overfitting, and a common method to avoid that is to evaluate a model by splitting a data set into two. The first one is the training set, on which the model is built and optimized. The second is the test set, on which the finished model is evaluated with unseen data [4].

For this research, 22,987 data points from the years 2001 to 2005 were separated for the training and validation set, and the remaining 7,505 data points from the years 2006, 2008, 2011, 2012, 2014 and 2015 were used as test data. This procedure was made in order to guarantee that the two batches of data have samples for all months, days of a month, days of a week and hours of a day.

3 METHODS AND MODELLING

The modelling of ozone fluctuations can be made through two types of models: deterministic or stochastic. Deterministic models use several equations to represent the atmosphere behaviour and thus forecast the ozone concentrations in a limited domain. Due to the complexity of this process, developing and maintaining them are expensive tasks and demands a large amount of computational power, since it has to process many chemical and physical interactions between diverse parameters like emissions, meteorology and land cover. Stochastic models, otherwise, have a simpler implementation because they try to formulate a mathematical relationship between the input and output variables based on the detection of some patterns [5]. Once such models are fitted to the data, the predictions are made using few computational resources.

Artificial neural networks are one type of stochastic models and, in the deep learning field, there are currently many different architectures available for implementation, being essential to examine which one best fits the problem that needs to be solved. Previous works used recurrent neural networks (RNN) to predict daily maximum concentrations of tropospheric ozone in the city of Palermo, Italy [6], and in the Mexicali (Mexico)-Calexico (USA) border area [7]. In Biancofiore et al. [8], RNN models were applied to predict O_3 concentration at time $t+\Delta t$, where Δt can be 1, 3, 6, 12, 24 and 48 h. A convolutional neural network (CNN) was employed in Eslami et al. [9] to predict the hourly ozone concentration on each day using parameters from the previous day. Eight separated multilayer perceptron (MLP) networks were used in Agirre et al. [10] to forecast the values of the variables $O_3(t+k)$, being $k = 1, 2, \ldots, 8$ h, at two rural stations located in the Autonomous Community of the Basque Country (North Central Spain). An MLP predictor was built in Tamas et al. [11] using one single output to forecast O_3 concentration 24 hours ahead in Corsica, France, in order to be able to anticipate pollution peaks formation. In Coman et al. [5] two MLP models were evaluated. The "dynamic" model used a cascade of 24 multilayer perceptrons arranged so that each MLP feeds the next one, and the "static" model was a classical single MLP with 24 outputs. For both configurations, the outputs were ozone concentrations for a 24 h horizon.

The present research focus on multilayer perceptron neural network models, due to its simplicity and large application for short-term forecasts. These networks have universal function approximation capabilities, with applicability in non-linear problems and ability to generalize to unseen data, being effective for prediction purposes [12]. However, prior studies that used this type of network to forecast hourly concentrations of ozone aim attention on MLPs with one single hidden layer, which can lead to models with limited representational power. Moreover, few of them uses a single model to forecast hourly ozone concentrations for all time steps ahead in a 24 h horizon. The proposed model employs this approach, since it can take advantage of a shared internal representation for all the forecasts and obtain a better generalization of the problem. Additionally, in the performed simulations the best results were achieved with deeper network topologies.

Multilayer perceptron networks have a flexible topology and among their main parameters are the number of layers and the number of neurons in each layer. At least three layers are required: an input layer, a hidden layer, and an output layer. The definition of the number of layers and neurons is variable, and the best composition is problem-specific [13].

Since the model objective is to predict tropospheric ozone levels hourly, up to twenty-four hours ahead, the output layer of the proposed model is composed by twenty-four neurons, one for each hour in advance. Several network designs were tested varying the number of inputs, hidden layers and nodes in each hidden layer. Besides, different nonlinear activation functions were experimented on nodes in the hidden layers, keeping the output neurons with linear activation function. In order to choose a good set of parameters for the training procedure, some optimization algorithms and values of learning rate, batch size and L2 regularization strength were tried.

The generated models were evaluated using the training data set with a 5-fold cross-validation. In this type of cross-validation, the data set is divided into parts of the same size. One part forms the validation set and the other parts form the training set. This process is repeated for each part of the data, and the combination of tests is used to make a reliable estimate of the model error [4].

The models' performances were measured based on statistics such as the mean absolute error (MAE), mean squared error (MSE), mean absolute percentage error (MAPE), Pearson's correlation coefficient (r), and regression coefficient (R^2). These metrics are described in the following equations:

$$MAE = \frac{1}{n}\sum_{i=1}^{n}|O_i - F_i|, \tag{1}$$

$$MSE = \frac{1}{n}\sum_{i=1}^{n}(O_i - F_i)^2, \tag{2}$$

$$MAPE = \frac{100}{n}\sum_{i=1}^{n}\left|\frac{O_i - F_i}{O_i}\right|, \tag{3}$$

$$r = \frac{\sum_{i=1}^{n}(O_i-\overline{O})(F_i-\overline{F})}{\sqrt{\sum_{i=1}^{n}(O_i-\overline{O})^2}\sqrt{\sum_{i=1}^{n}(F_i-\overline{F})^2}}, \tag{4}$$

$$R^2 = \frac{\left(\sum_{i=1}^{n}(O_i-\overline{O})(F_i-\overline{F})\right)^2}{\sum_{i=1}^{n}(O_i-\overline{O})^2\sum_{i=1}^{n}(F_i-\overline{F})^2}, \tag{5}$$

where O_i is the observed value, $\overline{O}$ is the mean of all observed values, F_i is the forecasted value, $\overline{F}$ is the mean of all forecasted values and n is the number of samples.

Table 2 summarizes the topology and characteristics of the MLP neural network that achieved the best results in the metrics evaluated. The training of this network used the Feed-forward Backpropagation algorithm with Adadelta optimizer, learning rate of 1.0, batch size of 4.0, L2 regularization strength of 2.10^{-5} and mean squared error as the loss function.

3.1 Data preparation

The best obtained model has eighteen inputs, composed by six temporal variables and by the twelve parameters measured in the air quality station. The wind representation was converted

Table 2: Multilayer perceptron neural network chosen topology.

Layers	Topology		
	Neurons	Activation function	Trainable parameters
Input	18	N/A	0
1st Hidden layer	15	ReLU	285
2nd Hidden layer	185	ReLU	2,960
3rd Hidden layer	250	ReLU	46,500
4th Hidden layer	200	ReLU	50,200
5th Hidden layer	225	ReLU	45,225
6th Hidden layer	185	ReLU	41,810
Output	24	Linear	4,464

from scalar values of direction and speed to vector components U and V, using eqns (6) and (7). This transformation was made because the scalar representation could mislead the model, since direction values close to 0° or to 360° indicate that the wind is blowing in the same direction, although these values are numerically distant

$$U = speed * \sin((270 - direction)(\frac{\pi}{180})) , \tag{6}$$

$$V = speed * \cos((270 - direction)(\frac{\pi}{180})) . \tag{7}$$

As indicated by other works [10], [11], the use of periodical variables, as sine and cosine functions, representing the time cycles, lead to better results in predict ozone concentrations. Therefore, the following temporal inputs were used:

- sin (2π (60 h + m)/1440), cos (2π (60 h + m)/1440), where h = 0, 1, 2, …, 23 is the hour of the day and m = 0, 1, 2, …, 59 is the minute of the hour;
- sin (2πd/7), cos (2πd/7), where d = 0, 1, 2, …, 6 is the day of the week with 0 representing Sunday and 6 representing Saturday;
- sin (2πy/12), cos (2πy/12), where y = 1, 2, …, 12 is the month of the year.

Before model fitting, all inputs and their ground truth targets were normalized between -1 and 1. This procedure changes the data to a common scale, avoiding that one input have excessive importance in consequence of its value range [11]. Once the output of the model is obtained, the variables are de-normalized.

3.2 Reference models

To measure the efficiency of the proposed neural network, two models were used for reference. The first one is called Persistence model and is commonly used as a baseline to evaluate the performance of a forecasting model. In this predictor, the forecasts for all time steps ahead are set as the current value, which can be expressed mathematically as $y(t+\Delta t) = y(t)$, where y is the forecast target and t is time [14].

The second reference model is composed by a group of twenty-four linear regressors, each one responsible to predict a different time step of the next 24 hours of ozone concentration. The regressors are based on multiple linear regression and were fitted using the same inputs and targets as the MLP model. Regularization of L1 type was used in the models training, technique also known as Lasso Regression [15]. Several regularization parameters, which defines the regularization strength, were evaluated using a 5-fold cross-validation over the training data set and the parameter that produced the best performance for each regressor was chosen.

3.3 Computational tools

All computational experiments were implemented using the *Python* language. Experiments with MLP neural networks were performed using the *TensorFlow* platform through its *Keras* high-level API (Application Programming Interface), and the *Scikit-learn* machine learning library was used to evaluate the linear regression model.

4 RESULTS

A comparison between the metrics of evaluated ozone forecasting models, using the test data set, is shown in Table 3. Values close to 0.0 are best for the MAE and MSE, values close to 0.0% are satisfactory for the MAPE, and values close to 1.0 are adequate for the R^2 and r. A Pearson's correlation coefficient of −1.0 implies a negative linear correlation between the forecasted and the ground truth values, and a value of 0.0 implies that there is no linear correlation between these variables [14].

Table 3: Comparison table between ozone forecasting models using performance metrics over the test data set.

Model	Performance metrics				
	MSE	MAE	r	R^2	MAPE
Persistence	295.19	12.90	0.384	0.148	113.08
Lasso	126.02	8.73	0.692	0.479	90.10
MLP	101.75	7.68	0.770	0.593	70.55

The values of the metrics introduced on Table 3 refers to the models as a whole, considering all the 24 predictions at the same time. Thus, the proposed multilayer perceptron is a very effective model, surpassing the reference models in all considered metrics. The persistence model obtained the worst performance, as expected due to its simplicity. Tables 4 and 5 presents the statistics for some forecasting horizons of the MLP model and of the Lasso linear model, respectively.

With exception of the predictions for the first hour ahead, where the Lasso model has slightly better results, the neural network outperforms the linear model in all other time horizons. This is shown on Fig. 2 using as reference the mean squared error and the Pearson's correlation coefficient (r). Besides, the prediction errors of the MLP have a more stable behaviour along the forecasted time steps.

The model accuracy is graphically shown in Figs 3 and 4, where ozone concentration forecasts are displayed for 1 hour ahead and 24 hours ahead in a period of approximately 7 days of measurements. The blue lines represent the actual ozone concentrations and the orange lines represent predictions of ozone concentrations. Fig. 3 demonstrates a period in the beginning of June 2006, which represents the end of autumn season in the south hemisphere. On the other hand, the Fig. 4 display a period in the end of December 2014 and beginning of January 2015, which represents the start of summer season.

Table 4: Multilayer perceptron performance metrics over the test data set for some forecasting horizons.

Forecast horizon	Multilayer perceptron performance metrics				
	MSE	MAE	r	R^2	MAPE
T+1	42.90	4.94	0.910	0.829	41.14
T+2	63.37	5.93	0.863	0.745	51.10
T+3	77.46	6.61	0.829	0.688	57.65
T+6	99.42	7.56	0.774	0.599	67.39
T+9	110.52	8.01	0.746	0.557	72.44
T+12	114.09	8.19	0.741	0.549	75.13
T+15	113.53	8.23	0.742	0.551	76.20
T+18	111.69	8.16	0.748	0.559	75.81
T+21	106.72	7.97	0.757	0.574	75.84
T+24	101.50	7.77	0.768	0.590	76.73

Table 5: Lasso linear model performance metrics over the test data set for some forecasting horizons.

Forecast horizon	Lasso linear model performance metrics				
	MSE	MAE	r	R^2	MAPE
T+1	34.90	4.26	0.924	0.855	34.23
T+2	74.28	6.54	0.831	0.691	61.31
T+3	103.40	7.94	0.755	0.571	79.04
T+6	141.71	9.43	0.641	0.411	98.89
T+9	138.04	9.23	0.653	0.426	93.60
T+12	125.95	8.80	0.695	0.483	88.08
T+15	137.99	9.25	0.656	0.431	97.88
T+18	147.52	9.58	0.626	0.392	102.67
T+21	134.72	9.12	0.669	0.448	98.68
T+24	105.80	7.92	0.755	0.570	78.69

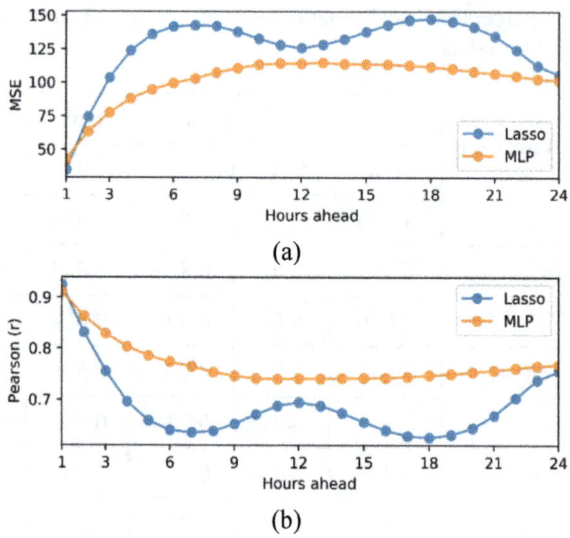

Figure 2: Comparison between MLP and Lasso predictions for each hour ahead. (a) Using mean squared error; and (b) Using Pearson's correlation coefficient.

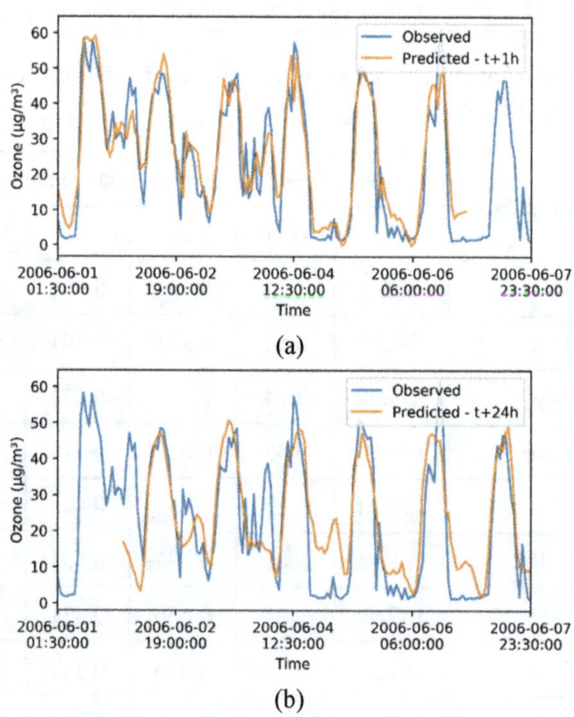

Figure 3: Multilayer perceptron predictions for a seven-day period in June 2006. (a) Predictions 1 hour ahead; and (b) Predictions 24 hours ahead.

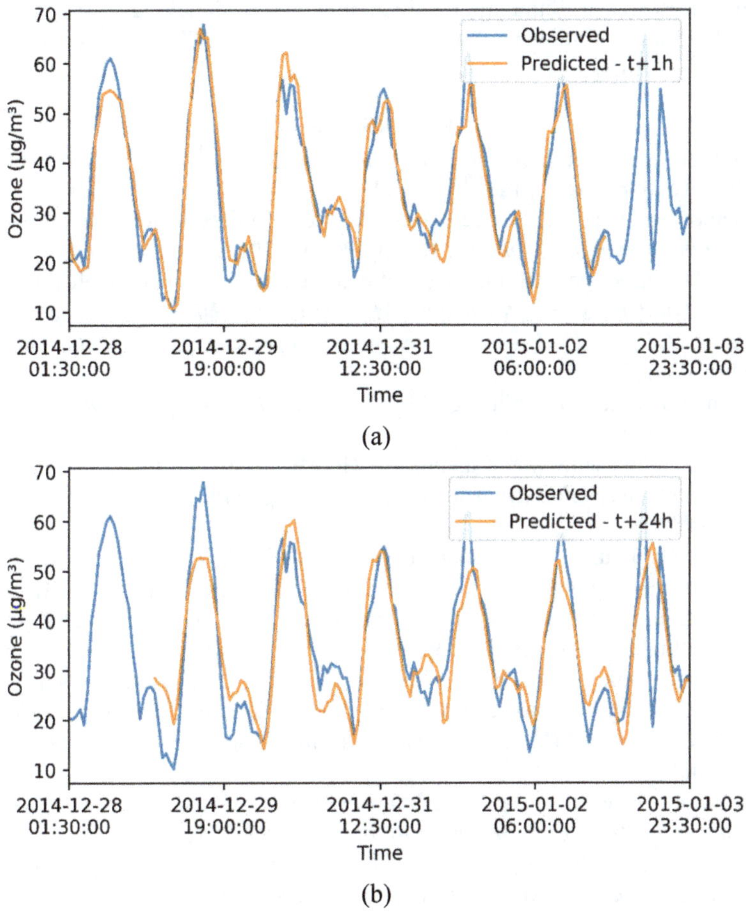

Figure 4: Multilayer perceptron predictions for a seven-day period between December 2014 and January 2015. (a) Predictions 1 hour ahead; and (b) Predictions 24 hours ahead.

5 CONCLUSIONS

The results indicated a reasonable performance for the proposed forecasting model, which can be used by authorities and citizens to take preventive measures that avoid imminent health risks due to O_3 exposure. Moreover, it has been shown that deep learning techniques can be successfully applied to hourly forecasting of ozone concentrations in urban areas. Once such models are trained and fit to the data, the inference process, i.e. the forecasting procedure, has a very low computational cost, meaning that it can be used as an alternative approach in comparison with numerical modelling systems, which require much more computational power.

ACKNOWLEDGEMENTS

This work was partially supported by the Bahia State Research Support Foundation (Fundação de Amparo à Pesquisa do Estado da Bahia - FAPESB), in the form of a Development of Technological Innovation grant for L. Alves in the project "Implantation of

Infrastructure Research in Simulation and Computational Modelling in the State of Bahia using High Performance Processing", developed at Senai CIMATEC Supercomputing Center for Industrial Innovation.

REFERENCES

[1] Filippidou, E.C. & Koukouliata, A., Ozone effects on the respiratory system. *Progress in Health Science*, **1**, pp. 144–155, 2011.

[2] Emberson, L.D. et al., Ozone effects on crops and consideration in crop models. *European Journal of Agronomy*, **100**, pp. 19–34, 2018.

[3] World Health Organization, Occupational and Environmental Health Team, *Air Quality Guidelines for Particulate Matter, Ozone, Nitrogen Dioxide, and Sulfur Dioxide: Global Update 2005: Summary of Risk Assessment*, World Health Organization, 2006.

[4] Reitermanová, Z., Data splitting. *WDS'10 Proceedings of Contributed Papers*, Part 1, pp. 31–36, 2010.

[5] Coman, A., Ionescu, A. & Candau, Y., Hourly ozone prediction for a 24-h horizon using neural networks. *Environmental Modelling and Software,* **23**(12), pp. 1407–1421, 2008.

[6] Brunelli, U., Piazza, V., Pignato, L., Sorbello, F. & Vitabile, S., Two-days ahead prediction of daily maximum concentrations of SO_2, O_3, PM_{10}, NO_2, CO in the urban area of Palermo, Italy. *Atmospheric Environment*, **41**(14), pp. 2967–2995, 2007.

[7] Salazar-Ruiz, E., Ordieres, J.B., Vergara, E.P. & Capuz-Rizo, S.F., Development and comparative analysis of tropospheric ozone prediction models using linear and artificial intelligence-based models in Mexicali, Baja California (Mexico) and Calexico, California (US). *Environmental Modelling & Software*, **23**(8), pp. 1056–1069, 2008.

[8] Biancofiore, F. et al., Analysis of surface ozone using a recurrent neural network. *Science of the Total Environment*, **514**, pp. 379–387, 2015.

[9] Eslami, E., Choi, Y., Lops, Y. & Sayeed, A., A real-time hourly ozone prediction system using deep convolutional neural network. arXiv preprint arXiv:1901.11079, 2019.

[10] Agirre, E., Anta, A., Barron, L.J. & Albizu, M., A neural network based model to forecast hourly ozone levels in rural areas in the Basque Country. *WIT Transactions on Ecology and the Environment*, vol. 101, WIT Press: Southampton and Boston, pp. 109–118, 2007.

[11] Tamas, W., Notton, G., Paoli, C., Voyant, C., Nivet, M.L. & Balu, A., Urban ozone concentration forecasting with artificial neural network in Corsica. *Mathematical Modelling in Civil Engineering*, **10**(1), pp. 29–37, 2014.

[12] Agirre, E., Anta, A. & Barron, L.J., Forecasting ozone levels using artificial neural networks. *Forecasting Models: Methods & Applications*, ed. J. Zhu, iConcept Press, pp. 208–218, 2010.

[13] Russell, S. & Norvig, P., *Artificial Intelligence: A Modern Approach*, 3rd ed., Pearson Education Inc., 2010.

[14] Zucatelli, P.J. et al., Short-term wind speed forecasting in Uruguay using computational intelligence. *Heliyon*, **5**(5), 2019.

[15] Tibshirani, R., Regression shrinkage and selection via the lasso. *Journal of the Royal Statistical Society: Series B (Methodological)*, **58**(1), pp. 267–288, 1996.

UNPAVED ROAD INFLUENCE AREAS IN HYDROCARBON EXPLORATION PROJECTS

MIGUEL ANTONIO DE LUQUE VILLA[1] & ALEXANDER VALENCIA CRUZ[2]
[1]Universidad de Cundinamarca, Colombia
[2]CAIA INGENIERIA SAS, Colombia

ABSTRACT

The influence area of a project includes environmental impacts caused by the development of the project on abiotic, biotic and socioeconomic components such as air, water, and soil. Knowing the location and extent of this area is essential for the environmental management of a project because it indicates where resources should be invested. This paper proposes a guide for the determination of areas of influence through the dispersion of particulate material in hydrocarbon projects. The present research looked at the unpaved road that leads from Puerto Gaitán (Meta, Colombia) to the Rubiales Field. The Gaussian dispersion model CALPUFF was used to determine the dispersion of particulate matter in the road section that connects Puerto Gaitán and the Rubiales field. 2000 discrete receptors were placed in a perpendicular line in three (3) sections of the road; these receptors were used to determine the concentration behaviour of particulate matter from the centre of the road every 2 m, up to 2000 m on each side. In order to determine the concentration profiles of the particulate matter, the modelling results were plotted in Excel, and a correlation analysis was carried out between the meteorology, the road sections, the emission factors and the particulate matter dispersion. This information was adjusted to different types of probability distributions in order to determine the representative function of the behaviour of the concentrations which included all meteorological data. The distributions with the best fit of the data corresponded to the Beta distribution. A methodology was generated for the calculation of the influence area in unpaved roads for hydrocarbon exploration projects. The methodology was used for the environmental licensing process of the LLA-66 exploration area of the company BC exploration and production of hydrocarbons SL branch Colombia, located in the city of Puerto Lopez in the Department of Meta, Colombia. It was found that the area of influence extended 6 m to each side of the axis of the road.

Keywords: air pollution, influence areas, unpaved roads, hydrocarbon exploration projects, modelling.

1 INTRODUCTION

The mining energy sector in Colombia is one of the pillars of the government's development plan, which is why it is important to know to what extent air quality is affected when devising energy development strategies [1]. Many industrial activities, particularly energy, could be highly polluting and directly affect air quality and, consequently, human health [1]. In developing countries, dispersion models are increasingly being used and can be very useful in environmental impact assessments, as well as supporting the development of local action plans that improve air quality [2]. A project's area of influence includes environmental impacts caused by the development of the project on abiotic, biotic and socioeconomic components, such as air, water, and soil [3].

Knowing the location and extent of this area is essential for the environmental management of a project because it indicates where resources should be invested. This paper proposes a guide for the determination of areas of influence through the dispersion of particulate material of hydrocarbon projects, in which the transit of vehicles on unpaved roads is presented with the CALPUFF model [4].

WIT Transactions on Ecology and the Environment, Vol 236,
www.witpress.com, ISSN 1743-3541 (on-line)
doi:10.2495/AIR190141

2 MATERIALS AND METHODS

2.1 Study area

The present research looked at the unpaved road that leads from Puerto Gaitán (Meta, Colombia) to the Rubiales Field (Fig. 1).

Figure 1: Puerto Gaitán – Rubiales field road.

2.2 Emissions factors

EPA guide AP-42, Chapter 13, miscellaneous sources (Section 13.2.2 – Unpaved roads) [5] was used to calculate the emission factors (eqn (1)).

$$E_F = k\left(\frac{s}{12}\right)^a \left(\frac{W}{3}\right)^b, \tag{1}$$

where k, a, and b are empirical constants; E_F = size-specific emission factor (g/VKT); s = surface material silt content (%), (calculated granulometry by a carved mechanic [6]); W = mean vehicle weight (tons) ; VKT = vehicle km travelled.

From the above, the values of the emission factors obtained are presented in Table 1.

Table 1: Unpaved road PST emission estimates parameters.

Section	s(6)	W	k	a	b	E_F (g/VKT)	Average daily traffic (vehicles/ day)	Distance (km)	VKT	Particulate matter emission (g/m^2 * s)
1	10.4	22	4.9	0.7	0.45	3063	980	40	33051.48	0.001160
2	10.4	22	4.9	0.7	0.45	3063	525	40	17706.15	0.000630
3	9.5	22	4.9	0.7	0.45	2875	14	40	472.16	0.000019

2.3 Air quality modelling

The Gaussian dispersion model CALPUFF was used to determine the dispersion of particulate matter in the road section that connects Puerto Gaitán and the Rubiales field. The simulations were carried out under different meteorological conditions and emissions factors.

2.3.1 Simulation scenarios

To define the simulation scenarios, the available meteorology of the area and the emission factors of the road were taken into account. The topography was constant. The available meteorology for the study included data from the Rubiales Morelia Airport station for the year 2013. It was decided to simulate the entire year, separated month by month, resulting in 12 meteorological data sets, in order to increase the scenarios. In total, 39 scenarios were used.

2.3.2 Topographic information
It is necessary to process the topography in the dispersion model, a completely flat terrain is not assumed, using the Digital Elevation Model (DEM) on the area to be studied, which obtains the altitude relative to sea level of each source and receptor. For our study, we used the file 10s090w_20101117_gmted_mea075 from The Advanced Spaceborne Thermal Emission and Reflection Radiometer (ASTER) Global Digital Elevation Model (GDEM) [7], defining the coordinates in which it was necessary to import the elevations (Fig. 2).

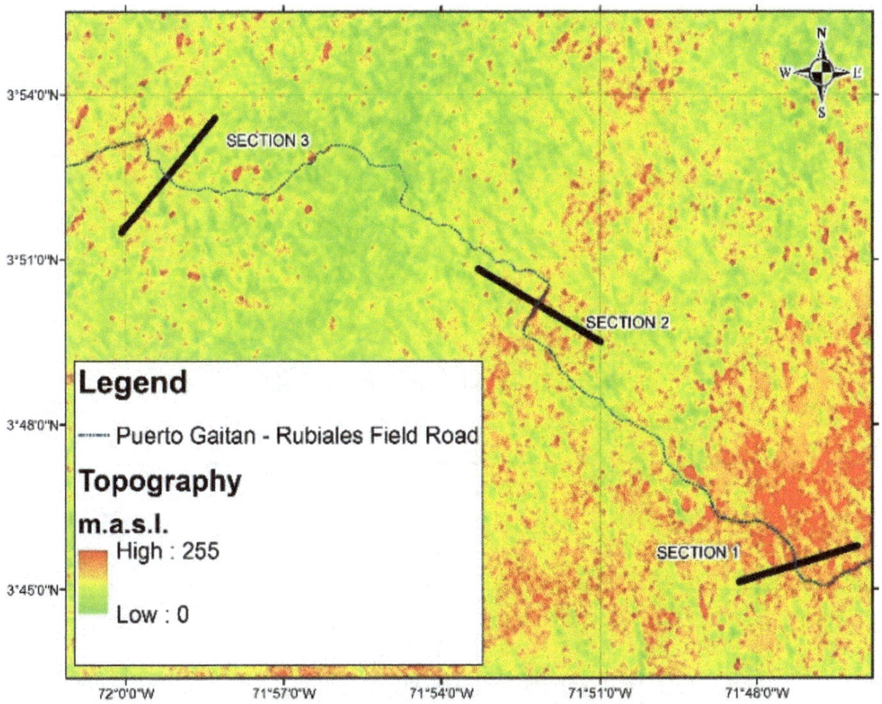

Figure 2: Study area topography.

2.3.3 Meteorology information
The wind rose representative of the analysed meteorology was obtained with the software WRPLOT View [8]; in this program, the values for wind speed and direction are set as input parameters. The result is shown in Fig. 3; the prevailing winds came from the east and northeast, and the predominant speeds were between 0.5 and 2.1 m/s.

2.3.4 Modelling grid
For the modelling domain, 10,000 receptors were distributed uniformly on a 40 km x 40 km grid every 400 m to determine the global particulate distribution. Fig. 4 shows the location of the modelling grid.

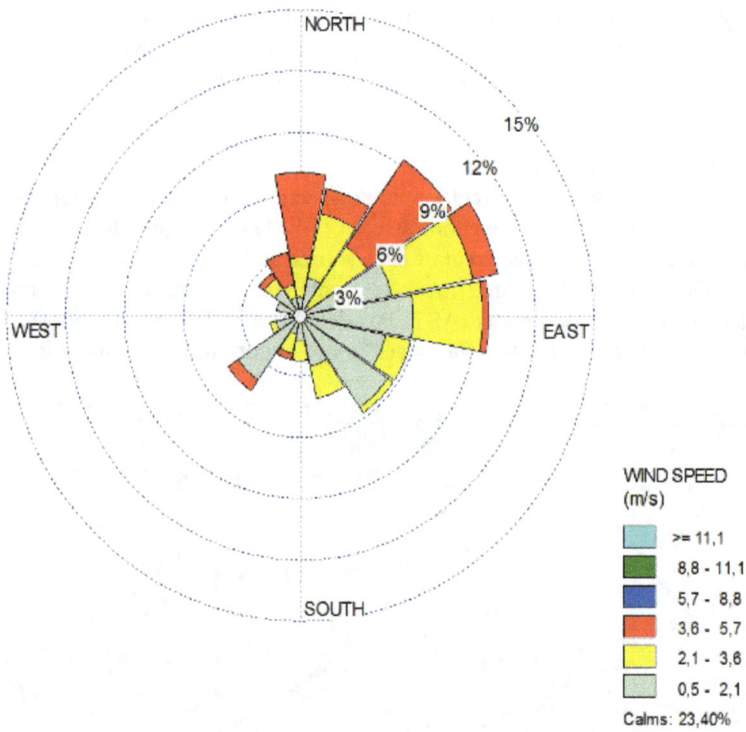

Figure 3: Wind rose Rubiales field.

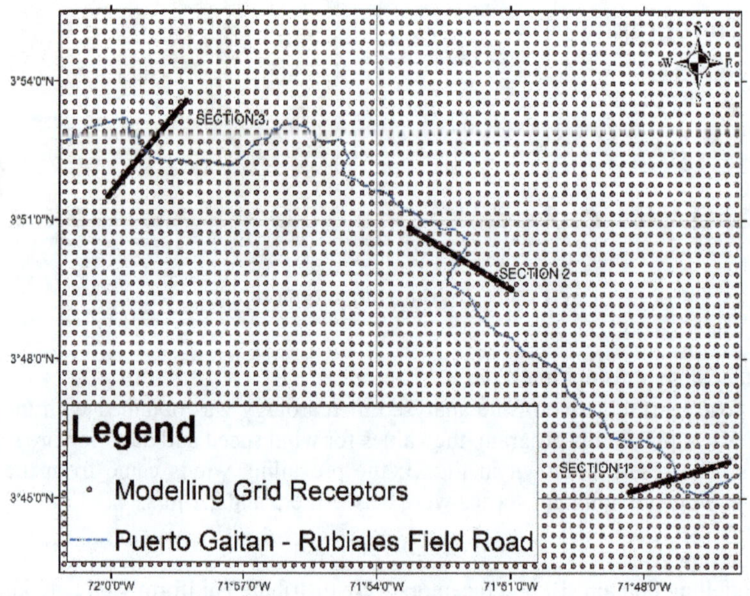

Figure 4: Modelling grid.

2.3.5 Discrete receptors

2000 discrete receptors were placed in a perpendicular line in three sections of the road; these receptors were used to determine the decay of the concentrations on the sides of the road every 2 m, up to 2000 m on each side. The location of the discrete receptors is shown in Fig. 5.

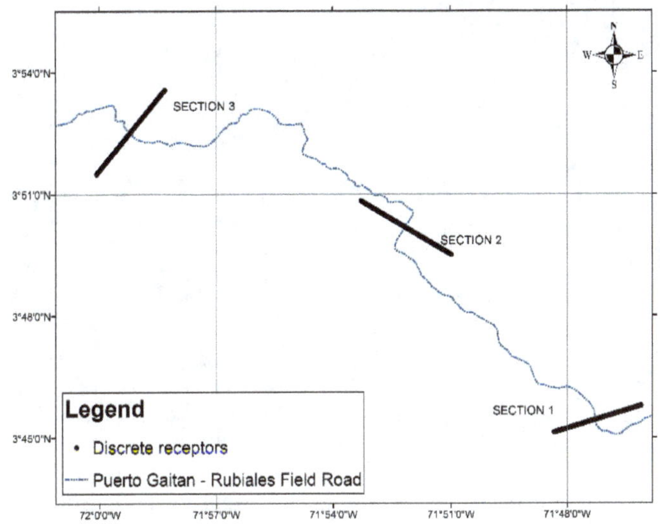

Figure 5: Discrete receptors Puerto Gaitán – Rubiales field road.

2.4 Identification of the probability density function adjusting to the behaviour of the particular matter

In order to determine the concentration profiles for the particulate matter, the modelling results were plotted in Excel, and a correlation analysis was carried out between the meteorology, the road sections and the emission factors and the particulate matter dispersion.

The results reported by the CALPUFF were processed to determine the average annual arithmetic concentrations. This information was adjusted to different types of probability distributions in order to generate a representative function of the behaviour of the concentrations, which included all meteorological data. This adjustment was made with the EasyFit distribution adjustment software [9], in which about 55 distributions were evaluated. EasyFit evaluates all available distributions in the literature using statistical goodness-of-fit tests (Chi-square, Kolmogorov-Smirnov and Anderson-Darling) and indicates the means of the p-value statistical parameter if the data conform to a given distribution. In this case, the goodness adjustment tests were defined with 95% confidence.

3 RESULTS

3.1 Modelling results

The modelling results in three different scenarios where the meteorology was similar, and the emission factors were different showed that the dispersion of particulate matter for the annual exposure period behaved similarly (Figs 6 and 7).

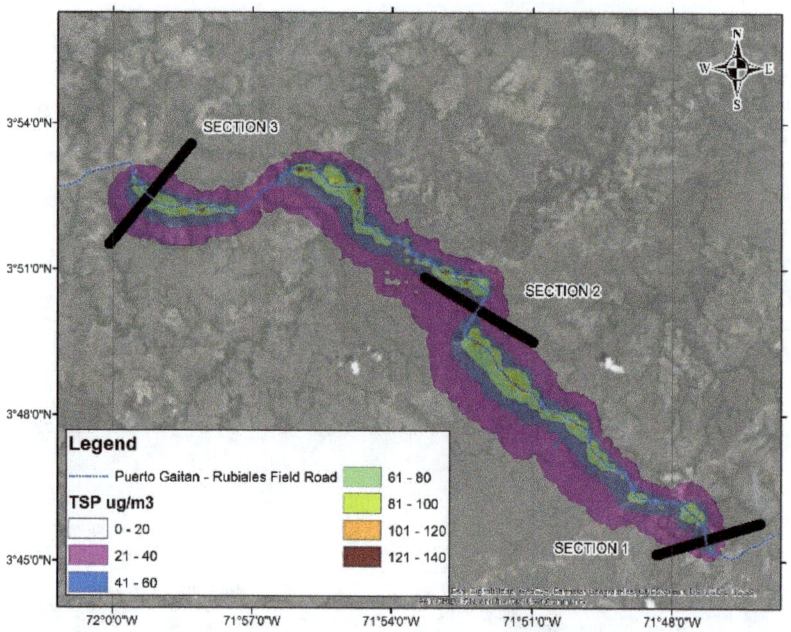

Figure 6: Concentration isopleths for the annual exposure period of particulate matter Scenario 1.

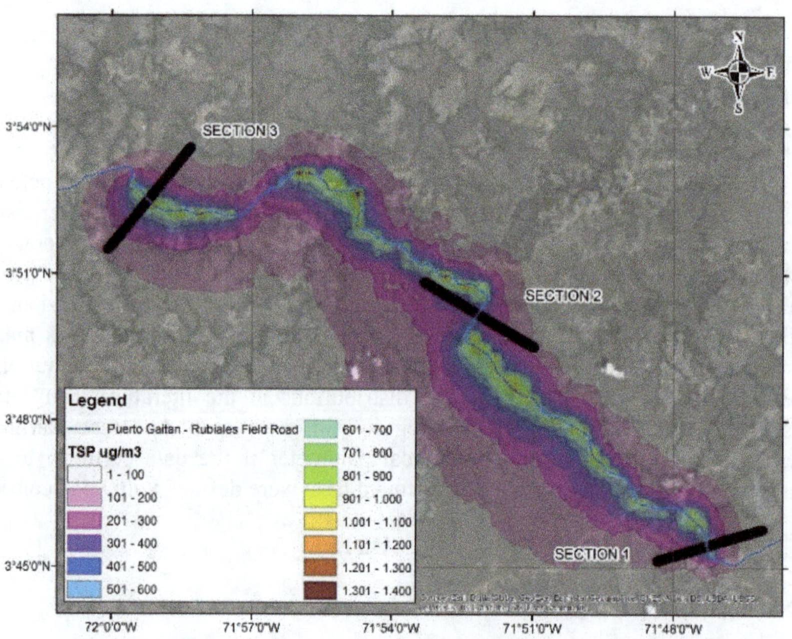

Figure 7: Concentration isopleths for the annual exposure period of particulate matter Scenario 2.

The behaviour of the dispersion of particulate matter was similar in every section of the road, independent of the meteorology of the study area, and the air quality was directly proportional to the emission factor (Figs 8 and 9).

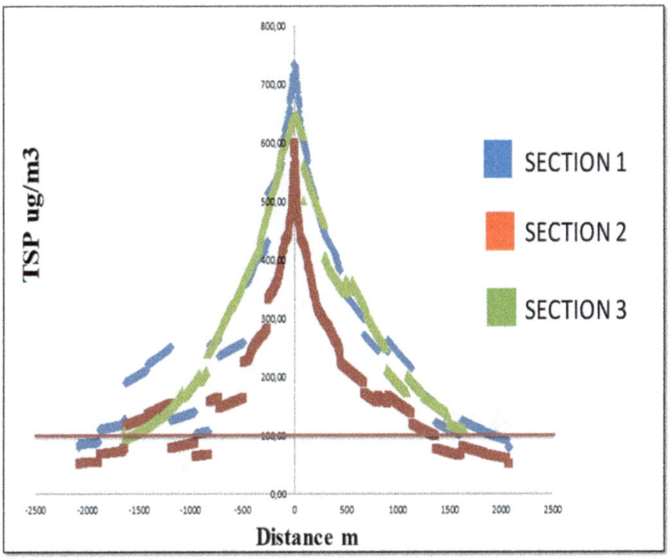

Figure 8: Average particulate matter concentration profiles for 2000 m on the side of the road Scenario 2.

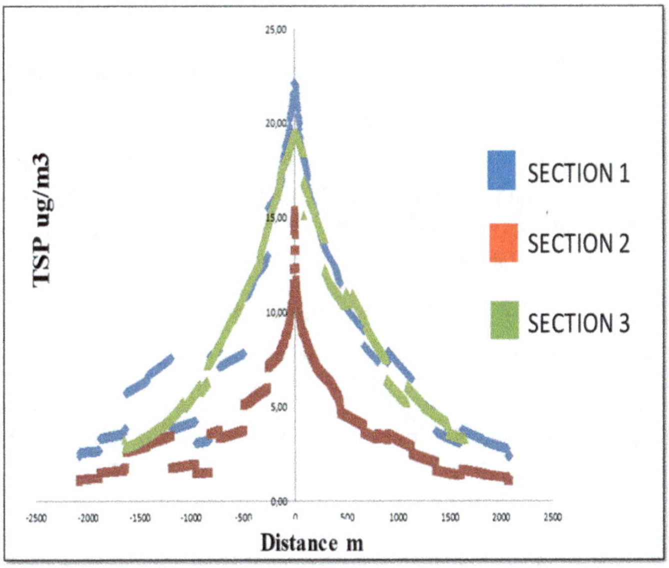

Figure 9: Average particulate matter concentration profiles for 2000 m on the side of the road Scenario 3.

3.2 Adjustment of the probability density function for the behaviour of particular material

The results of the EasyFit Software indicated that the distributions with the best fit of the data corresponded to the Beta distribution (Fig. 10). Similar studies obtained the same distribution [10].

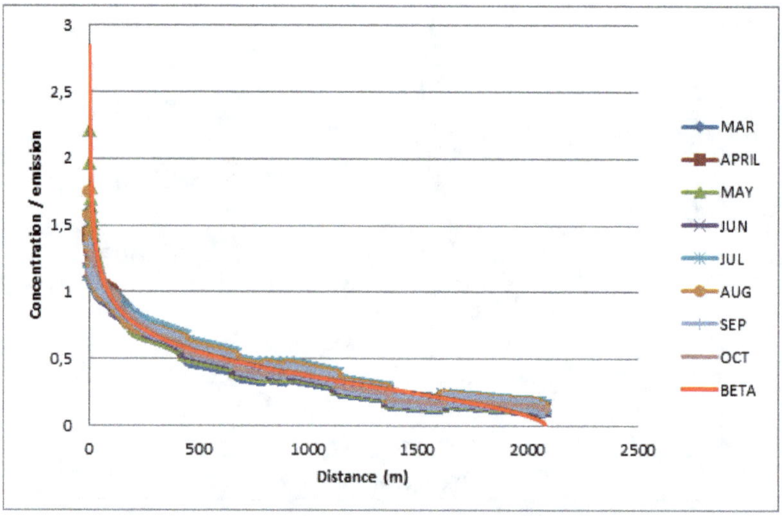

Figure 10: Adjustment of the beta distribution for concentrations of particulate matter.

3.3 Methodology to find the area of influence in unpaved roads in hydrocarbon exploration projects

Eqn (2) is proposed to determine the annual concentration of particulate matter.

$$Ca = Fa \text{ Beta } (\alpha 1, \alpha 2, A, B), \tag{2}$$

where, Ca is the annual concentration in µg / m$^{3;}$ Fa is the annual dispersion factor. Beta is the distribution function with parameters α1, α2. A and B represent minimum and maximum dispersion distances, respectively. Table 2 provides values for these parameters for TSP dispersion around unpaved roads.

Table 2: Beta distribution parameters.

Parameter	Dispersion factor	α1	α2	A	B
Particulate matter annual concentration	834.74	0.726	1.567	0	2,085

3.4 Case study

The methodology was used for the environmental licensing process of the LLA-66 exploration area of the company BC exploration and production of hydrocarbons SL branch Colombia, located in the city of Puerto Lopez in the Department of Meta, Colombia. It was found that the area of influence extended 6 m to each side of the axis of the road (Fig. 11).

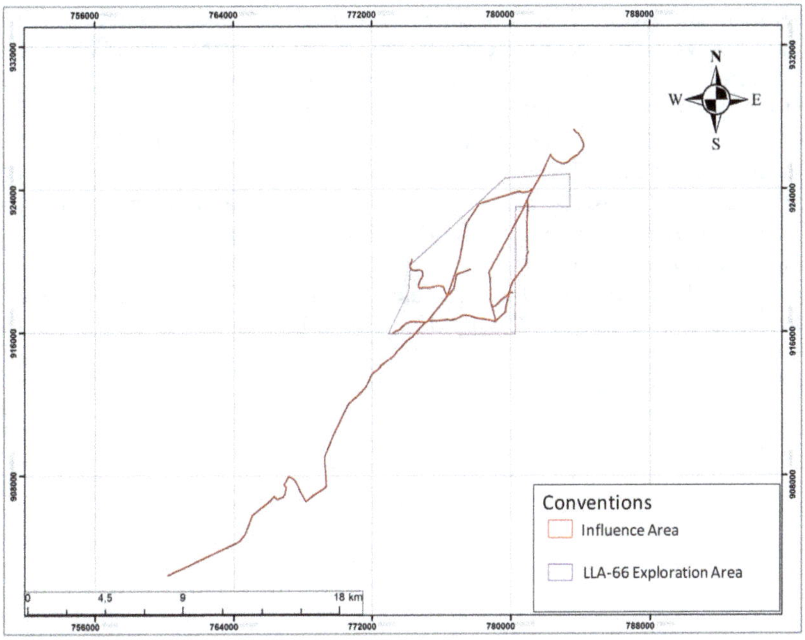

Figure 11: Area of influence on the atmospheric component LLA-66 exploration area.

4 CONCLUSIONS

According to the modelling results and distribution analysis performed in this study, we concluded that the decay of the particulate material on unpaved roads is adjusted to a beta distribution with the following parameters:

ACKNOWLEDGEMENTS

The authors thank the companies CAIA engineer and BC exploration and production of hydrocarbons SL branch Colombia for all of their support and information that they provided for the project.

REFERENCES

[1] Turtós, L., Modeling guide for the local dispersion of gaseous pollutants and particles with the AERMOD Model System, 2010.

[2] Conde, A.P., *Local Action Plans to Improve Air Quality and Environmental Impact Studies Supported by Atmospheric Dispersion Models*, National Congress of the Environment: Madrid, p. 21, 2008.

[3] ANLA. *Autoridad Nacional de Licencias Ambientales*. www.anla.gov.co/contenido/contenido.aspx?catID=1324&conID=7956. Accessed on: 26 May 2014.

[4] Scire, J., Strimaitis, D. & Y.R.J., CALPUFF Dispersion Model, Earth Tech, Inc. 196 Baker Avenue Concord, MA 01742, no. January, 2000.

[5] US EPA, Emission Factor for Unpaved Road, US Environmental Protection Agency: Washington, 2001.

[6] Instituto Nacional de Vías, Granulometric analysis of soils by size. I.N.V. E – 123 – 07, 2007.

[7] NASA LP DAAC, *ASTER Level 1 Precision Terrain Corrected Registered At-Sensor Radiance*. Version 3. NASA EOSDIS Land Processes DAAC, USGS Earth Resources Observation and Science (EROS) Center, Sioux Falls, South Dakota. https://lpdaac.usgs.gov. Accessed on: 1 Jan. 2016.
[8] Thé, J.L., Thé, C.L. & Johnson, M.A., WRPLOT View Wind and Rain Rose Plots for Meteorological Data, 2016.
[9] Schittkowski, K., EASY-FIT: A software system for data fitting in dynamical systems EASY-FIT: A Software System for Data Fitting in Dynamic Systems. *Structural and Multidisciplinary Optimization*, **23**(2), pp. 153–169, 2002.
[10] Huertas, M.E., Huertas, J.I., & Valencia, A., Vehicular road influence areas. *Atmospheric Environment*, **151**, pp. 108–116, 2017.

REGRESSION MODELLING FOR PREDICTING VISIBILITY IN A TROPICAL URBAN ENVIRONMENT

SAVITHA ULAVI & SARAGUR MADANAYAK SHIVA NAGENDRA
Department of Civil Engineering, Indian Institute of Technology, India

ABSTRACT

Air pollution and visibility reduction is a major concern in cities. The present study focused on understanding the impact of air quality and meteorology on visibility impairment for a tropical coastal city (Chennai, India). Visibility measurements were carried out for 176 days representing summer, winter and post-monsoon season. The meteorological and air quality parameters for the study period were taken from the nearest continuous monitoring station supervised by the Central Pollution Control Board of India (located within 2 km of the visibility monitoring site). It was found that the average visibility experienced by the city is 11.24 ± 5.47 km. The seasonal variations in the data monitored showed maximum average visibility (15.06 ± 2.87 km) during summer, followed by winter (7.64 ± 3.53) and post-monsoon season (5.80 ± 2.72). The stepwise linear regression model was developed to predict visibility. The model developed for winter showed $PM_{2.5}$ is the significant predictor with reasonable R^2 value of 0.68. It was observed that there is a strong negative correlation between fine particulate matter ($PM_{2.5}$) and visibility. During summer ozone, $PM_{2.5}$, wind speed, relative humidity, oxides of nitrogen (NO_x) turned out to be the significant predictors for visibility and Pearson correlation showed inverse correlation between visibility and ozone (-0.683), as well as $PM_{2.5}$ (-0.587). Thus the model developed for summer was able to capture the photochemistry of the summer and explained 71% variation in visibility. The number of low visibility events (< 2 km) were estimated for different seasons excluding fog events (RH > 90). The results showed large numbers of low visibility events in winter (348) followed by post-monsoon (205). The results also indicated that haze events increased with an increase in particulate matter concentration (> 60 µg/m³), RH (> 65) and decrease in wind speeds (< 1 m/s).

Keywords: air pollution, regression, meteorology, visibility reduction.

1 INTRODUCTION

Poor air quality due to fine particulate matter (aerodynamic diameter < 2.5 µm) pollution and visibility reduction is a major concern in cities. The phenomenon of urban visibility reduction is complex and manifests as a result of interaction of air emissions and meteorology. The basic mechanism attributable to low visibility is scattering and absorption by gases and particles in the atmosphere [1], [2]. Meteorological parameters like wind speed, wind direction, ambient temperature, relative humidity influence visibility indirectly by influencing sources and sinks of gases and particulate matter [3]. There could exist a several combinations of meteorological parameters say low wind speed and high humidity which could lead to increased residence time of pollutants, there by higher pollutant build up leading to inversion/stable atmospheric conditions. This increased aerosol loading could further reduce solar radiation reaching earth's surface, leading to depressed planetary boundary layer which intern progress towards weakened diffusion [2], [4]. In addition particulate mass concentration, particle number concentration, size, chemical composition, mixing state and hygroscopicity have a significant impact on visibility impairment [5], [6].

Quantification of relationship between visibility, air quality and meteorology is critical for understanding physical or chemical processes which impair visibility and to forecast poor visibility events. Such studies on urban visual air quality as well play important role in shaping environmental policy and management response system. The present study focused on understanding the following aspects: (i) monitoring and characterizing of visibility for

WIT Transactions on Ecology and the Environment, Vol 236, © 2019 WIT Press
www.witpress.com, ISSN 1743-3541 (on-line)
doi:10.2495/AIR190151

tropical coastal urban environment; (ii) to quantify a relation between visibility, air quality and meteorology through a stepwise linear regression model; and (iii) to explore the possible mechanism or underlying theory which may contribute for visibility reduction in different seasons and to analyse the low visibility events.

2 STUDY AREA DESCRIPTION AND MEASUREMENTS

The monitoring site is located in Chennai, a coastal city situated on Coromandel coast of the Bay of Bengal (13.0827°N, 80.2707°E). It is a fast growing Indian metropolitan city with a population of 8.6 million [7]. Rapid urbanisation, industrial and vehicular growth has resulted in a greater environmental stress over city. The earlier studies carried out which aimed at understanding the prevailing air quality in a city (between 1991 to 2005) have shown exceedance of PM_{10}, NO_x, surface ozone concentrations at several pockets of Chennai [8], [9]. In addition the source apportionment studies carried out post 2005 in Chennai have reported presence of increased particulate phase sulphate, gaseous phase oxides of nitrogen and EC/OC ratio in particulate matter [10]. This type of atmospheric composition indicate potential for increased particulate hygroscopicity, enhanced light absorption and scattering there by posing greater risk of atmospheric visibility deterioration.

With orientation to climate of city, Chennai is hot and humid throughout the year. It experiences summer (March–May), winter (January and February). The monsoon period is classified as pre-monsoon (south-west monsoon) June to September and post-monsoon (north-eastern monsoon) from October to December [11]. Chennai experiences tropical wet and dry climate as per Koppen climate classification. Visibility monitoring was carried out on a roof top (20 meters above ground level) of Indian Institute of Technology Madras (IITM) Engineering Design building block (shown in Fig. 1). IITM campus is a beautiful wooded land spread over 250 acres and the campus is surrounded by residential area, educational and research institutes and commercial centres thereby heavy traffic roads. Thus the monitoring site serves as a perfect urban background. Belfort model 6000 visibility sensor was used to monitor visibility and it works on the principles of forward light scattering (used in earlier studies [12], [13]). Visibility range the instrument can detect is 6 m to 80 km with accuracy of ± 10%. Visibility data was captured for every minute and monitoring was carried out for 61 days representing summer (April and May 2018), 56 days representing post-monsoon (November and December 2018) and 59 days representing winter (January and February 2019). Meteorological (relative humidity, wind speed, wind direction) and air quality parameters ($PM_{2.5}$, SO_2, NO_x, ozone) were taken from CPCB continuous monitoring site located within IITM campus [14].

2.1 Methods

The data collected from CPCB was examined for outliers using Quartile method. Every one minute average visibility data, monitored and every 15 minute average air quality and meteorological data, obtained from CPCB was converted into daily average values. IBM SPSS statistics Version 25 was used to build season-wise regression model and to carry out statistical analysis. Origin Pro 2015 was used as plotting software. MatLab version R2015a was used as a coding platform to find planetary boundary height from radiosonde observations. The data points corresponding to RH > 90 were eliminated (as to exclude fog events contributing for low visibility) and haze events were examined.

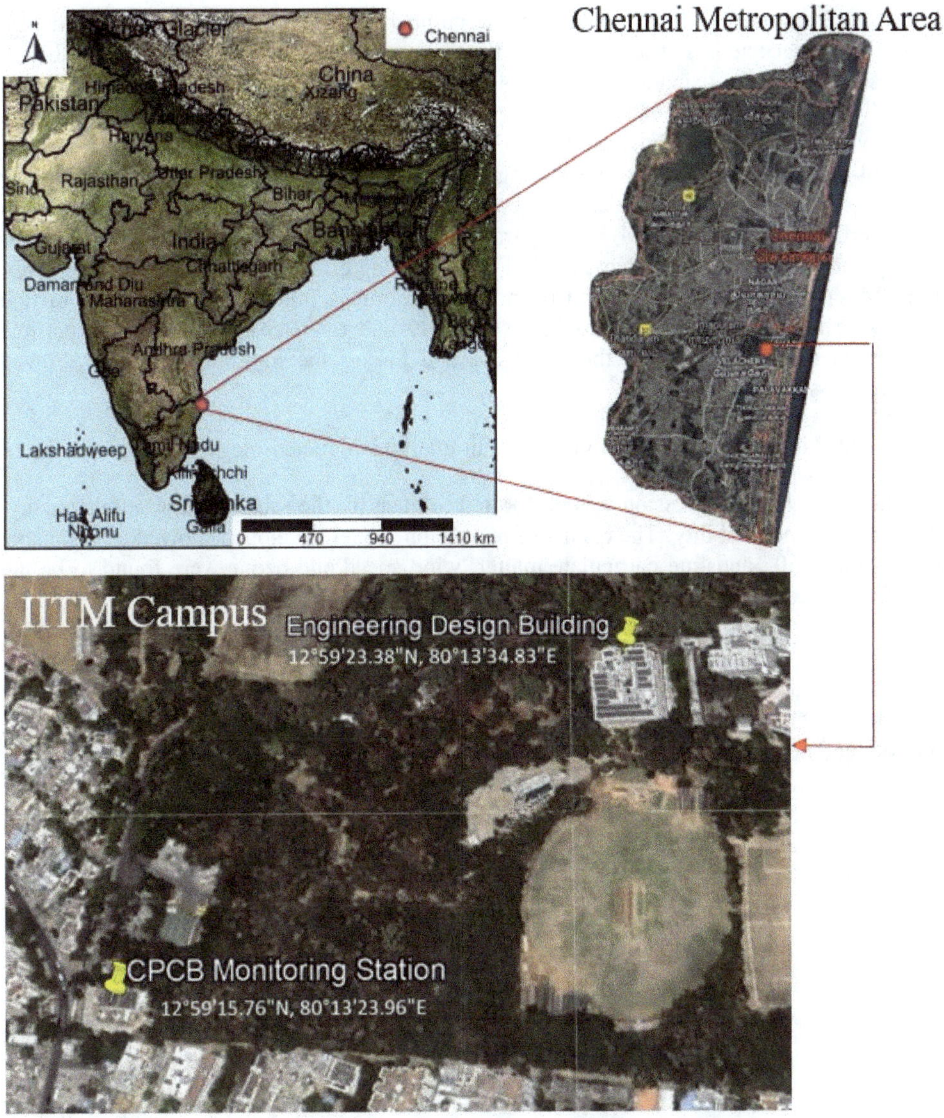

Figure 1: Satellite view of India highlighting the geographical location of monitoring site in Chennai.

3 RESULTS AND DISCUSSION

3.1 Characteristics of visibility monitored and seasonal variations

The average visibility experienced by the city is 11.24 ± 5.47 km with a modal value being 14 km (for 176 days representing summer, post-monsoon and winter season). The average extinction coefficient is 0.40 km^{-1}(at 880nm). The five number summary (refers to descriptive measures in statistics) minimum < Q_1 < median < Q_3 < maximum is estimated to be

2.44 < 5.17 < 8.96 < 13.67 < 21.70. The five number summary was calculated considering the daily average and it is clear that 50% of the data set is below 8.96 km (indicative of haze). The test for normality using normal Q–Q plot and Kolmogorov–Smirnov test showed data is normally distributed (N = 30847) with a significance value of 0.0. The seasonal variations in data have been observed (shown in Fig. 2). Summer marks the better visibility with a mean value of 15.60 ± 4.16 km. However during both post-monsoon and winter season visibility is below 10 km indicating the occurrence of hazy days. The impaired visibility days were characterized by increased $PM_{2.5}$ concentration shown in time series plot of visibility, air quality and meteorological parameters in Fig. 3. The lag effect was observed between NO_x and $PM_{2.5}$ (increase in oxides of nitrogen concentration is followed by sharp rise in fine particulate concentration). This could be possibly due to heterogeneous gas to particle conversion reactions. 57% of the daily average during the monitoring period showed visibility less than 10 km.

3.2 Quantifying relation between visibility, air quality and meteorology

The step wise linear regression was performed to quantify the relation between air quality, meteorology and visibility. The results of regression model developed showed a reasonable R^2 value of 0.502 and fine particulate matter, wind speed and ozone were found to be the significant predictors (given in eqn (1))

$$\text{Visibility} = 12.091\text{-}0.111(PM_{2.5}) + 0.8817\ (WS)\text{-}0.73\ (\text{ozone}). \quad (1)$$

A further attempt was made to know if there is any improvement in the model by considering the seasonal data. Thus steps were repeated and model was rebuilt for winter, summer and post-monsoon season.

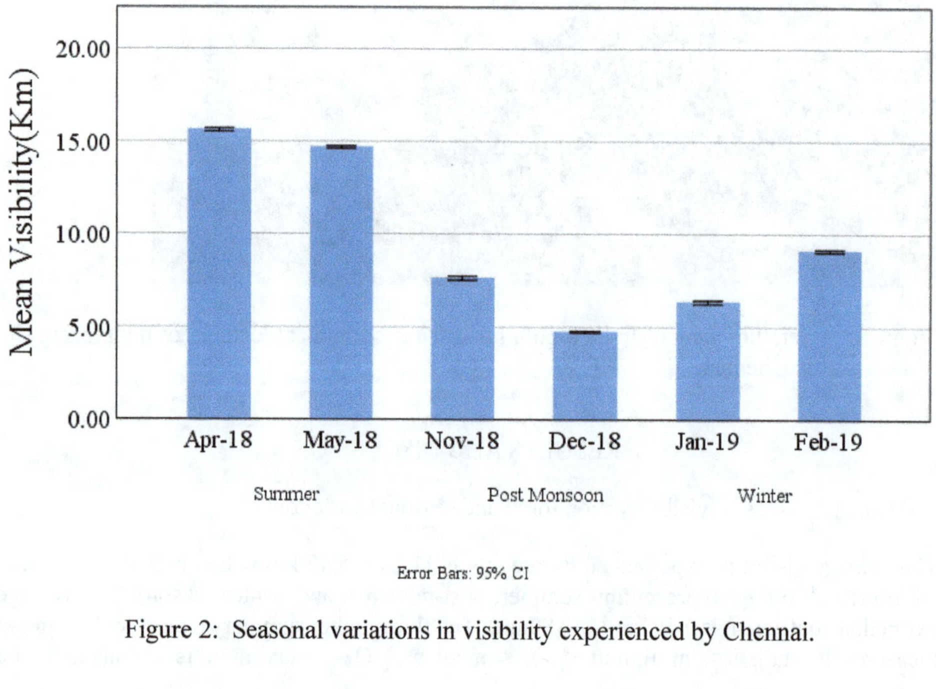

Figure 2: Seasonal variations in visibility experienced by Chennai.

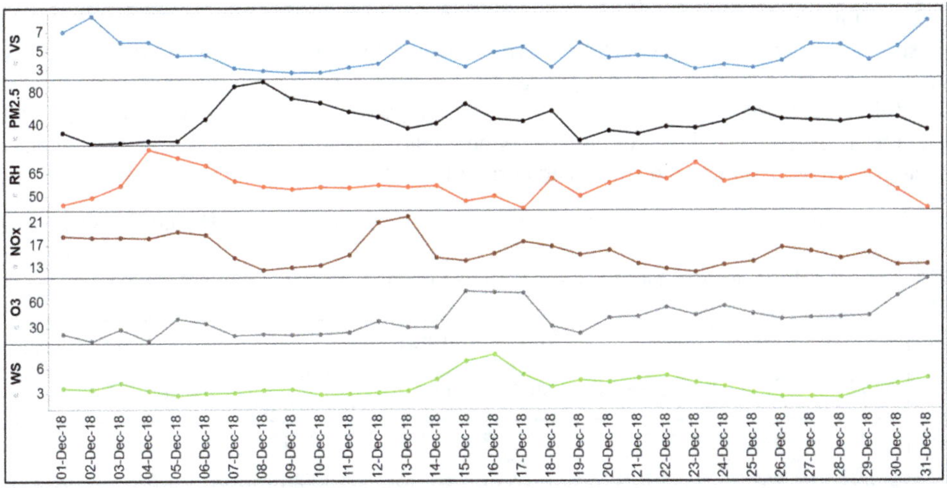

Figure 3: Time series plot showing variations in visibility (VS = blue), particulate matter ($PM_{2.5}$ = black), relative humidity (RH = red), oxides of nitrogen (NO_X = brown), ozone (O_3 = grey) and wind speed (WS = green) for December 2018.

3.2.1 Winter

It was interesting to know that $PM_{2.5}$ turned out to be the significant predictor variable during winter and 68.2% of variations in visibility could be explained by fine particulate matter alone ($R^2 = 0.682$) given in eqn (2). The Pearson correlation study showed there exits strong negative or inverse correlation between visibility and $PM_{2.5}$ (-0.826) and ozone (-0.52). Positive correlation was observed between visibility and wind speed (shown in Table 1). Stepwise regression equation obtained for winter is presented in eqn (2)

$$\text{Visibility} = 14.304\text{-}0.202(PM_{2.5}). \tag{2}$$

Table 1: Pearson correlation between visibility, air quality and meteorological parameters.

Parameters	Visibility	$PM_{2.5}$	NO_x	Ozone	RH	WS
Visibility	1.00	-0.826	-0.256	-0.52	-0.078	0.329
$PM_{2.5}$		1.00	0.429	0.475	-0.003	-0.330
NO_x			1.00	-0.125	-0.122	0.170
Ozone				1.00	-0.326	-0.319
RH					1.00	-0.68
WS						1.00

This could be possibly due to radiation inversion conditions, suppressed planetary boundary layer which generally prevail during winter season. These stable weather conditions trap the pollutants close to the earth's surface and affect optical properties of the atmosphere (scattering and absorption) there by reducing the ambient visibility. The distribution fit between visibility and $PM_{2.5}$ showed exponential decay relation during winter (with R^2 0.80).

3.2.2 Summer

In summer ozone, $PM_{2.5}$, WS, RH, NO_x turned out to be the significant predictors for visibility with R^2 value of 0.709. Pearson correlation showed strong negative correlation between visibility and ozone (-0.683), $PM_{2.5}$ (-0.587). Stepwise regression model summary for summer is given in Table 2 and the regression equation in eqn (3). Fig. 4 depicts frequency histogram for standardized residual and normal P–P plot.

$$\text{Visibility} = 32.23-0.126(\text{ozone})-0.057(PM_{2.5})-0.199(WS)-0.14(RH)-0.037(NO_x). \quad (3)$$

Table 2: Model summary for stepwise linear regression – summer data.

Model summary					
Model	R	R square	Adjusted R square	Standard error of the estimate	Durbin–Watson
1	0.683[a]	0.467	0.458	2.11612	
2	0.769[b]	0.592	0.577	1.86806	
3	0.806[c]	0.650	0.632	1.74376	
4	0.825[d]	0.680	0.657	1.68247	
5	0.842[e]	0.709	0.682	1.61946	1.102

a. Predictors: (Constant), Ozone
b. Predictors: (Constant), Ozone, $PM_{2.5}$
c. Predictors: (Constant), Ozone, $PM_{2.5}$, WS
d. Predictors: (Constant), Ozone, $PM_{2.5}$, WS, RH
e. Predictors: (Constant), Ozone, $PM_{2.5}$, WS, RH, NO_x
f. Dependent variable: Visibility

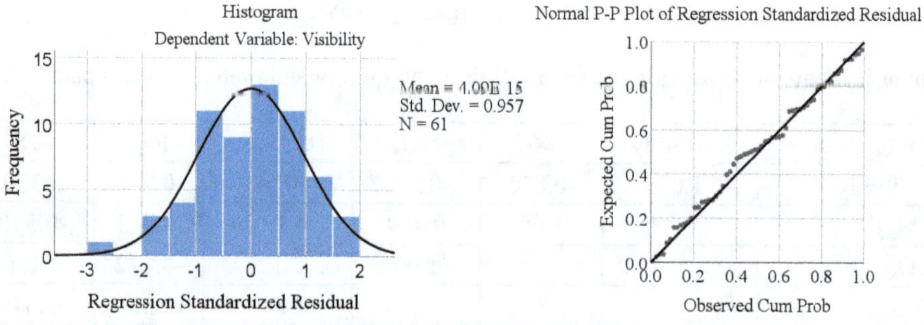

Figure 4: Frequency histogram for standardized residual and normal P–P plot.

Ground level ozone and secondary particulates formation through complex photochemical reactions during peak summer hours has been studied and reported in multiple urban environments [15], [16]. The presence O_3 and several other radicals (like Hydroxyl radical-OH, Peroxy radicals-RO_2 etc.) increase the oxidizing capacity of the atmosphere which intern increase the secondary fine aerosol particle (0.1 µm to 1 µm) by gas to particle conversion process. This secondary fine particulate matter formed ultimately lead to visibility reduction

[17], [18]. Thus the model developed for summer (eqn (3)) is able to capture the photochemistry of summer (smog) and explains approximately 71% of variation in visibility.

3.2.3 Post-monsoon

During post-monsoon season $PM_{2.5}$, ozone (-0.414 and -0.321 respectively) showed negative correlation with visibility and RH showed no significant correlation with visibility. The variation in visibility explained by the model (eqn (4)) was found to be low (29.1%).

$$\text{Visibility} = 14.18-0.061(PM_{2.5})-0.051(\text{ozone})-0.069(RH). \tag{4}$$

Post-monsoon season is the period where Chennai city receives major portion of its rainfall. The low R^2 value could be possibly due to unaccounted factors in the model like washout or precipitation events, seasonal changes in wind pattern, cloud cover, etc.

During this season the lowest visibility value observed was 2.65 km and corresponding $PM_{2.5}$ concentration was 73 µg/m³ (on 9/12/2018) and the highest visibility value witnessed was 14 km (on 6/11/2018) with corresponding $PM_{2.5}$ concentration of 39 µg/m³. However there were 12 days with $PM_{2.5}$ concentration lower than 30 µg/m³ which did not witness better visibility. The reduction in fine particulate matter due to wet deposition as a result of drizzle or precipitation events. However though particle concentration was reduced, precipitation events would have simultaneously increased the local relative humidity leading to increased extinction of light and there by visibility reduction [19].

3.3 Examination of low visibility occurrences

During visibility monitoring period severe low visibility (< 2 km) events were observed. Most often low visibility episodes occur due to fog, haze and fog–haze mixed events. However in the present study RH > 90 were not considered, thus fog events contributing to visibility impairment are excluded. Haze has been often reported as weather phenomena where visibility < 10 km at RH below 90% [19] and haze episodes are characterized by excessive aerosol loading. The occurrence of severe low visibility events were highest during winter (348) followed by post-monsoon season (205). As depicted in Fig. 5, the possibility of haze events increase with increase in RH and particulate matter concentration and decrease in wind speed. Most of these events have occurred during early morning and mid night hours. The reason for such events during night and early morning hours would be the formation of stable-nocturnal boundary layer. Thus to justify this planetary boundary layer (PBL) data for Chennai city was analysed from radiosonde observations [20]. The data is available for morning (5:30 am) and evening (5:30 pm). The Pearson correlation performed between visibility and PBL for dataset showed positive correlation during January (0.578), February (0.210) during winter. This means that as the PBL increases visibility increases and vice versa, however for post-monsoon season no significant correlations were observed. PBL varied between 0.6–2 km during low visibility events.

The national ambient air quality standard for $PM_{2.5}$ is 60 µg/m³ (24 hour average). It is clear from the Fig. 5 that large number of haze episodes have occurred even when the $PM_{2.5}$ concentrations are below the standard, when RH exceeds 70 and wind speeds are lower than 0.5 m/s during winter. However no severe low visibility events were observed during summer though RH > 70 at several instances. The lowest visibility recorded during summer is 4.35 km. This could be possibly due to decrease in the strength of land sea breeze effect during post-monsoon and winter periods (as sea-breeze circulation occurs when 5°C or

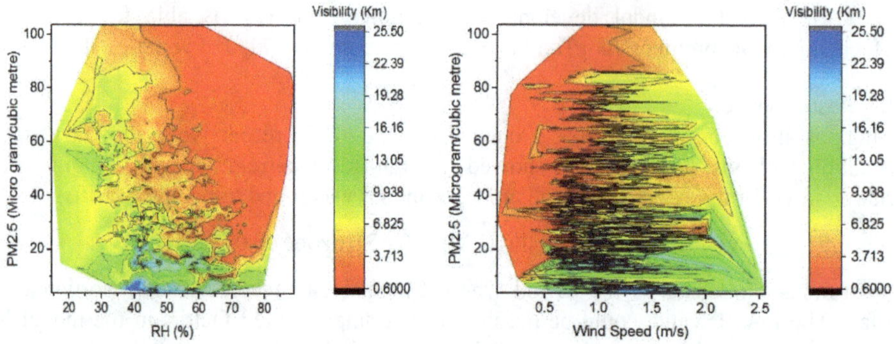

Figure 5: Occurrence of low visibility events during winter period (January 2019) and relation with $PM_{2.5}$, relative humidity and wind speed.

greater temperature difference between land and water body [21]. The lower wind speeds favour accumulation of moisture in the ambient environment, which further favours hygroscopic growth of particles leading to excessive extinction of light. Thus it clearly indicates that there is need for season specific ambient air quality standards to be practiced or more stringent $PM_{2.5}$ emission reduction norms during winter for experiencing better visibility.

4 CONCLUSION

Visibility monitoring campaign was carried out for 176 days representing summer, winter and post-monsoon season. From the results obtained it is found that Chennai city experiences average visibility of 11.24 ± 5.47 km with modal value of 14 km. Further analysis of seasonal variations in visibility infer, visibility is better in summer (15.60 km) followed by winter (7.64 km) and post-monsoon season (5.80 km). City experiences hazy days (< 10 km) during winter and post-monsoon season. Stepwise linear regression for winter showed fine particulate matter ($PM_{2.5}$) is the significant predictor explaining 68% variation in visibility. However for summer it is observed that ozone, $PM_{2.5}$, WS, RH, NO_x are the significant predictors, which together explain 71% variation in visibility and indicated that model captures the photochemistry of the summer. The occurrence of low visibility events are found to be highest during winter followed by post-monsoon season and positive correlation was observed between PBL and visibility during winter.

REFERENCES

[1] Zhao, H. et al., Characteristics of visibility and particulate matter in an urban area of northeast China. *Atmospheric Pollution Research*, **4**(4), pp. 427–434, 2013. DOI: 10.5094/APR.2013.049.

[2] Quan, J., Tie, X., Zhang, Q., Liu, Q., Li, X. & Gao, Y., Characteristics of heavy aerosol pollution during the 2012–2013 winter in Beijing, China. *Atmospheric Environment*, **88**, pp. 83–89, 2014. DOI: 10.1016/j.atmosenv.2014.01.058.

[3] Singh, A., Bloss, W.J. & Pope, F.D., 60 years of UK visibility measurements: Impact of meteorology and atmospheric pollutants on visibility. *Atmospheric Chemistry and Physics*, **17**(3), pp. 2085–2101, 2017. DOI: 10.5194/acp-17-2085-2017.

[4] Luan, T., Guo, X., Guo, L. & Zhang, T., Quantifying the relationship between $PM_{2.5}$ concentration, visibility and planetary boundary layer height for long-lasting haze and

fog–haze mixed events in Beijing. *Atmospheric Chemistry and Physics*, pp. 203–225, 2018. DOI: 10.5194/acp-18-203-2018.

[5] Titos, G., Cazorla, A., Zieger, P., Andrews, E. & Lyamani, H., Effect of hygroscopic growth on the aerosol light-scattering coefficient: A review of measurements, techniques and error sources. *Atmospheric Environment*, **141**, pp. 494–507, 2016. DOI: 10.1016/j.atmosenv.2016.07.021.

[6] Seinfield, J.H. & Pandis, S.N., *Atmospheric Chemistry and Physics: From Air Pollution to Climate Change*, 3rd ed., John Wiley & Sons, 2016.

[7] Population Census 2011. www.census2011.co.in/.

[8] Gupta, I. & Kumar, R., Trends of particulate matter in four cities in India. *Atmospheric Environment*, **40**, pp. 2552–2566, 2006.

[9] Pulikesi, M., Baskaralingam, P., Elango, D., Rayudu, V.N., Ramamurthi, V. & Sivanesan, S., Air quality monitoring in Chennai, India in the summer of 2005. *Journal of Hazardous Materials*, **B136**, pp. 589–596, 2006.

[10] CPCB, Report on air quality monitoring, emission inventory and source apportionment study for Chennai. http://cpcb.nic.in/displaypdf.php?id=Q2hlbm5haS5wZGY.

[11] Attri, S.D. & Tyagi, A., Climate Profile of India: Indian Meteorological Department (IMD), Ministry of Earth Sciences, 1–5, 2010. www.indiaenvironmentportal.org.in/files/climate_profile.pdf.

[12] Cheng, Z. et al., Long-term trend of haze pollution and impact of particulate matter in the Yangtze River Delta, China. *Environmental Pollution*, **182**, pp. 101–110, 2013. DOI: 10.1016/j.envpol.2013.06.043.

[13] Yang, Y.R. et al., Characteristics and formation mechanism of continuous hazes in China: A case study during the autumn of 2014 in the North China Plain, pp. 8165–8178, 2015. DOI: 10.5194/acp-15-8165-2015.

[14] CPCB, Continuous monitoring station data. https://app.cpcbccr.com/ccr/#/caaqm-dashboard-all/caaqm-landing.

[15] Ji, D. et al., Investigating the evolution of summertime secondary atmospheric pollutants in urban Beijing. *Science of Total Environment*, **572**, pp. 289–300, 2016. DOI: 10.1016/j.scitotenv.2016.07.153.

[16] Kim, H., Zhang, Q. & Heo, J., Influence of intense secondary aerosol formation and long-range transport on aerosol chemistry and properties in the Seoul metropolitan area during spring time: Results from KORUS-AQ. *Atmospheric Chemistry and Physics*, pp. 7149–7168, 2018. DOI: 10.5194/acp-18-7149-2018.

[17] Aneja, V.P., Brittig, J.S., Kim, D. & Carolina, N., Ozone and other air quality-related variables affecting visibility in the southeast United States. *Journal of Air and Waste Management Association*, **54**, pp. 681–688, 2004.

[18] Fabian, P. & Dameris, M., *Ozone in the Atmosphere: Basic Principles, Natural and Human Impacts*, Springer Heidelberg: New York, 2014.

[19] Fu, X. et al., Changes in visibility with $PM_{2.5}$ composition and relative humidity at a background site in the Pearl River Delta region. *Journal of Environmental Science*, **40**, pp. 10–19, 2015. DOI: 10.1016/j.jes.2015.12.001.

[20] University of Wyoming, Atmospheric soundings. http://weather.uwyo.edu/upperair/sounding.html.

[21] Stull, R.B., *An Introduction to Boundary Layer Meteorology*, Kluwer Academic Publishers, pp. 3–20, 1999.

CHARACTERIZATION OF ODOROUS EMISSIONS FROM A CIVIL WASTEWATER TREATMENT PLANT IN ITALY

MARCO RAVINA[1], DEBORAH PANEPINTO[1], JHEYSON MEJIA ESTRADA[2], LUCA DE GIORGIO[3], PIETRO SALIZZONI[2], MARIA CHIARA ZANETTI[1] & LORENZA MEUCCI[3]
[1]Department of Environment, Land and Infrastructure Engineering, Politecnico Di Torino, Italy
[2]Laboratoire de Mécanique des Fluides et d'Acoustique, University of Lyon, France
[3]SMAT Research Centre, Italy

ABSTRACT

The characterization and reduction of odour emissions represents an open debate among the scientific community. Odour nuisances are connected to a large number of substances, mostly detectable at low concentrations. Direct estimation of odour impacts through olfactometry is not always applicable, as this approach requires air sampling and a pool of trained panellists. Measuring the concentration of odorous substances provides support to the characterization of emission sources and the design of odour monitoring systems. Civil wastewater treatment plants (WWTPs) are known sources of odours. The objective of this project is the design and development of an integrated odour emission monitoring system at the Castiglione Torinese WWTP in Italy. In this paper, the preliminary characterization of the emission sources and the odour emitting components are presented. The characterization of the emission sources and tracers was obtained by mean of a number of site inspections and measurement campaigns held between 2017 and 2019. In the last campaign, held in January 2019, chemical odour tracers (H_2S, NH_3, VOC) and dynamic olfactometry measurements were performed simultaneously. The screening of VOC species through gas chromatograph/mass spectrometer analysis of air samples was also performed. Odour emitting components were ranked in terms of odour activity value (OAV). Results show that VOC is the only group of compounds that is always detectable on the site. NH_3 and H_2S may, in some cases, be present at considerable concentrations. Results of OAV calculations show that a number of VOCs are detected on the site with a high spatial and temporal frequency. Additional considerations are reported on the site-specific correlation between chemical species and odour measurements. This detailed characterization of the emission sources and tracers results in the design of the final integrated monitoring system, which will be based on continuous measurement of H_2S, NH_3 and VOC and advanced dispersion modelling.

Keywords: odour monitoring, wastewater treatment, odour dispersion modelling, odour activity value, odour tracers, olfactometry.

1 INTRODUCTION

Odours procure nuisance among the population and have been reported to increase health effects even with no toxic concentration [1]. Odour nuisances are connected to a large number of chemical substances, mostly detectable at low concentration thresholds. Civil wastewater treatment plants (WWTPs) are known sources of odours. The analysis of odour impacts in civil WWTPs is a challenging task, due to the design of this kind of plants.

Many factors influence whether an emission will produce disturbing impacts: odorants are characterized by specific sensorial properties, and co-existing odorants may be subject to interactive effects. The need to regulate odour impacts requires specific methods for odour measurement and assessment. Odour control strategies depend on the knowledge of the interactions between its chemical components and their dilution in the atmosphere after being emitted [2]. At today, the application of olfactometry methods present some limitations due to (i) the high costs of air sampling and subsequent laboratory analysis; and (ii) the impossibility of continuous measurements [3]. The application of electronic noses, although promising, is still limited, due to technological problems, as sensor drift over time and

WIT Transactions on Ecology and the Environment, Vol 236, © 2019 WIT Press
www.witpress.com, ISSN 1743-3541 (on-line)
doi:10.2495/AIR190161

undesired sensor sensitivity to variable atmospheric conditions [4]. In addition, electronic noses are still not representative of the odour impacts in case a complex mixture of odorants is assessed. Using analytical techniques, i.e. concentrations measurements of odorous substances, is not directly correlated to the determination of odour properties. Nevertheless, the chemical identification of odorous compounds is necessary for odour monitoring and abatement. With the purpose of relating a chemical composition of an odorous mixture to its odour concentration, different methods were proposed. These methods include the odour activity value (OAV [5]), the sum of the individual odour intensities (SOI [6]) and the equivalent odour concentration (EOC [7]). Odour activity value (OAV) represents an established methodology for obtaining significant information about odours based on the results of chemical analyses in the field [8]. OAV, being a sort of total concentration weighted by the odour thresholds of the single compounds contained in an odorous mixture, does account for the different relative contribution of each compound to the mixture total odour concentration, and therefore is able to better describe the odour properties of an odorous mixture than just the total VOC. The main critical aspects of this method are those related to concentration measurement of odorant species, i.e. the detection limit of the instruments and the complete characterization of all possible odorants.

In order to obtain a continuous and complete monitoring of odour impacts, the methods based on analytical measurements must be connected to dispersion modelling. Odour dispersion modelling allows the evaluation of spatial and temporal distribution of odour impacts. For this reason, modelling is essential to characterize the possible sources of odour nuisances and take relevant remedies. Modelling odours emitted by WWTPs follows the general approach of other types of sources. In general, non-stationary Lagrangian puff and particle models have to be preferred to Gaussian models, due to the inability of these latter to handle calm and stagnation conditions, lack of three-dimensional meteorology and steady-state assumption [4]. The main issue with odour modelling is that to fully predict odour annoyance, the fluctuations of concentrations at a small time scale (second or minute) at the receptor are critical.

Owing to these considerations, the characterization of odorous emissions from WWTPs requires an integrated approach, in which analytical, olfactory and modelling methodologies are applied and linked. The objective of this research project is to propose the design and development of such an integrated odour emission monitoring system. The SMAT's wastewater treatment plant (WWTP), the largest chemical, physical and biological treatment plant in Italy, was used as a case study.

The monitoring system is composed by a network of fixed continuous measurement stations on the site, and the development of a modelling chain simulating the local dispersion of odour emissions in the area. The system must be designed in ways to analyse the possible contribution of SMAT's WWTP to odour nuisances in case of complaints of the citizens. This paper starts from presenting the results of the emission characterization phase. Based on this characterization, the methodology and features of the continuous monitoring system design were defined and discussed.

2 MATERIALS AND METHODS

The research project has been developed through the following phases:

- characterization of the emission sources and the chemical species used as indicators of odour emissions;
- selection and design of the continuous monitoring system; and

- connection of the continuous monitoring system to the modelling chain for the analysis and evaluation of odour nuisance episodes.

2.1 Site description

SMAT's centralized civil WWTP is the largest chemical, physical and biological treatment plant in Italy. The plant treats municipal and industrial wastewater with a capacity of more than 2,000,000 of equivalent inhabitants. It consists of four lines devoted to the wastewater treatment and one line for sludge treatment. The water line, with an average flow rate of about 25,000 m^3/h, is made up of the following processes: grid screens, grit and grease removal, primary sedimentation, pre-denitrification, biological treatment, secondary sedimentation, phosphorous removal and final filtration (Fig. 1).

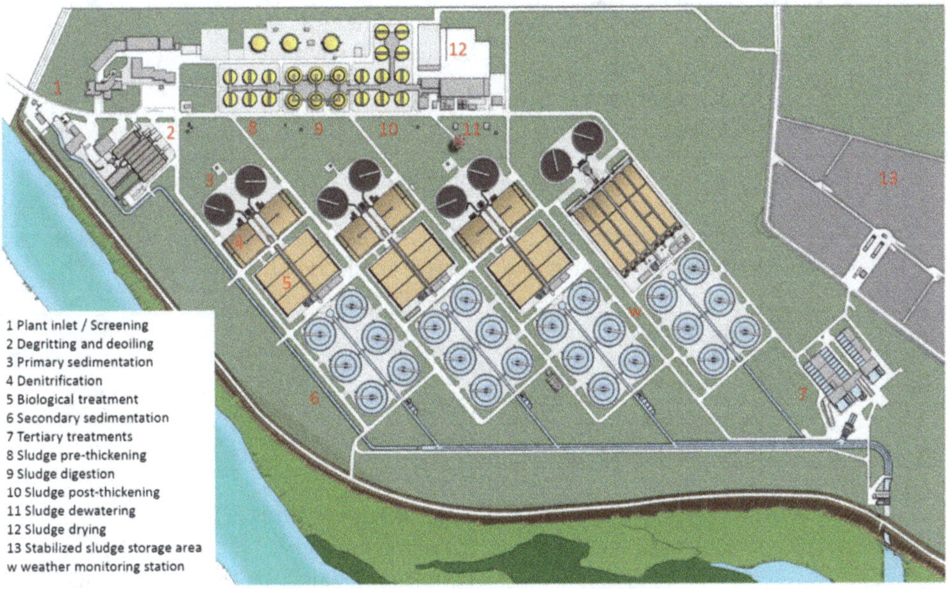

Figure 1: Plant of SMAT's WWTP [9].

In September 2017, a weather monitoring station was installed onsite. The location was selected in way to avoid any potential disturbance of air circulation generated by structural elements.

The datalogging system provides average values of the weather variables over 10 minutes.

The average wind distribution recorded in the period September 2017–March 2019 show different wind distribution between night time and day time. Night time is characterized by low wind (around 0.5–1 m/s) with main direction SE. Day time is characterized by slightly higher wind (up to 5 m/s) with main direction NE. No significant seasonal variation is evidenced, although higher variability of wind distribution appears during spring and winter.

2.2 Emission sources characterization

A preliminary characterization phase was conducted to evaluate the odour emission sources. This phase had the objective of defining the areas of the WWTP with a major influence on odour impacts. To this end, five monitoring campaigns were held onsite between October 2017 and January 2019. Each campaign aimed at characterizing different aspects. A brief description of the objectives, methods and chemical species considered for each campaign is reported in Table 1.

Table 1: Measurement campaigns description.

ID	Date	Objective	Description	Measures
1	Oct-17	Preliminary screening of emission sources	All components of the process monitored in three separate days	VOC, NH_3
2	Nov-17	Spatial and temporal analysis of main emission sources selected after Campaign 1	Concentrations were measured in different days at different times; close to the emission source and at a distance of 10 m	VOC, NH_3
3	Mar-18	Characterization of volatile compounds emitted by the plant	Contemporary concentrations measurement and samples collection for GC–MS analysis	VOC, H_2S
4	May-18	Analysis of endogenous/exogenous contribution of emission sources to concentrations	Concentrations were measured close to the emission sources and at the site boundaries	VOC
5	Jan-19	Analysis of the relationship between concentrations and odours	Contemporary concentrations measurement and samples collection for olfactometry and GC–MS analysis	VOC, NH_3, H_2S, olfactometry

The analysis focused on three group of chemical species, i.e. H_2S, NH_3 and VOC. VOC concentration was measured with the use a portable photo ionization detector (PID, model Tiger, ION Science, 0.1 ppb of resolution). NH_3 concentration was measured with the use a portable electrochemical device (model Gas Alert Extreme, BW Technologies, 0.1 ppm of resolution). H_2S concentration measurements were contracted to an external laboratory, and executed according to NIOSH 6013-1994 method.

In the last monitoring campaign, measurements of VOC, H_2S and NH_3 concentrations were done in parallel with air sampling and subsequent olfactometry analysis by an external laboratory. A total of 20 samples were collected, 17 samples of ambient air and three from diffused sources. Ambient air samples were collected in nalophan bags. Samples from diffused sources (degritting tank, primary sedimentation tank and stabilized sludge storage area) were collected by wind tunnel sampling. Odour concentrations were determined in a ODOURNET TO8 olfactometer according to standard EN 13725:2004.

2.3 VOC screening and data processing

A VOC screening was performed during Campaigns 3 and 5. This phase had the objective of characterizing the chemical species emitted by the WWTP and detecting possible odours tracers. The data were subsequently used for evaluating a possible correlation between VOC and odour concentrations. VOC screening in Campaign 3 was done according to EN ISO 16017-1:2002 method. Six air samples were collected in the following areas of the plant: plant inlet, grit removal, primary settler, sludge drying, stabilized sludge external storage, sludge thickening. Air samples were collected on adsorption tubes, then analysed in a laboratory with a gas chromatograph/mass spectrometer (GC–MS). VOC screening in Campaign 5 was done according to EPA-TO-15 1999 method. In this latter campaign, three air samples were collected, two from ambient air (plant inlet and sludge drying) and one by means of wind tunnel sampling (external sludge storage).

Data collected from the monitoring campaigns (concentrations and VOC screening) were used to calculate the odour activity value (OAV) of single species. OAV is defined as the ratio of the concentration of a specific odorant to its odour threshold (OT) value. The sum of single OAVs (SOAV) was then calculated. Odour thresholds of single substances were taken by Nagata [10] and integrated with additional sources [11], [12].

For Campaign 5, a specific OU/SOAV ratio for each measurement point was calculated. The following methodology was used:

- A VOC profile was assigned to each point depending on its location on the area. The VOC profile was selected from the VOC screening performed in Campaign 3 (six samples).
- Total VOC concentration measured with the portable instrument was divided following the VOC profile assigned, obtaining concentrations of single species.
- OAV and SOAV were calculated in each point.
- Measured odour concentration was divided by the SOAV, obtaining a specific OU/SOAV ratio for each sampling point.

3 RESULTS

Results of measurement campaigns and data elaboration, as well as the main features of the integrated odour monitoring system, are presented in this section.

3.1 Results of monitoring campaigns

The results of the monitoring campaigns are resumed in Fig. 2.

During Campaign 1, on average, VOC concentrations were higher next to preliminary treatment section, primary settling and sludge de-watering section. There is agreement in the scientific literature that these treatment stages represent the most important contributions to odour emissions in civil WWTPs [13], [14]. Concentrations were lower downstream of the primary treatment (nitrification tanks and secondary settlers). NH_3 concentration was below the detection limit of the instrument (0.1 ppm), with the only exception of two areas: 2 ppm in correspondence of the plant inlet (degritting stage), and 2 ppm in correspondence of Module 2 secondary settling stage.

During Campaign 2, measurements were restricted to the areas with higher concentration in Campaign 1. In these areas, VOC concentrations were measured close to the emission source or in the immediate proximity (around 5 m). Results of Campaign 2 showed that concentration values quickly decay by moving away from the emission sources, indicating a

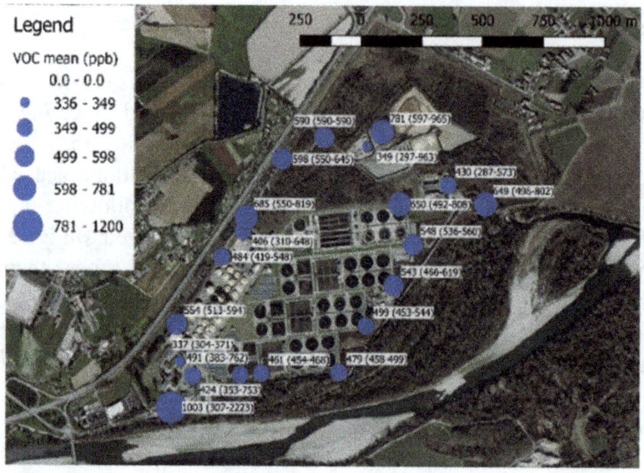

Figure 2: Mean, maximum and minimum VOC concentrations measured in monitoring campaigns.

rapid dispersion of the contaminants. Concentrations measurements of NH_3 confirmed the results of Campaign 1. Results of Campaign 3 confirmed the results of previous campaigns, evidencing a higher VOC concentration in correspondence of the plant inlet and the sludge drying area. Considering H_2S measurements of Campaign 3, the concentration was below the detection limit of the method (0.0556 μg/m^3) in four of the six sampling points. H_2S concentration of 1.2 μg/m^3 was measured next to the grit removal stage and a value of 16 μg/m^3 next to the primary settling stage. These values show that H_2S may, in some circumstances, be present at values above its odour threshold, that is around 0.6 μg/m^3. VOC measurements of Campaigns 4 and 5 were extended to the whole plant area (Fig. 2). These latter campaigns evidence a significant variability of VOC concentration around the plant, with high background values in the order of 350–680 ppb. It is expected that the presence of VOC in the area was also due to the possible contribution of external sources. These external sources may be represented by the adjacent motorway, by the industrial site located 3 km north of the plant and by the surrounding agricultural activities. Results of parallel measurements of VOC, H2S, NH3 and OU of Campaign 5 are reported in Table 3. Odour concentration was higher in correspondence of the degritting tank, primary settler and (mostly) in the stabilized sludge external storage area. A significant difference in concentration is found between wind tunnel and ambient air samples.

3.2 VOC screening and SOAV calculation

In VOC screening, 132 species were detected in Campaign 3 and 40 species in Campaign 5. Provided that a higher number of species and chemical groups was analysed, results of Campaign 3 were considered more precise and used in the subsequent stage of analysis.

VOC screening of Campaign 3 showed that the main family of compounds are aromatics, aldehydes and halogenated aliphatics. Few species were detected in all sampling points. These are: tetrachlorethylene (average concentration on all sampling points of 2.73 μg/m^3), toluene (6.33 μg/m^3), m + p-xylene (2.88 μg/m^3), limonene (4.36 μg/m^3) and dimethyl disulfide (DMDS, 0.46 μg/m^3). The same compounds are reported in similar studies on odour emissions from civil WWTPs found in bibliography [15], [16].

The list of VOC species with highest OAV and the SOAV for Campaigns 3 and 5 is reported in Table 2. For Campaign 3, a SOAV between 6.10 and 30.39 is found. Major contribution to SOAV can be assigned to four chemical groups or species in order of importance: aldehydes, sulphur compounds (H_2S and DMDS), ethylbutyrate and propionic acid. H_2S contributes to the 87% of the SOAV in correspondence of the primary settler, and to the 26% in correspondence of the grit removal stage. These results show that, if present, H_2S contributes significantly to odour emission at the SMAT's WWTP. The results of Campaign 5 show that aldehydes, octanal and decanal in particular, give the major contribution to SOAV. DMDS is also present in correspondence of the external sludge storage area. SOAV at the plant inlet and sludge de-watering area of Campaign 5 is higher than that of Campaign 3, due to the relevant contribution of octanal to total odour activity (55–68%).

Table 2: SOAV and chemical species with higher OAV in Campaign 3 and Campaign 5.

Plant section (composition profile)	Campaign 3		Campaign 5	
	SOAV	Species with higher OAV	SOAV	Species with higher OAV
Plant inlet (P1)	6.10	Acetaldehyde 1.090; Ethylbutyrate 0.990; Butirraldehyde 0.721; Propionic acid 0.592; Octane 0.434	38.77	Octanal 21.359 Decanal 8.766 Acetic acid 3.620 Nonanal 1.610 Hexanal 1.465
Grit removal (P2)	7.71	Hydrogen sulphide 2.000 Acetaldehyde 1.371 Butirraldehyde 0.914 Propionic acid 0.751 Propionaldehyde 0.440	–	–
Primary settler (P3)	30.39	Hydrogen sulphide 26.664 Acetaldehyde 1.090 Butirraldehyde 0.673 Propionic acid 0.462 Propionaldehyde 0.336	–	–
Sludge thickening (P4)	19.08	Ethylbutyrate 9.036 Propionic acid 1.954 4-isopropyltoluene 1.354 3-isopropyltoluene 1.279 Phenol 0.964	–	–
Sludge drying (P5)	7.22	dimethyl disulfide 2.228 Acetaldehyde 1.336 Butirraldehyde 1.154 Ammonia 0.684 dimethyl trisulfide 0.462	44.32	Octanal 30.437 Decanal 4.127 Heptanal 3.153 Hexanal 2.352 Nonanal 2.515
Stabilized sludge external storage (P6)	9.47	Propionic acid 4.593 Acetaldehyde 1.195 Butirraldehyde 0.914 Ammonia 0.620 4-isopropyltoluene 0.461	16.13	Disulfide,dimethyl 7.143 Hexanal 5.321 Toluene 1.476 1-Butanol 0.800 Benzene,1,2,4-trimethyl-0.635

3.3 OU/SOAV ratio

Starting by the parallel olfactometry and chemical measurements of Campaign 5, a OU/SOAV ratio was calculated for each of the 20 sampling points. To calculate SOAV, a specific composition profile between those of Campaign 3 was assigned to each point (Table 3). Total VOC concentration was divided into concentration of single species depending on the composition profile assigned. Composition profiles were selected from Campaign 3 because of the higher precision of the VOC screening performed in this campaign. OAV of single species was calculated and then summed up to obtain SOAV. For each sampling point, OU concentration was then divided by the SOAV and a OU/SOAV ratio was obtained. The results are reported in Table 3. Different values of OU/SOAV ratio were found if the odour concentration was determined through wind tunnel or ambient air sampling. In the first case, a OU/SOAV between 1.04 and 10.31 was found. In the second case, the OU/SOAV ratio ranged between 0.11 and 0.46.

Table 3: Calculation of OU/SOAV ratio for the sampling points of Campaign 5.

ID	Plant area	VOC concentration (ppb)	SOAV profile	SOAV calculated	OU concentration (OU/m^3)	OU/SOAV
1	Plant inlet	1512	P1	357.1	47	0.13
2	Grit removal*	753	P2	172.8	180	1.04
3	Primary settler*	756	P3	114.9	540	4.70
4	Sludge drying	648	P5	127.9	33	0.26
5	Stabilized sludge external storage – position 1*	963	P6	348.5	1100	3.16
6	Stabilized sludge external storage – position 2*	964	P6	349.2	3600	10.31

*Wind tunnel sampling.

3.4 Integrated odour monitoring system

The results of preliminary monitoring campaigns of VOC, NH_3 and H_2S may be resumed with the following considerations:

- VOC is the only group of compounds that was always detectable on the site. Nevertheless, it could not be excluded that NH_3 and H_2S may be present at considerable concentration and contribute significantly to odour impacts.
- A number of areas in the site may be detected and evidenced as primary emission sources of VOC to be potentially linked to odour emission. These areas are the preliminary treatments, primary settling, sludge drying and sludge thickening. The stabilized sludge external storage area should also be considered as a primary emission source, as characterized by the highest odour concentration. Other areas may be defined as secondary emission sources. These are aeration tanks, secondary settling tanks and sludge thickening area.

- If measurements are taken moving away from the emission source, concentrations show a significant variability in space and magnitude. Information collected is not sufficient to directly evaluate the possible contribution of external sources to total concentration.

The VOC screening of Campaigns 3 and 5 provided information on the chemical composition of the compounds emitted by the WWTP or present in the area. A number of species or groups may be found in the whole area and associated with the emission sources. Future monitoring campaigns on the emission sources surrounding the area of study could provide additional information in order to define possible chemical tracers for odour nuisances. Results of VOC screening were also elaborated to study the link between chemical and odour concentrations. The resulting values of OU/SOAV ratios were used to link chemical measurements to the modelling of odour emissions.

The resulting general scheme of the integrated odour monitoring system is reported in Fig. 3. Starting from VOC, H_2S and NH_3 concentration measurements, odour concentration is estimated with the method described in Section 3.3. Odour concentrations are used to calculate odour emission rates for primary and secondary sources. Odour emission rates are thus introduced into a dispersion model to evaluate the magnitude and spatial distribution of odours. The system is designed to operate in near-real time, i.e. receiving weather data and concentrations from the monitoring stations, processing them and running the dispersion model without a significant delay (in the order of minutes to hours). This is achieved by mean of a set of algorithms that (i) manages and synchronizes data acquisition from weather and concentration sensors; (ii) converts chemical concentrations into odour concentration; (iii) calculates odour emission flow; and (iv) prepares the input and runs the dispersion models.

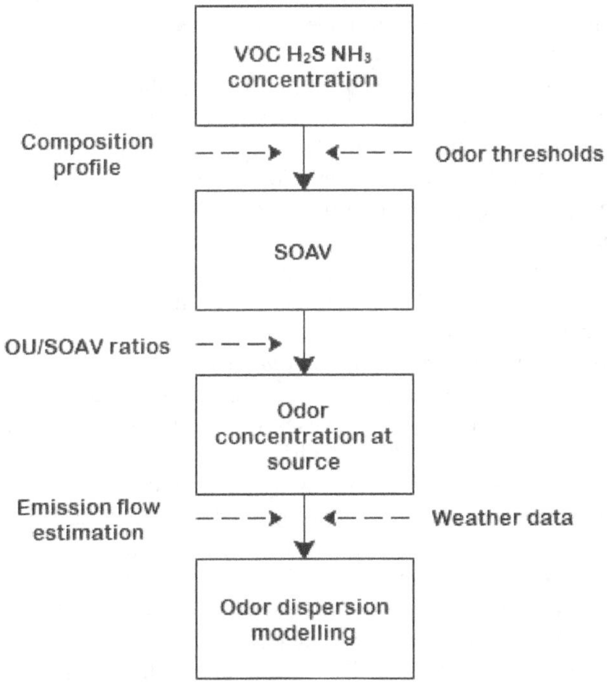

Figure 3: Operating scheme of the continuous odour monitoring system.

The system for the continuous measurement of chemical species is composed by a monitoring station for every emission source. Each unit is equipped with sensors for VOC, NH_3 and H_2S measurement. The sensors have low detection limits, in the order of 10 ppb for VOC, 0.5 ppm for NH_3 and 10 ppb for H_2S respectively. Similarly, the measuring range of the sensors must be wide enough to cover possible concentration peaks (in the order of 0–15 ppm for VOC, 0–25 ppm for NH_3 and 0–1000 ppb for H_2S respectively). The sampling frequency is 10 minutes, adequate to the analysis of peak emission episodes.

Specific gas sensors commercially available comprise a broad variety of chemical, electrochemical, catalytic, and optical detectors with a high sensitivity and selectivity for some target odorants. They present low response times and low cost. Their main drawback is that they are susceptible of interferences from humidity [17]. To address this problem, some of these sensors have been equipped with a specific filter (e.g. the CairClip sensors, EnveaTM [18]).

The modelling system is going to be based on the Safety Atmospheric Lagrangian Model (SLAM) developed at Laboratory of Fluid Mechanics and Acoustics of Lyon University, France [19]. SLAM is composed by two main modules: the first is devoted to meteorological data processing and extraction of the wind field, the second to pollutant dispersion. The meteorological pre-processor is based on Monin–Obukhov similarity theory. The dispersion module simulates the trajectories of a large number of particles from an increased average velocity field at every point and at each iteration of a random component representing the turbulent fluctuation.

4 CONCLUSION

This paper had the objective of defining the design of an integrated system for the continuous monitoring of odour emissions at SMAT's WWTP. The preliminary scientific and experimental characterization phase provided the relevant information to overcome this objective. The integrated monitoring system will be managed by a novel set of algorithms that (i) collect and synchronize data acquisition from weather and concentration sensors; (ii) convert chemical concentrations into odour concentration; (iii) calculate odour emission flow; and (iv) prepare the input and runs the dispersion models. Given its high level of automation, the system will allow a fast reproduction of odour dispersion in case of odour nuisance episodes, providing information on the contribution of SMAT's WWTP to odour impacts in the surroundings.

In the study of odour quantification and minimization, several aspects remain unresolved, due to the complexity of the topic. These aspects include the measurement methods (chemical analyses, olfactometry, electronic noses), the characterization of odorants (in particular, odour thresholds), the characterization of additive or synergistic effects and the dispersion modelling phase (definition of odour emission rates and peak concentrations). The sensitivity of population to odour problems is constantly increasing, since odours have been demonstrated to affect citizens' health and not only cause nuisance. Seen this, a trade-off between scientific knowledge and the necessity of providing technological solutions is needed. The system presented herein attempts to provide such a solution. It is expected that its design and features shall be improved after entering into operation. In this sense, additional field measurements will provide further details. Also, the implementation of alternative methods for i) the conversion of chemical concentrations into odour units and ii) the estimation of concentration fluctuation shall be tested on the system. As a final result of this research project, SMAT's WWTP will be equipped with a complete and integrated monitoring system of odour emissions. The same methodology may, in principle, be extended to other WWTPS, as well as other typologies of odour emitting sources.

ACKNOWLEDGEMENT

The authors acknowledge SMAT company for the support and the data provided.

REFERENCES

[1] Hayes, J.E., Stevenson, R.J. & Stuetz, R.M., The impact of malodour on communities: A review of assessment techniques. *Sci. Total Environ.*, **500–501**, pp. 395–407, 2014. DOI: 10.1016/j.scitotenv.2014.09.003.

[2] Carrera-Chapela, F., Donoso-Bravo, A., Souto, J.A. & Ruiz-Filippi, G., Modeling the odour generation in WWTP: An integrated approach. *Rev. Water Air Soil Pollut.*, **225**, p. 1932, 2014. DOI: 10.1007/s11270-014-1932-y.

[3] Gostelow, P., Parsons, S.A. & Stuetz, R.M., Odour measurements for sewage treatment works. *Water Res.*, **35**, pp. 579–597, 2001.

[4] Capelli, L., Sironi, S., Del Rosso, R. & Guillot, J.M., Measuring odours in the environment vs. dispersion modelling: A review. *Atmos. Environ.*, **79**, pp. 731–743, 2013. DOI: 10.1016/j.atmosenv.2013.07.029.

[5] Rincon, C.A. et al., Odour concentration (OC) prediction based on odour activity values (OAVs) during composting of solid wastes and digestates. *Atmos. Environ.*, **201**, pp. 1–12, 2019. DOI: 10.1016/j.atmosenv.2018.12.030.

[6] Kim, K.H. & Park, S.Y., A comparative analysis of malodor samples between direct (olfactometry) and indirect (instrumental) methods. *Atmos. Environ.*, **42**, pp. 5061–5070, 2008. DOI: 10.1016/j.atmosenv.2008.02.017.

[7] Wu, C., Liu, J., Zhao, P., Piringer, M. & Schauberger, G., Conversion of the chemical concentration of odorous mixtures into odour concentration and odour intensity: A comparison of methods. *Atmos. Environ.*, **127**, pp. 283–292, 2016. DOI: 10.1016/j.atmosenv.2015.12.051.

[8] Lebrero, R., Bouchy, L., Stuetz, R. & Muñoz, R., Odour assessment and management in wastewater treatment plants: A review. *Crit. Rev. Env. Sci. Tec.*, **41**(10), pp. 915–950, 2011. DOI: 10.1080/10643380903300000.

[9] SMAT (Società Metropolitana Acque Torino). www.smatorino.it/. Accessed on: 6 May 2019.

[10] Nagata, Y., Measurement of odor threshold by triangle odor bag method. Odor Measurement Review, Japan Ministry of the Environment, pp. 118–127, 2003.

[11] van Gemert, L.J., *Odour Thresholds. Compilations of Odour Threshold Values in Air, Water and Other Media.* Oliemans Punter & Partners BV: The Netherlands, 2011.

[12] Dunlop, M.W., Blackall, P.J. & Stuetz, R.M., Odour emissions from poultry litter: A review litter properties, odour formation and odorant emissions from porous materials. *J. Environ. Manage.*, **177**, pp. 306–319, 2016. DOI: 10.1016/j.jenvman.2016.04.009.

[13] Talaiekhozani, A., Bagheri, M., Goli, A. & Khoozani, M., An overview of principles of odour production, emission, and control methods in wastewater collection and treatment systems. *J. Environ. Manage.*, **170**, pp. 186–206, 2016. DOI: 10.1016/j.jenvman.2016.01.021.

[14] Lin, J., Li, L., Liu, J. & Li, M., Odours and volatile organic compounds emission from main processing units of wastewater treatment plant. *Chinese J. Environ. Eng.*, **10**, pp. 2329–2334, 2016.

[15] Sivret, E.C., Wang, B., Parcsi, G. & Stuetz, R.M., Prioritisation of odorants emitted from sewers using odour activity values. *Water Res.*, **88**, pp. 308–321, 2016. DOI: 10.1016/j.watres.2015.10.020.

[16] Fisher, R.M., Le-Minh, N., Sivret, E.C., Alvarez-Gaitan, J.P., Moore, S.J. & Stuetz, R.M., Distribution and sensorial relevance of volatile organic compounds emitted

throughout wastewater biosolids processing. *Sci. Total Environ.*, **599–600**, pp. 663–670, 2017. DOI: 10.1016/j.scitotenv.2017.04.129.

[17] Brattoli, M., de Gennaro, G., de Pinto, V., Demarinis Loiotile, A., Lovascio, S., Penza, M., Odour detection methods: Olfactometry and chemical sensors. *Sensors*, **11**, pp. 5290–5322, 2011. DOI: 10.3390/s110505290.

[18] EnveaTM, CairClip sensors. http://cairpol.com/en/home/. Accessed on: 6 May 2019.

[19] Vendel, F., Soulhac, L., Mejean, P., Donnat, L. & Duclaux, O., Validation of the Safety Lagrangian Atmospheric Model (SLAM) against a wind tunnel experiment over an industrial complex area. *14th International Conference on Harmo, within Atmos. Disp. Modell. for Regul. Purposes*, Kos, Greece, 2011.

NANOTECHNOLOGY-BASED CONTROL OF HAZARDOUS AIR POLLUTANTS EMISSION: PILOT SCALE TRIALS FOR SIMULTANEOUS CAPTURE OF H_2S, NH_3, AND ODOURS FROM LIVESTOCK FACILITIES

GUADALUPE VALDES LABRADA[1], SURAJ KUMAR[1], BERNARDO PREDICALA[2] & MEHDI NEMATI[1]
[1]Department of Chemical and Biological Engineering, University of Saskatchewan, Canada
[2]Prairie Swine Centre Inc., University of Saskatchewan, Canada

ABSTRACT

Generation and emission of hazardous gases such as hydrogen sulphide (H_2S), and ammonia (NH_3 from industrial settings and livestock production facilities represent one of the major air pollution challenges. Traditional approaches like physicochemical processes used in industrial settings, and diet manipulation, manure confinement, and addition of inhibitors commonly employed in livestock operations are associated with technical drawbacks, excessive cost, inability to completely eliminate pollutants, and difficulty of implementation on a small scale. Thus, there is a need for development of more effective and feasible technologies. One such innovative approach is the use of nano-based adsorbents to mitigate the emission of these air pollutants. Effective capture of individual H_2S and NH_3 using metal oxide nanoparticles (ZnO and TiO_2) have been reported in our earlier work. To evaluate the potential for wider applications, we have now investigated simultaneous capture of NH_3, H_2S, and odours by a mixture of ZnO and TiO_2 nanoparticles and a tailor-made composite adsorbent (ZnO and TiO_2 nanoparticles deposited on activated carbon). Laboratory work with pre-mixed gases of various compositions revealed that an increase of H_2S and NH_3 concentrations led to higher adsorption capacities with both nano-adsorbents. Higher temperatures enhanced the adsorption of H_2S but led to lower adsorption capacities for NH_3. Characterization of adsorbents through CNHS analyses, thermogravimetry, FT-IR and XRD revealed that ZnO and TiO_2 both adsorbed NH_3 and H_2S. While ZnO had a much higher affinity for H_2S through chemisorption, TiO_2 was more effective in adsorption of NH_3 by physisorption. Results from trials conducted in a semi-pilot scale adsorption system fed with swine manure gases, and those in a livestock research facility where a nano-based circulation–filtration system was deployed in a real situation confirmed the effectiveness of nano-adsorbents in capture of NH_3, H_2S, and odours from representative gases in real settings.

Keywords: ammonia, hydrogen sulphide, odour, nano-based emission control, pilot trial.

1 INTRODUCTION

Hazardous air pollutants such as ammonia (NH_3) and hydrogen sulphide (H_2S) are generated and emitted as part of industrial processes. Agricultural activities aiming at livestock production and associated facilities are other major contributors to emission of these hazardous gases [1]–[4]. Ammonia is an odorous air pollutant with serious impact on human and animal health. Ammonia inhalation could irritate nose and throat and cause nausea and respiratory tract problems [5]. Formation of ground level ozone and fine ammonium nitrate particulates are some of the environmental challenges associated with the emission of ammonia [6], [7]. Like ammonia, hydrogen sulphide emission poses serious health and environmental risks due to its toxic and corrosive nature. Hydrogen sulphide also contributes to formation of other air pollutants such as sulphur oxides and atmospheric acidic depositions [3], [4], [8].

Traditional approach such as physicochemical processes used in industrial settings and those like diet manipulation, manure confinement, and addition of inhibitors to manure that are commonly employed in livestock operations are associated with technical drawbacks,

www.witpress.com, ISSN 1743-3541 (on-line)
doi:10.2495/AIR190171

excessive cost, inability to completely eliminate these hazardous air pollutants, and difficulty of implementation in the small scale. Thus, there is a need for development of more effective and feasible technologies. One such innovative approach is the use of nano-based adsorbents to mitigate the emission of NH_3, H_2S, and odours. Our research effort aiming at application of nanotechnology to tackle the emission of NH_3, H_2S, and odours included several phases as depicted in Fig. 1. In summary we have investigated the application of pure metal oxide nanoparticles (e.g. TiO_2 and ZnO), as well as composite nano-adsorbents (TiO_2 nanoparticles deposited on activated carbon) for the removal of individual NH_3 and H_2S from gaseous streams, with the results reported elsewhere [5], [8], [9].

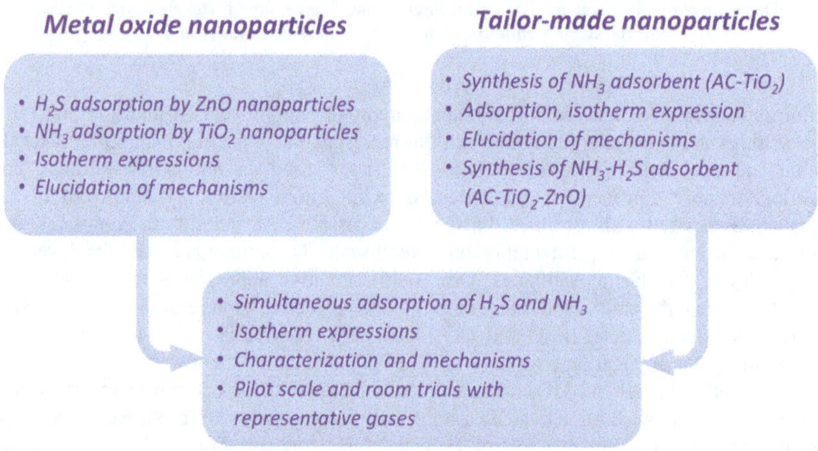

Figure 1: Overview of research strategy for tackling emission of hazardous gases.

The current paper focuses on other phases of this research that aimed at simultaneous capture of NH_3 and H_2S from gaseous streams and includes an overview of our findings from the laboratory work conducted with pre-mixed gases, semi-pilot scale tests with representative gases (e.g. gases emitted from swine manure), and trial in livestock production rooms.

2 EXPERIMENTAL SYSTEMS AND PILOT TEST FACILITY

2.1 Evaluation of emission control in laboratory scale system

The laboratory experimental system for simultaneous capture of ammonia and hydrogen sulphide was a modification of the systems used for the capture of individual NH_3 and H_2S and consisted of feed gas tanks, an adsorption column with the nano-based adsorbent, mass flow meters, differential pressure transducer, thermocouple, stainless steel tubing including the required sampling ports and an online gas chromatograph [5], [8], [9]. The nano-based adsorbents used in the experimental runs were either a binary mixture of commercial ZnO and TiO_2, or a tailor-made composite adsorbent that consisted of activated carbon with deposited ZnO and TiO_2 nanoparticles.

Using the devised mass flow controller, the premixed gases (1000 ppmv NH_3 – balanced with He, and 1000 ppmv H_2S – balanced with He) were diluted with He to achieve the desired ammonia and hydrogen sulphide concentrations in the mixture. The feed gas was then passed through the adsorption column packed with the designated adsorbent to generate the breakthrough curves at various gas compositions and temperatures. The flow rate of the influent mixed gas was controlled at 100 mL min^{-1}. The outlet of the adsorption column was directed to the online gas chromatograph to measure the concentration of NH_3 and H_2S in real time. The temperature of the column was controlled at the designated level using a heating tape connected to a temperature controller. Conducting the experiments with mixed gases containing various levels of NH_3 and H_2S (50–500 ppmv of each hazardous gas) at various temperatures (22–280°C) allowed us to assess the impacts of gas composition and temperature on the effectiveness of each nano-based adsorbent in simultaneous capture of NH_3 and H_2S. The generated data was then used to determine the adsorption capacity of each nano-based adsorbent under various operating conditions and also to identify suitable isotherms to describe the adsorption process. The laboratory scale experimental system and adsorption column are shown in Fig. 2.

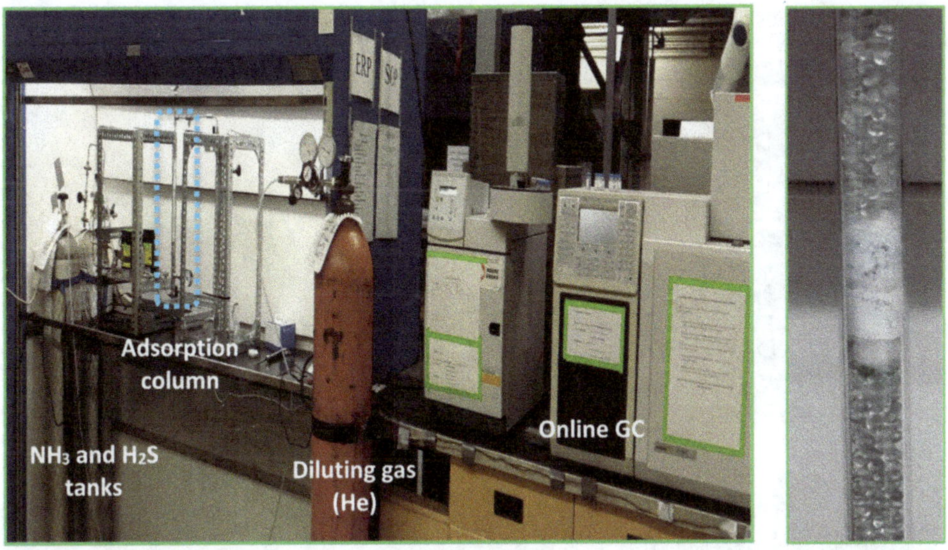

Figure 2: Experimental set-up (left). Adsorption column packed with ZnO and TiO_2 nanoparticles and glass beads (right).

2.2 Simultaneous capture of NH_3 and H_2S from swine manure gas in semi-pilot scale system

The effectiveness of ZnO and TiO_2 nanoparticles in removing NH_3, H_2S, and odours from representative gases was assessed using gases emitted from the stored swine manure. Swine manure was collected from the manure pit of a grow-finish pig production room in an actual pig barn. Before using in the experimental runs, the collected manure was transferred to several containers with tight covers and stored for three weeks at room temperature to allow anaerobic digestion and production of manure gases. The semi-pilot scale set-up consisted of a centrifugal fan, an adsorption column, NH_3 and H_2S sensors, rubber tubings, and galvanized ducts. The adsorption column was made of transparent PVC cylinder with an internal diameter of 10 cm and a height of 25 cm. The nano-adsorbent consisting of both ZnO and

TiO_2 nanoparticles was placed in the column. Mesh pad with glass wool were used at the bottom and the top of the nanoparticle bed to support the particles and to prevent their carry over with the effluent gas. The bottom of the adsorption column was connected to the manure container headspace using a flexible rubber tubing, while the top of the column was connected to the centrifugal fan by galvanized ducts. The centrifugal fan generated the flow necessary to withdraw the gases from the manure containers headspace and to pass them through the adsorption column. Ammonia and hydrogen sulphide sensors were installed before and after the adsorption column to determine the concentration of NH_3 and H_2S in the influent and effluent gasses. During the experimental run, manure containers were used in sequence whereby each container was agitated intermittently to release manure gases from the slurry for a 20-minute cycle (i.e. 2 minutes agitation at the start of the run and then every 5 minutes). Once the 20-minute cycle was completed for the first container, the flexible tubing connection was moved and connected to the next manure container and the procedure was repeated until the 140 minutes overall trial time was completed. Fig. 3 shows various components of the semi-pilot scale set-up.

Figure 3: Experimental set-up including Adsorption column packed with ZnO and TiO_2 nanoparticles.

2.3 Emission control trials in a research pig production facility

To assess the effectiveness of ZnO and TiO_2 nanoparticles in mitigating NH_3 and H_2S emissions from livestock facilities, room-scale trials were conducted in two fully instrumented and identical pig production rooms at the Prairie Swine Centre Inc. The dimensions of each chamber were 4.2 m × 3.6 m × 2.7 m and each housed a pen with approximate dimensions of 2 m × 1.25 m × 0.3 m. Pens were surrounded by plastic matrix flooring for easy access to the collection tubs underneath the slatted floor. Chambers were maintained at a negative pressure through the ventilation system. Fresh air was forced through a filtration unit by a centrifugal fan before entering the chambers. The air in the room was exhausted from the chamber through a sidewall exhaust fan. To maintain the rooms at consistent temperature, air conditioning unit and an electric heater had been devised to cope with seasonal temperature variations. To investigate the effectiveness of developed nano-based filter, two air circulation–filtration system were made and installed in each environmental chamber. Each unit consisted of a filtration compartment, an axial fan and the required ducts and tubings. The inlet duct was placed near the manure pit and the treated air (passed through the filter) was distributed back to the room through the outlet tubing. During the tests the axial fan drew the contaminated air near the surface of the manure pit and passed it through the duct where the filter housing was installed. Filtered air was then distributed back to the room through another duct that was connected to the fan outlet. The duct had 8 equally-spaced holes which allowed the treated air to flow back into the chamber. The filtration compartment of the air circulation–filtration system that was used in the treatment room was loaded with approximately 200 g of each ZnO and TiO_2 nanoparticles, while in the control room a commercial filter pad was used in the filtration compartment. The filter housing in the treatment room was made of plastic styrene with honeycomb structure to ensure uniform distribution of the nanoparticles across the filter area. The upstream and downstream faces of the filter housing were covered with a commercially-available filter pad and a layer of glass wool to confine the nanoparticles within the plastic housing.

Evaluation of nano-based filtration system was conducted in several trials each lasting 30 days. The first 15 days of each trial were used to accumulate the manure in the pit. The circulation–filtration system was then tested on days 20, 25, and 30 of each trial. Prior to each test pigs were moved from the treatment and control rooms to an adjacent room and were only returned to the rooms after completion of tests and once sufficient ventilation was achieved. During each test, manure slurry in the collection tub was agitated using a steel rake and a recirculating pump. The mixing and recirculation were done simultaneously for 5 minutes. This allowed to mimic a situation that occurs during the periodic cleaning and/or drainage of the accumulated manure in underfloor pits of swine production rooms (i.e. highest level of H_2S, NH_3, and odours are usually experienced during the clearing of manure pit and drainage of manure). Air circulation–filtration system was operated as soon as the mixing started for 20 minutes. Concentrations of NH_3 and H_2S were monitored over the entire period of test (20 minutes), using gas sensors that were installed at the filter inlet and outlet. An additional set of NH_3 and H_2S sensors were installed at the human level (approximately 1.6 m above the floor) to evaluate NH_3 and H_2S concentrations within the chamber air space. A similar procedure was followed in the control room.

3 REPRESENTATIVE RESULTS

3.1 Simultaneous adsorption of NH_3 and H_2S in laboratory experiments

This section presents highlights of the results obtained with a binary mixture of commercial TiO_2 and ZnO. As indicated earlier adsorption experiments were conducted with gas mixtures containing 50 to 500 ppmv of each NH_3 and H_2S at 22, 70, 140 and 280°C. Based on the experimental results, ammonia breakthrough curves shifted to the left and breakthrough time became shorter as NH_3 concentration in the mixture was increased. At a constant NH_3 concentration, the breakthrough time decreased due to increase of temperature. In a similar fashion, increase of H_2S concentration in the mixture led to shorter breakthrough times. Contrary to what observed with ammonia, at a constant H_2S concentration the increase of temperature prolonged the breakthrough time. Fig. 4 shows typical breakthrough curves generated with the lowest and highest evaluated concentrations of 50 and 500 ppmv of each NH_3 and H_2S at 22 and 140°C. The breakthrough curves generated at other concentrations and temperatures showed similar pattern.

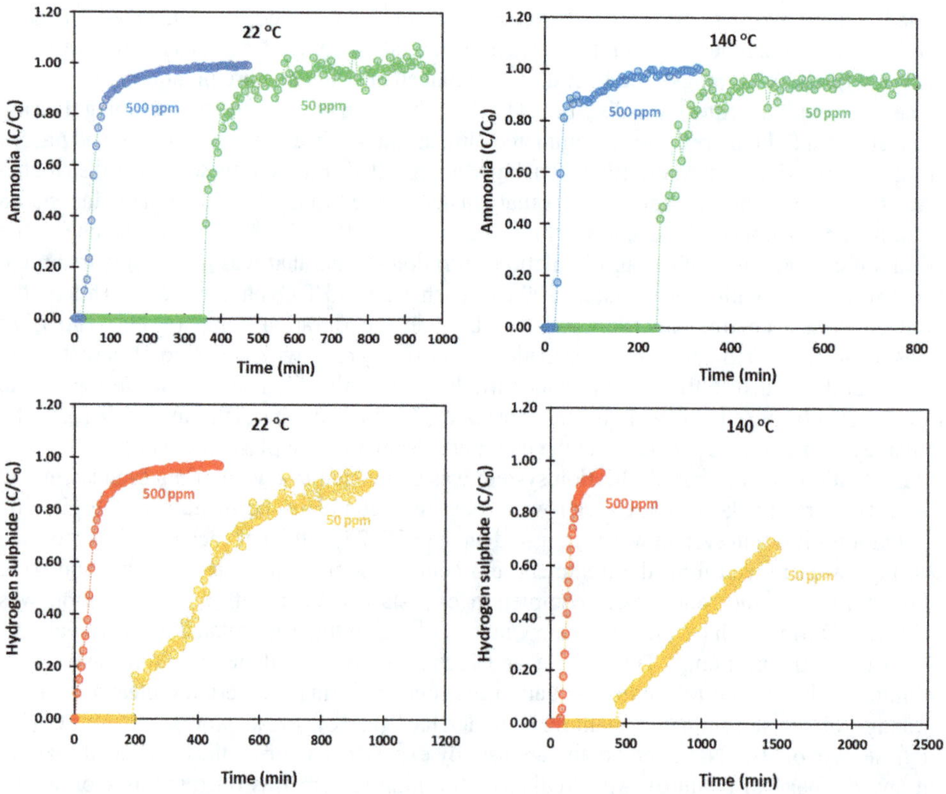

Figure 4: Representative breakthrough curves obtained with gas mixtures containing 50 ppmv or 500 ppmv of each NH_3 and H_2S at 22 and 140°C.

3.2 Capture of NH_3 and H_2S from swine manure gases in the semi-pilot scale system

As described earlier representative gases containing NH_3 and H_2S were generated by storing swine manure in several containers and allowing sufficient time for anaerobic digestion and accumulation of NH_3 and H_2S. The headspace gas from each container then provided the feed gas (influent) to the semi-pilot adsorption system for a period of 20 minutes during which the manure slurry in the container was subjected to intermittent mixing. Monitoring the concentrations of NH_3 and H_2S in the influent gas revealed high NH_3 and H_2S concentrations at the beginning of each 20-minute cycle (~200 and 300 ppmv NH_3 and H_2S, respectively) which then started to decrease as mixing continued. This pattern that was observed in all seven cycles is somewhat similar to the NH_3 and H_2S concentration profiles observed in production facilities during the cleaning and drainage of manure from the pits. Interestingly no ammonia or hydrogen sulphide was detected in the treated gas (effluent of semi-pilot scale adsorption system) during the entire trial (140 minutes), even when the influent concentrations of NH_3 and H_2S were at their maximum values. This revealed the effectiveness of nano-based filtration system in removal of NH_3 and H_2S from representative gases.

3.3 Emission control trials in research production facility

Fig. 5 shows the maximum concentrations of NH_3 and H_2S recorded by gas sensors located at the inlet (emitted gas from the manure pit) and outlet of the installed air circulation–filtration system (treated gas). These concentrations were recorded during the mixing of manure pits in the chambers. As shown the maximum NH_3 concentrations in the emitted gas on days 15, 20, 25 and 30 were 63, 68, 76 and 42 ppmv, respectively. These were then decreased to 4, 21, 34 and 20 ppmv as a result of the treatment in the air circulation–filtration system (removal percentage: 52–69%). A similar pattern is also seen for H_2S whereby the maximum H_2S concentrations in the emitted gas on days 15, 20, 25 and 30 were recorded as 51, 47, 56 and 18 ppmv, respectively, with the corresponding concentrations in the effluent gas being 19, 14, 17 and 11 ppmv (removal percentage: 39–70%).

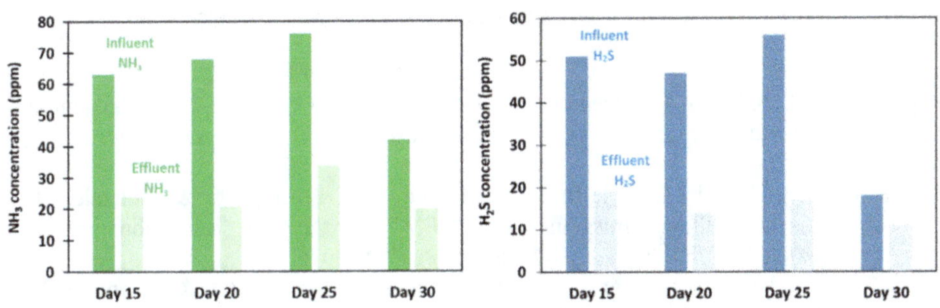

Figure 5: Maximum concentrations of NH_3 (left panel) and H_2S (right panel) recorded by gas sensors located at the inlet and outlet of the installed air circulation–filtration system.

In the control chamber where only a commercial filter pad was used in the air circulation–filtration system (i.e. no nanoparticles), no substantial differences in concentrations of NH_3 and H_2S in the influent and effluent gases were observed, indicating that the marked decrease in the level of NH_3 and H_2S in the treatment room trials was due to the use of nanoparticles in the air circulation–filtration system. It is important to point out that modest modifications to the air circulation–filtration system such as increasing the depth of the nanoparticles bed by increasing the quantity of nanoparticles and modifying the design of the filter compartment (e.g. decreasing the cross-sectional area of the filter compartment) to ensure that the entire gas stream passes through the filter bed completely are highly likely to lead to complete capture of NH_3 and H_2S from the gases emitted from the manure pit.

4 SUMMARY OF FINDINGS

The results of the present study in the laboratory system with pre-mixed gases and in semi-pilot scale with gases emitted from the stored manure revealed that metal oxide nanoparticles (TiO_2 and ZnO) were effective in simultaneous capture of ammonia and hydrogen sulphide from both pre-mixed and representative gases. The equilibrium adsorption capacities of both ammonia and hydrogen sulphide increased as concentration of these gases were increased in the mixture. Equilibrium adsorption capacity of hydrogen sulphide increased with the increase of temperature in the range 22°C to 280°C, while a decrease in the adsorption capacity of ammonia due to the increase of temperature was seen. Characterization of the exposed TiO_2 and ZnO nanoparticles by various techniques revealed that ZnO and TiO_2 both adsorbed NH_3 and H_2S. However, ZnO had a much higher affinity for H_2S and TiO_2 was more effective in adsorption of NH_3. The results of characterization analyses together with the contrasting patterns with regard to the dependency of adsorption capacity on temperature revealed the dominance of physical adsorption in case of NH_3 and chemisorption in case of H_2S. Application of the devised air filtration–circulation system with the nano-based filter showed the effectiveness of ZnO and TiO_2 nanoparticles in the capture of NH_3 and H_2S from a livestock production facility, though modest modification to the design of air filtration–circulation system and increase in the quantity of applied nanoparticles might be required for complete elimination of these hazardous emissions.

ACKNOWLEDGEMENT

This work was made possible by an Agriculture Development Fund grant (ADF 20140246) from the Ministry of Agriculture, Government of Saskatchewan, Canada.

REFERENCES

[1] Clarisse, L., Clerbaux, C., Dentener, F., Hurtmans, D. & Coheur, P.-F., Global ammonia distribution derived from infrared satellite observations. *Nature Geosciences*, **2**, pp. 479–483, 2009.

[2] Morán, M., Ferreira, J., Martins, H., Monteiro, A., Borrego, C. & González, J.A., Ammonia agriculture emissions: From EMEP to a high resolution inventory. *Atmospheric Pollution Research*, **7**(5), pp. 786–798, 2016.

[3] Ni, J.Q., Heber, A.J., Diehl, C.A., Lim, T.T., Duggirala, R.K., & Haymore, B.L., Summertime concentrations and emissions of hydrogen sulfide at a mechanically ventilated swine finishing building. *Transactions of the ASAE*, **45**(1), pp. 193–199, 2002.

[4] Chénard, L., Lemay, S.P. & Laguë, C., Hydrogen sulfide assessment in shallow-pit swine housing and outside manure storage. *Journal of Agricultural Safety and Health*, **9**(4), pp. 285–302, 2003.

[5] Rezaei, E., Schlageter, B., Nemati, M. & Predicala, B., Evaluation of metal oxide nanoparticles for adsorption of gas phase ammonia. *Journal of Environmental Chemical Engineering*, **5**(1), pp. 422–431, 2017.

[6] Bejan, D., Graham, T. & Bunce, N.J., Chemical methods for the remediation of ammonia in poultry rearing facilities: A review. *Biosystem Engineering*, **115**(3), pp. 230–243, 2013.

[7] Webb, J., Thorman, R.E., Fernanda-Aller, M. & Jackson, D.R., Emission factors for ammonia and nitrous oxide emissions following immediate manure incorporation on two contrasting soil types. *Atmospheric Environment*, **82**, pp. 280–287, 2014.

[8] Awume, B., Tajallipour, M., Nemati, M. & Predicala, B., Application of ZnO nanoparticles in control of H_2S emission from low temperature gases and swine manure gas. *Water, Air and Soil Pollution*, **228**, pp. 147–162, 2017.

[9] Rezaei, E., Azar, R., Nemati, M. & Predicala, B., Gas phase adsorption of ammonia using nano TiO_2-activated carbon composites: Effect of TiO_2 loading and composite characterization. *Journal of Environmental Chemical Engineering*, **5**(6), pp. 5902–5911, 2017.

INTEGRATED ENVIRONMENTAL HEALTH RISK ASSESSMENT FRAMEWORK FOR FIREWOOD-INDUCED INDOOR AIR POLLUTION

KHOMOTSO SEMENYA & FANNIE MACHETE
College of Agriculture and Environmental Sciences, University of South Africa, Republic of South Africa

ABSTRACT

This paper presents an integrated environmental health risk assessment framework to assess risks associated with firewood-induced indoor air pollution as an environmental health hazard. The combustion of firewood is associated with the emission of particulate matter, volatile organic and inorganic compounds or gases that are hazardous to human health and the environment. To date, there have been numerous studies and environmental health risk assessment approaches used to measure, monitor and assess environmental health risks associated with firewood-induced and related pollutants. However, many of the aforementioned risk assessment methods followed a narrow approach, thus rendering these studies methodologically one-sided. Consequently, some of these studies drew inferences about the air pollutants and the associated human health risks without identifying the tree species from which such firewood was harvested, as well as the different properties of wood and combustion characteristics including the dynamics of the fireplace where such combustion took place. Some of these studies failed to identify a specific community where particular firewood is used, since different firewood species grow in different areas and are therefore commonly used in communities where the wood is easily found. Most experimental studies on firewood-induced air pollution examined firewood emissions from undefined or unknown wood species. It is for these reasons that the current study aimed to close these methodological gaps by developing a comprehensive integrated environmental health risk assessment framework that considers the firewood user households, common types of wood and conditions of their harvesting, combustion and emissions in the laboratory environment. This study integrated observations, ethnobotanical meta-analysis and experimental study designs into one comprehensive integrated environmental health risk assessment (IEHRA) framework to assess the risks associated with exposure to volatile organic compounds (VOCs) from firewood combustion. The IEHRA was applied in the Senwabarwana villages to assess environmental health risks associated with selected firewood-induced volatile organic compounds. The results were sufficiently comprehensive to identify common ethnobotanical plant species used for firewood in the study area, the frequency of fire making, the conditions in which fire was made and common health problems experienced by firewood user households. Finally, different species of firewood were burnt in a laboratory under simulated kitchen conditions and VOCs were collected for analysis. The results of the VOCs samples differentiated high- and low-risk firewood species.

Keywords: indoor air pollution, environmental health, environmental health risk assessment, exposure assessment, risk characterisation, toxicity assessment, volatile organic compounds.

1 INTRODUCTION

The scientific credibility of research and its outcomes depend on the strength and rigor of the research methods followed. Environmental health risk assessment (EHRA) is a complex, comprehensive and scientific method of estimating environmental induced health risks associated with probable exposure to potential environmental hazards, as applied in the current study. The above resonates with a range of EHRA definitions from myriad other authors. For example, according to The Institute of Environmental Medicine [1], EHRA is a "multidisciplinary field of environmental health practice that is focused around the methods used to evaluate exposure, predict health risks and outcomes". National Health and Medical Research Council [2] refers to EHRA as a process of estimating the "probability that, within

WIT Transactions on Ecology and the Environment, Vol 236, © 2019 WIT Press
www.witpress.com, ISSN 1743-3541 (on-line)
doi:10.2495/AIR190181

a certain timeframe an adverse outcome will occur into a population exposed to chemical pollutants (in air, water, soil or food) under specific conditions". Clearly, a scientifically credible, replicable and methodologically sound framework is required to qualitatively and quantitatively estimate such environmental health risks.

There have been numerous studies that have used different approaches to assess indoor and outdoor induced environmental health risks, including domestic firewood induced air pollution, for example Kapwata et al. [3], Olave et al. [4], Mitchell et al. [5], Parajuli et al. [6], Jung et al. [7] and Joon et al. [8], to name a few. It is evident from that these studies inferred on air pollution induced health risk from a one sided and narrow methodological base. Several methodological gaps from most air pollution induced health risk assessments or EHRA identified and addressed in the current EHRA framework are as follows.

1.1 Survey as environmental health risk assessment tool

Survey is one of the most frequently used study designs used in air pollution studies. Survey is a list of questions aimed at extracting specific data from a group of people, [9], [10]. Surveys may be conducted by phone, mail, via the internet, and sometimes face-to-face, Agency for Toxic Substance and Disease Registry [11]. Most employed tools used in survey to collect data are questionnaires and interviews [12]. In air pollution study's authors, Kapwata et al. [3], Parajuli et al. [6] and Rumchev et al. [13] use survey to obtain information about cooking patterns, exposure to potential risk factors, smoking history, occupation and exposure to biomass fuels used in the home for cooking, heating and lighting, details of the house construction, demographic data and an assessment of socioeconomic status. Survey may be structured or semi-structured with open-ended or closed-ended questions, followed by response options. Kothari [10] noted that the use of closed questions leads to ease of processing answers; enhanced comparability of answers; easier to show relationship between variables; easier to make comparisons between variables and easier to make comparisons between respondents. However, Gratt [14] noted that closed-ended questions to be easily analysed statistically but seem to limit the respondent's response. In contrary Open-ended questions lead to a greater variety of responses (able to explore the answer) from participants but are difficult to analyse statistically because the data must be reduced in some manner, [12]. Survey have different modes of administration ranging from self-administration, telephonically or via email. Creswell [12] finds self-administered surveys to be cheaper compared to other modes of administration and having an advantage or reaching a large sample size, cover wide geographical area and excellent for capturing sensitive topics. Other modes of administration have disadvantage of survey being insufficiently completed (questions left unanswered) and in some cases may not be returned to the researcher or be returned late [10]. For example Langbein [15] experienced massive lack of response on questionnaire questions on stove usage, cooking fuels and cooking location and therefore omitted this information in analysis.

1.2 Observation as environmental health risk assessment tool

In an observational study, the sample population being studied is measured, or surveyed, as it is [12]. The researcher observes the variables but does not influence the population in any way or attempt to intervene in the study [16]. There is no manipulation by the researcher. Instead, data is simply gathered, and correlations are investigated. Since observational studies do not control any variable, the results can only allow the researcher to claim association, not causation. Chakraborty et al. [17] used observation to check ventilation pattern of the houses by considering the total number of windows in kitchen and living room.

1.3 Experiment as health risk assessment tool

Survey and observations cannot reflect the mass or concentration of a pollutant that is inhaled by an individual or community or households. The primary challenges with such methodological approach are that surveys and observations are perceptive than factual or concrete. Without experimental measurements, actual air quality problems and the resulting health burden cannot be assessed. Some researchers use experiments to get information on concentrations of pollutants in the smoke formed during indoor fuel combustion or the environment, for example Mitchell et al. [5] and Wright [18]. Dai et al. [19] and Azuma et al. [20] followed experimental methods to assess potential health risks associated with domestic firewood use. Unlike an observational study, an experimental study has the researcher purposely attempting to influence the results (Mouton [16]). The goal is to determine what effect a particular treatment has on the outcome. Researchers take measurements or surveys of the sample population. The researchers then manipulate the sample population in some manner. Since variables are controlled in a designed experiment, the results allow the researcher to claim causation [12]. Other studies (Mitchell et al. [5], Nolte et al. [21], Williams et al. [22] and Plejdrup et al. [23]) used analyses firewood-induced air pollution through experimental methods at a laboratory without prior understanding of the tree species in use or conditions from which such tree species were harvested.

2 CHALLENGES WITH SURVEY, OBSERVATION AND EXPERIMENT AS TOOL IN ENVIRONMENTAL HEALTH RISK ASSESSMENT

Some of the studies discussed above do not employ exposure methods to find the linkages between adverse health effects and air pollutants. These studies drew general inferences about the emissions and human health risks without relating the plant species, their use to a specific community. The studies also did not confirm the real use of such wood by a real community. Other studies (Ramos and De Albuquerque [24], Tabuti et al. [25] and Evtyugina et al. [26]) survey method in determining the different firewood species used for domestic fire making. These studies identified tree species through their indigenous ethnobotanical names. This approach was replicated with improvement (adapted) in the current study. The improvements were based on the limitations associated with the verification challenges experiences by Díez and Pérez [27]. This study found that participants mentioned plants that could not be seen by researchers making it impossible to identify such plants. The limitations identified from Venter and Venter [28] and De Winter et al. [29] studies were, namely:

a) The ethnobotanical names from communities are often inconsistent in both writing and pronunciation, thus leading to grammatical errors in the naming of the plant species.
b) South African communities are more ethnically heterogeneous than before; these communities often speak more than one language, leading to a cultural mix. Thus, affecting the ethnobotanical knowledge and use.
c) There is insufficient ethnobotanical literature that clearly differentiates indigenous plants according to various classes and subclasses of species and their characterization features.

3 ENVIRONMENTAL HEALTH RISK ASSESSMENT THEORIES

Several studies have applied human health environmental risk assessment (HHRA) to assess risk of chemicals [30]–[35]. The risk assessment method can be direct or indirect. The indirect method as applied in these studies [30]–[35] uses data from fixed monitoring stations to estimate exposure to pollutants while direct methods uses biological measurements [36]. Indirect methods have an advantage when the focus is on larger population exposure, which can be done at a shorter time and with minimal resources. Direct assessments are time consuming and expensive, and they are also not feasible for measuring more than one pollutant due to the inconvenience of attaching several samplers close to person's breathing zone. In this study, the environmental health risk assessment is adapted and given a new name Integrated Environmental Health Risk Assessment (IEHRA). It is used to address the concern of potential impact of volatile organic compounds on human health by examining exposures resulting from firewood combustion and the effects of such volatile organic compounds on human health. The framework is based on identifying the emission from firewood combustion in Senwabarwana villages.

The IEHRA framework was therefore applied in Senwabarwana to assess the potential environmental health risks from the volatile organic compounds formed during the combustion of firewood. For the purposes of this study, the Environmental Health Risk Assessment (Machete [30]) was adopted and adapted to a new name – the IEHRA framework (see Fig. 1).

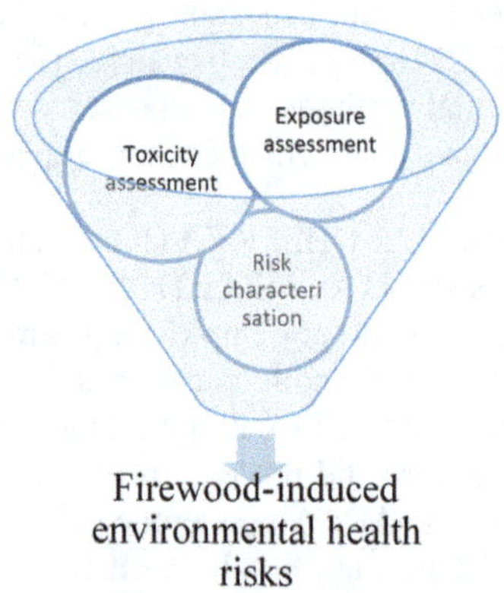

Figure 1: Integrated Environmental Health Risk Assessment framework.

The IEHRA framework was used to assess concerns about the potential impact of VOCs on human health, by examining exposure resulting from firewood combustion and the effects of such VOCs on human health. This framework is based on identifying the emissions from firewood combustion, as measured in Senwabarwana villages. As presented in subsequent sub-sections, the framework encompasses three stages. As shown through Fig. 1, each of the three pillars (components) of the IEHRA framework adopted in this study are discussed in detail below.

3.1 Toxicity assessment

To assess toxicity, this stage combines hazard identification and dose response assessment, [31]. There is no single measure of toxicity – different procedures are used, depending on the forms of toxicity (carcinogenic or non-carcinogenic effects). For toxicity assessment, this study integrated four study designs and approaches (survey, observation, ethnobotanical analysis, experimentation).

3.1.1 Surveys

Surveys were conducted by means of structured interviews to assess and analyse the circumstances in which firewood is used in the study area. The survey also laid the groundwork for hazard identification. The current research identified and analysed the different types of trees from which households harvested their firewood. Through descriptive statistical analysis and frequency analysis, priority tree species were identified.

3.1.2 Observation

Observation checklist was used to capture qualitative information such as the delivery methods of collected wood, storage area, harvesting tools, kitchen characteristics, building height, floor area, roof type, wall material, stove type, ventilation (number and size of windows, doors and chimney). During this process, the voucher specimens of each plant were photographed, collected, preserved and kept in the Horticulture Centre of the University of South Africa (UNISA), Florida Campus.

3.1.3 Ethnobotanical analysis

The names of the plants were given in Sepedi, which is the vernacular language predominantly spoken in Senwabarwana. Books, publications and different websites on vernacular names of plants, were used to assist in identifying the plants by comparing and checking the features. However, the researcher had difficulties in identifying some plants, as they were not appearing in any of the books used. Notwithstanding, plants were arranged in a list with correct common name, botanical name, vernacular name, family name and description of each plant.

3.1.4 Experimentation

This stage followed the guidelines set by the Department of Health [37] as well as the manufacturer's guidelines [38], which are based on the World Health Organization (WHO) guidelines. According to Innovative Occupational Hygiene Solutions [39], South Africa does not have legislation and methodologies for monitoring indoor air pollution. Therefore, most companies employ international assessment standards for collecting samples from building surfaces and air, examining the effects of pollutants on human occupants, and determining life-sustaining indoor air quality [40]. Depending on the types of pollutants to be monitored.

3.1.5 Sample preparation

The five most commonly used tree species (Red bush willow, Sickle bush, Leadwood, Black monkey thorn) were tested during the experimental phase. According to Creswell [12], many factors affect the performance of a fuel during combustion, including the heat value index, the heat potential, the durability of the embers, its elemental composition, its physical properties (e.g. density, size), its ash residue and the availability of oxygen. All these factors should be taken into consideration when combusting wood fuel. However, this study is not concerned with the combustion properties of firewood, but rather with the emissions. Therefore, the physicochemical properties of the commonly preferred species were not tested. Amongst the participants it was found that the wood was intended for residential heating and cooking and had been air-dried. The wood was cut in equal logs (1 kg) prior to combustion. According to Inkoom and Crentsil [41], the combustion of 1 kg of firewood produces harmful levels of emissions. The combustion methods were designed to reflect the Senwabarwana kitchen and stove setup of cooking using a three stone open fire method. A monitor was placed in the cooking area at a level equivalent to the breathing zone of an individual engaged in cooking (0.5–1 m above the floor, and 0.5–1 m from the source). For the purposes of this study, it was assumed that factors affecting emissions (such as moisture, ash content, amount of gasifying substances, shape and materials, flue gas outlet dimensions, air supply, stack height, size and shape, natural draft, air supply and mixing, combustion rate, operation habits), from residential wood combustion are constant. This is because residents do not scientifically test for these factors prior to combustion, and this process must reflect the true combustion processes.

3.1.6 Sampling of pollutants

Sampling includes the collection of pollutants (analyte) from a known volume of air and the stabilisation of the analyte on the sampling medium [26]. Sampling is usually the most important step in any analytical procedure. Errors committed at this stage cannot be corrected later during the analysis, therefore care was taken during the sampling methods selection process. This study used 1 L Teflon Tedlar Bags (plastic) to sample a volume of air. These bags are simple to use, affordable, reusable and available in various sizes (normally from 500 mL to 1.00 L). This method enables the measurement of short-term exposures for many substances (sampling duration of a few minutes). A disadvantage of using Tedlar bags is that compounds may not remain stable for more than 24–48 hrs [42]. Some bags are also permeable to certain chemicals, and losses of significant amounts of sample have been observed when they are stored for prolonged periods. To avoid this, samples were taken to the laboratory immediately after sampling. Moreover, Tedlar bags can allow humidity to diffuse when relative humidity levels differ between inside and outside Gawryś et al. [43]. A double-layer Tedlar bag has been designed with a drying agent between the two films, to limit the impact of external humidity on a low-humidity sample Scientific Kit Corporation (SKC) [38].

The bag was connected to a pump, which drew out the volume of air at a flow rate of 5 mL/min for 1 h. The bag was filled to less than 80% of its maximum volume, in accordance with the manufacturer's instructions. The sampling period was 1 h, which is equivalent to the cooking period. For accurate measurements, three replicate tests were performed for each wood species.

3.1.7 Sample storage

Samples collected in Tedlar bags were placed in a clean and cool environment (at room temperature) out of direct sunlight to prevent photo degradation. The bag samples were taken to the laboratory with the valve closed, and an identification tag attached. For best results, samples were analysed within 12 hrs of collection, and no later than 24–48 hrs after collection. The samples were sent to the SKC Safety Health and Environment laboratory immediately after sampling. The samples were marked by indicating numbers, the date collected, and the tree species used, to avoid discrepancies.

3.2 Exposure assessment

To comprehensively assess potential human exposure to firewood-induced VOCs, the current study used numerous data from two of the four study designs discussed earlier: observation and experimentation. During the observation phase, the researcher assessed the physical structure of the kitchen designs, the floor area (size) of the kitchen in relation to the family size, the number of windows/vents and their ratio to the floor area, and the potential for cross-ventilation, among other things. Lastly, through experimentation, the burning of different wood species within defined standards and the collection of ambient indoor emissions enabled the researcher to simulate the potential emissions that can be inhaled and that can enter the human body via different routes of exposure.

3.3 Risk characterisation

The risk characterisation step combines information on toxicity and exposure, to describe what is likely to happen to those who are exposed. Therefore, risk characterisation, which is the last step in the IEHRA process, summarises all data from the previous two stages/steps (Woolfenden [42], USEPA [44]). In addition, risk characterisation is the stage of IEHRA at which conclusions are drawn, based on the strengths, weight and limitations of the evidence or available data about the environmental hazards resulting from domestic firewood use [11], [31].

4 RESULTS AND DISCUSSION

The results are presented according to the three IEHRA stages, namely toxicity assessment, exposure assessment and rich characterisation

4.1 Toxicity assessment

Firewood is the dominant fuel type used in the study area. The types of firewood species used as fuel, being preferred for their quality as firewood are Red bush willow (Mohweleri), Sickle bush (Moretshe), Leadwood (Motswiri), Black monkey thorn (Mokgwa), Umbrella thorn (Mushu). Communities have preferences in respect of the type of wood they choose for combustion, as different woods exhibit different burning characteristics [25]. The firewood is collected dry. In cases where wet wood is collected, it is stored in the backyard to allow it to dry before use. It was observed that all the respondents used open-fire three-stone stoves (without chimneys) for cooking. The use of these stoves' releases smoke into the kitchen. Black soot was observed on the walls of some kitchens. This exposes the person responsible for cooking, along with family members and/or neighbours to pollutants in the smoke, which might have affected their health.

During site visits and observation of household kitchens, none of the kitchens was found to show signs of potential presence of firewood sources of BTEX such as glues, paints, furniture wax etc. Thus, except for potential ambient BTEX sources, the above-mentioned domestic BTEX sources were excluded as potential sources of the BTEX found in these kitchens. It was observed that all the respondents used firewood in open-fire three-stone stoves for cooking and this practice has the potential to cause adverse health effects through the inhalation of smoke released from combustion. Through experimentation, toxicological assessment revealed that there are several pollutants released during combustion of firewood species. It was revealed that VOCs such as benzene, toluene, ethylbenzene and xylene are released as pollutants from the combustion of the firewood used as fuel in Senwabarwana. Fig. 2 shows the selected VOCs emitted per firewood species and their concentration. Higher concentrations could be expected in winter months when the area becomes prone to pollution accumulation due to climatic conditions. The results indicate that Mushu had the highest benzene concentration followed by Moretshe, exceeding the detection limit of one micron per m^3 ($1 \mu g/m^3$). The lowest benzene concentration was detected in Motswiri, Mokgwa and Mohwiliri.

These measured concentrations are treated as intake rather than dose because it is known how much of the inhaled pollutant is absorbed by the body. The legal limits on emissions of, and exposure to, VOCs vary from country to country and are set by authorities such as the European Union and the United States Occupational Safety and Health Administration (OHSA). At present, there are no official regulations for VOC emissions for South Africa,

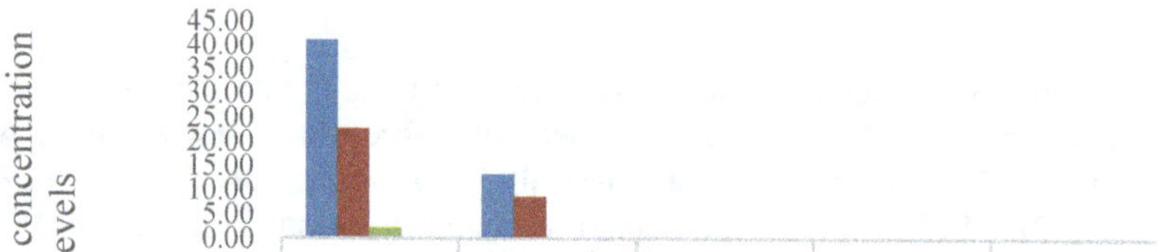

	Mushu	Moretshe	Motsiri	Mokgwa	Mohwiliri
Benzene	41.00	13.30	0.01	0.01	0.01
Toluene	22.65	8.68	0.01	0.00	0.00
Ethylbenzene	2.15	0.01	0.00	0.00	0.00
Xylene	0.01	0.01	0.00	0.00	0.00

Figure 2: Detected levels of VOCs concentration per selected tree species (µg/m^3).

apart from benzene for protection of human health and the environment despite VOCs playing an important role in the tropospheric chemistry and potentially causing serious effects on human health. The National Environmental Management: Air Quality Act 2004 (Act No. 39 of 2004) of the Republic of South Africa gives the national ambient air quality standards for pollutants such as sulphur dioxide, nitrogen dioxide, particulate matter, lead, ozone and benzene, but nothing for other VOC, hence international standards are often used (Republic of South Africa, 2004). The legal limits assume that toxic effects will not occur until a threshold dose is exceeded [2], [9].

According to the Agency for Toxic Substance and Disease Registry [11], benzene is a genotoxic carcinogen in humans, and no safe level of exposure can be recommended. It is therefore necessary to reduce indoor exposure levels as far as possible. From parameters published by the Department of Environmental Affairs, the national ambient air quality standard for benzene, per year, is 10 µg/m^3. The health effects associated with benzene are that it causes cancer and damage to the immune system, as well as neurological problems and possibly reduced fertility.

As noted in Fig. 2, high concentrations of toluene were emitted from Mushu, followed by Moretshe, and low concentrations from Motswiri, and no toluene was detected for Mokgwa and Mohwliri. Exposure to toluene causes headaches, fatigue, nausea and drowsiness. The World Health Organization non-cancer 30 mins guideline value is 1000 µg/m^3 based on odour annoyance. This study found toluene concentrations to be below the WHO standards, when participants were cooking with firewood.

Ethylbenzene and xylene were only emitted from Mushu and were not detected in the other firewood species. Exposure to ethlybenzene causes irritation of the eyes and throat, as well as liver disease. The odour threshold for ethylbenzene is 10,000 µg/m^3 (ATSDR [11]). The WHO ambient air guidelines for xylene are 4 800 for 24 hrs µg/m^3 and 870 µg/m^3 as an annual value [9]. This study found the concentrations of ethylbenzene and xylene to be lower than those recommended by the World Health Organization. A pollutant at a concentration below its toxicological reference value is not considered to represent a health risk [11]. The comparison between the intake value (dose) and the dose-response value indicates the level of exposure to be below the toxicological reference value, which it is unlikely that even sensitive individuals will experience adverse health effects.

4.2 Exposure assessment

The pollutants listed above can cause various health effects once exposure has occurred. Such health effects vary, depending on the length of time for which an individual has been exposed, and the concentrations s/he was subjected to. The respondents in Senwabarwana indicated that they cook once and/or twice a day (morning and/or evening). It is assumed that cooking takes approximately 1 h [44]. The kitchens are used mainly for cooking and body warming in winter. It was therefore necessary to assess the kitchen structures as they may influence exposure to firewood smoke during cooking. The structural design of kitchens in the study area is characterised by a limited number of windows, the size of which is usually small, thereby limiting the exchange of indoor air with outdoor air. It was observed that respondents opened windows and doors when cooking, to allow ventilation. However, pollutants cannot disperse properly due to poor kitchen structures putting residents at risk.

4.2.1 Reported health effects

Each of the products of firewood combustion has certain health effects on the human body. 46% of respondents self-reported headaches more frequently, followed by eye problems (33%) – sore, red and teary eyes; the burning of biomass fuel produces smoke that irritates the eyes [11]. According to the International Agency for Research on Cancer (IARC) [9], VOCs are irritants to the eyes and respiratory tract, and are carcinogenic. Although relying on self-reported diseases may make the study unreliable, a case-control study of indoor cooking smoke exposure and cataract prevalence in Nepal and India found that the use of solid fuel in unfuelled indoor stoves is associated with an increased risk of eye problems with cataract development in women [11]. That study therefore supports the findings of the current study. In many households, everyday exposure to air pollution may contribute to an increasing prevalence of asthma and cancer [17]. In this study, 12% of respondents self-reported asthma, and 3% self-reported cancer. Pneumonia was reported by only 1%, whereas 9% reported heart problems and 4% reported incidents of strokes. Literature indicates that short-term exposure to high levels of the BTEX can cause symptoms like eye, nose and throat irritation, headaches, nausea and vomiting, dizziness, and the worsening of asthma symptoms. Scientific studies suggest that long-term, chronic exposure at high levels can cause an increased risk of liver damage, kidney damage, cancer, and central nervous system damage.

4.2.2 Link between health effects and firewood species

Fig. 3 presents the relationship between the firewood species used and health effects. This is a test of statistical significance to compare frequency of reported health effects by type of firewood species. It is observed that respondents who uses mushu more often reported highest health effects complains followed by those using mohwiliri, moretshe, motswiri and mokgwa respectively. Mushu was found to emit more volatile organic compounds compared to other firewood species. It is therefore not surprising that there are more complaints regarding health effects for mushu as compared to other firewood species.

5 RISK ASSESSMENT

This step combines the information from the two previous steps to provide an indication of the nature and expected frequency of adverse health effects in exposed populations. The fundamental assumption of the sampling strategy consists of the fact that measured concentrations represent maximum concentrations to which all individuals could be exposed

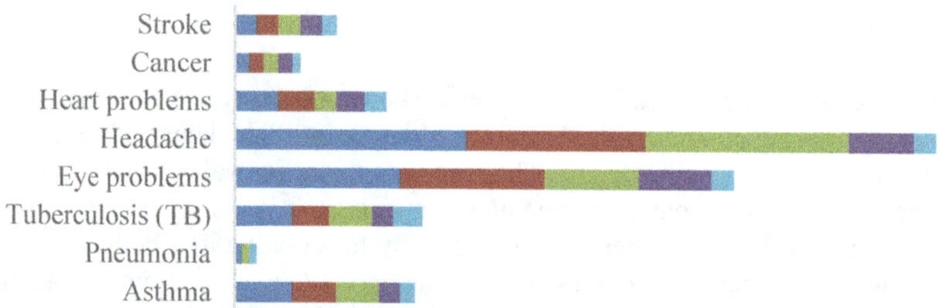

	Asthma	Pneumonia	Tuberculosis (TB)	Eye problems	Headache	Heart problems	Cancer	Stroke
Mushu (N=51)	8	1	8	23	32	6	2	3
Mohweleri (N=41)	6	0	5	20	25	5	2	3
Moretshe (N=33)	6	1	6	13	28	3	2	3
Motswiri (N=16)	3	0	3	10	9	4	2	3
Mokgwa (N=11)	2	1	4	3	3	3	1	2

Figure 3: Link between health effects and firewood species.

in the kitchen [44]. If this assumption is true, then the risk of developing health effects due to the presence of the studied volatile organic compounds can be assessed as negligible. This holds for all Benzene, Toluene, Ethylbenzene and Xylene (BTEX) (concentrations remain below the risk levels).

Benzene poses a health risk in households where mushu and moretshe are used. There is, however, no health risk associated with the use of mokgwa, mohwiliri and motswiri. Similarly, toluene poses a health risk when mushu and moretshe are used, while there is no health risk associated with the use of mokgwa, mohweleri and motswiri. Ethylbenzene poses a health risk when mushu is used as firewood. Xylene has no health risk associated with the use of all five-tree species. Due to a limitation of this study monitoring volatile organic pollutants over a limited period, it was therefore not possible to estimate risks over longer periods of exposure (chronic health effects).

6 CONCLUSION

This study reviewed existing air pollution related risk assessment frameworks and research methods used. During such reviews, it emerged that despite the existing of numerous risk assessment frameworks and research methods followed, a huge gap was identified in both the frameworks and methods used. Consequently, an opportunity of developing an improved and Integrated Environmental Health Risk Assessment (IEHRA) framework was identified. Thus, one of the major contributions of the current study was to improve from known EHRA frameworks by developing the current IEHRA and ultimately conduct the study in Senwabarwana Villages using this improved framework.

It was found that firewood is the main source of energy. Volatile organic compounds emitted from the selected firewood were assessed. Mushu was found to emit the highest concentrations of BTEX. Ethylbenzene and xylene were not detected in Motswiri, Mohwiliri and Mokgwa, while benzene and toluene were detected below the detection limit of 1 $\mu g/m^3$.

The study found the VOC concentration to be below the WHO standards for those cooking with firewood. Each of the VOCs derived from firewood combustion has certain health effects on the human body. However, lower limits were detected in this study, suggesting that the community of Senwabarwana – which relies on firewood – is at lower risk of the ill effects associated with burning firewood. The risk assessment of health effects reported in Senwabarwana leads to the conclusion that reported health complaints are not due to the presence of the measured compounds. It is recommended that further studies be conducted on other pollutants emitted by these firewoods.

REFERENCES

[1] The Institute of Environmental Medicine, Environmental health risk assessment. https://ki.se/en/imm/environmental-health-risk-assessment. Accessed on: 5 Oct. 2018.

[2] National Health and Medical Research Council (NHMRC), *Guidelines for Managing Risk in Recreational Water*, Commonwealth of Australia: Canberra, 2008.

[3] Kapwata, T., Language, B., Piketh, S. & Wright, C.Y., Variation of indoor particulate matter concentrations and association with indoor/outdoor temperature: A case study in rural Limpopo, South Africa. *Atmosphere*, **9**(4), p. 124, 2018.

[4] Olave, R.J., Forbes, E.G.A., Johnson, C.R. & Relf, J., Particulate and gaseous emissions from different wood fuels during combustion in a small-scale biomass heating system. *Atmospheric Environment*, **157**, pp. 49–58, 2017.

[5] Mitchell, E.J.S., Lea-Langton, A.R., Jones, J.M., Williams, A., Layden, P. & Johnson, R., The impact of fuel properties on the emissions from the combustion of biomass and other solid fuels in a fixed bed domestic stove. *Fuel Processing Technology*, **142**, pp. 115–123, 2016.

[6] Parajuli, I., Lee, H. & Shrestha, K.R., Indoor air quality and ventilation assessment of rural mountainous household of Nepal. *International Journal of Sustainable Built Environment,* **5**(2), pp. 301–311, 2016.

[7] Jung, C.H., Matsuto, T. & Tanaka, M., Behavior of metals in ash melting and gasification melting of municipal solid waste (MSW). *Waste Management*, **25**(3), pp. 301–310, 2005.

[8] Joon, V., Kumar, K., Bhattacharya, M. & Chandra, A., Non-invasive measurements of carbon monoxide in rural Indian women exposed to different cooking fuel smoke. *Aerosol and Air Quality Research*, **14**(6), pp. 1789–1797, 2014.

[9] The National Academic Press, Toxicity testing. www.nap.edu/read/11523/chapter/3. Accessed on: 5 Oct. 2018.

[10] Kothari, C.R., *Research Methodology, Methods and Techniques,* 2nd ed., Prakashan: New Delhi, 1990.

[11] Agency for Toxic Substance and Disease Registry, The toxic guide. www.atsdr.cdc.gov/substances/toxsubstance.asp?toxid=14. Accessed on: 3 Oct. 2018.

[12] Creswell, J.W., *Research Design: Qualitative, Quantitative and Mixed Methods Approaches*, 4th ed., SAGE, 2014.

[13] Rumchev, K., Brown, H. & Spickett, J., Volatile organic compounds: Do they present a risk to our health? *Reviews on Environmental Health*, **22**(1), pp. 39–55, 2007.

[14] Gratt, L.B., *Air Toxic Risk Assessment and Management: Public Health Risks for Normal Operations*, Van Nostrand Reinhold, 1996.

[15] Langbein, J., Firewood, smoke and respiratory diseases in developing countries – the neglected role of outdoor cooking. *PLoS ONE,* 2017.

[16] Mouton, M., *How to Succeed in your Masters and Doctoral Studies*, Van Schaik Publishers: Pretoria, 2001.

[17] Chakraborty, D., Mondal, N.K. & Datta, J.D., Indoor pollution from solid biomass fuel and rural health damage: A micro-environmental study in rural area of Burdwan, West Bengal. *Sustainable Built Environment*, **3**(2), pp. 262–271, 2014.
[18] Wright, C., Air pollution monitoring and evaluation framework for South Africa: prioritizing vulnerable communities. *WIT Transactions on Biomedicine and Health*, vol. 14. WIT Press: Southampton and Boston, pp. 55–64, 2009.
[19] Dai, H. et al., VOC characteristics and inhalation health risks in newly renovated residences in shanghai, China. *Science of the Total Environment*, **15**(577), pp. 73–83, 2017.
[20] Azuma, K., Uchiyama, I., Uchiyama, S. & Kunugita, N., Assessment of inhalation exposure to indoor air pollutants: Screening for health risks of multiple pollutants in Japanese dwellings. *Environmental Research*, **145**, pp. 39–49, 2016.
[21] Nolte, C.G., Schauer, J.J., Cass, G.R. & Simoneit, B.R.T., High polar organic compounds present in wood smoke and in the ambient atmosphere. *Environmental Science Technology*, **35**, pp. 1912–1919, 2001.
[22] Williams, A., Jones, J.M., Ma, L. & Pourkashanian, M., Pollutants from the combustion of solid biomass fuels. *Progress in Energy and Combustion Science*, **38**, pp. 113–137, 2012.
[23] Plejdrup, M.S., Nielsen, O. & Brandt, J., Spatial emission modelling for residential wood combustion in Denmark. *Atmospheric Environment*, **144**, pp. 389–396, 2016.
[24] Ramos, M.A. & De Albuquerque, U.P., The domestic use of firewood in rural communities of the Caatinga: How seasonality interferes with patterns of firewood collection. *Biomass & Bioenergy*, **39**, pp. 147–158, 2012.
[25] Tabuti, J.R.S., Dhillion, S.S. & Lye, K.A., Firewood use in Bulamogi County, Uganda: Species selection, harvesting and consumption patterns. *Biomass and Bioenergy*, **25**, pp. 581–596, 2003.
[26] Evtyugina, M. et al., Emission of volatile organic compounds from the residential combustion of Pyrenean oak and black poplar. *International Journal of Environmental and Ecological Engineering*, **7**(6), pp. 289–293, 2013.
[27] Díez, H.E. & Pérez, J.F., Physicochemical characterization of representative firewood species used for cooking in some Colombian regions. *International Journal of Chemical Engineering*, **2017**(1), pp. 1–13, 2017.
[28] Venter, F. & Venter, J., *Making the Most of Indigenous Trees*, Briza Publications: Pretoria, 1996.
[29] De Winter, B., De Winter, M. & Killick, D.J.B., *Sixty-Six Transvaal Trees*. Botanical Research Institute: Pretoria, 1966.
[30] Machete, F., Environmental health risks associated with e-waste exposure in Badplaas, Carolina and Elukwatini landfills, Republic of South Africa. *African Journal of Science, Technology, Innovation and Development*, **9**(6), pp. 679–684, 2017.
[31] Morakinyo, O.M., Adebowale, A.S., Mokgobu, M.I. & Mukhola, M.S., Health risk of inhalation exposure to sub-10μm particulate matter and gaseous pollutants in an urban-industrial area in South Africa: An ecological study. *Occupational and Environmental Medicine Research*, **7**(3), p. e013941, 2017.
[32] Oosthuizen, M.A., Wright, C.Y., Matooane, M. & Phala, N., Human health risk assessment of airborne metals to a potentially exposed community: A screening exercise. *The Clean Air Journal*, **25**(1), pp. 51–57, 2015.
[33] Thabethe, N.D.L., Engelbrecht, J.C., Wright, C.Y. & Oosthuizen, M.A., Human health risks posed by exposure to PM_{10} for four life stages in low socio-economic community in South Africa. *PanAfrican Medical Journal*, **18**(206), 2014.

[34] Habeebullah, T., Risk assessment of exposure to volatile organic compounds in the holy city of Makkah. *Environment Impact*, **162**, pp. 625–635, 2012.

[35] Muller, E., Diab, R.D., Binedell, M. & Hounsome, R., Health risk assessment of kerosene usage in an informal settlement in Durban, South Africa. *Atmospheric Environment*, **37**, pp. 2015–2022, 2003.

[36] Cattaneo, A. et al., Comparison between personal and individual exposure to urban air pollutants. *Aerosol Science and Technology*, **44**(5), pp. 370–379, 2010.

[37] Department of Health, *Guideline for the Management of Indoor Air Quality: A Guide for Environmental Health Practitioners in South Africa*, Pretoria, 2018.

[38] Scientific Kit Corporation, Application guide (Sampling train-air sample bags). www.skcltd.com. Accessed on: 14 Feb. 2018.

[39] Hopkins, D. & Williams, D., Guidance on human health risk assessment for environmental impact assessment in Alberta. Edmonton. https://open.alberta.ca/publications/guidance-on-human-health-risk-assessment-for-environmental-impact-assessment-in-alberta. Accessed on: 8 Oct. 2018.

[40] Cheng, Z., Li, B., Yu, W., Wang, H., Zhang, T. & Bu, Z., Risk assessment of inhalation exposure to VOCs in dwellings in Chongqing, China. *Healthy Buildings 2017 Europe*, 2017.

[41] Inkoom, D.K.B. & Crentsil, A.O., Estimation of indoor air pollution and health impacts due to biomass burning in rural northern Ghana, in case studies for developing globally responsible engineers. *Global Dimension in Engineering Education,* ed. GDEE, Barcelona. http://gdee.eu/index.php/resources.html. Accessed on: 5 Oct. 2018.

[42] Woolfenden, E., Sorbent-based sampling methods for volatile and semi-volatile organic compounds in air – Part 1: Sorbent-based air monitoring options. *Journal of Chromatography A*, **1217**(16), pp. 2674–2684, 2010.

[43] Gawryś, M., Fastyn, P., Gawłowski, J., Gierczak, T. & Niedzielski, J., Prevention of water vapour adsorption by carbon molecular sieves in sampling humid gases. *Journal of Chromatography A*, **933**(1–2), pp. 107–116, 2001.

[44] United States Environmental Protection Agency (USEPA), Framework for human health risk assessment to inform decision making. www.epa.gov/sites/production/files/2014-12/.../hhra-framework-final-2014.pdf. Accessed on: 26 Sept. 2018.

PROSPECTS FOR THE USE OF ALTERNATIVE FUELS AND ENERGY BY ROAD TRANSPORT IN RUSSIA

ALLA GOLUBEVA & ELENA MAGARIL
Department of Environmental Economics, Ural Federal University, Russia

ABSTRACT

Road transport is one of the main consumers of petroleum products and sources of air pollution. In the conditions of Russia's fast-growing vehicle fleet, it is necessary to develop and implement effective measures to reduce the consumption of scarce oil fuels and stimulate the use of alternative eco-friendly fuels. The aim of this work is to assess the prospects for the use of alternative fuels and energy by cars in Russia. This paper outlines the main ways of reducing the gasoline and diesel fuel consumption by road transport, including the use of the motor fuel alternative types. As the most promising direction of alternative energy development in the road transport sector regarding Russia's conditions, a transition to gas-cylinder fuel is proposed. This study presents the main problems of road transport gasification and the ways to solve them. The effective actions in the selected fields will reduce the consumption of scarce fuel and energy resources and enable improvement of the environmental situation in megalopolises.

Keywords: alternative fuels and energy, motor transport, liquefied petroleum gas, compressed natural gas, liquefied natural gas, traditional motor fuel, transport gasification.

1 INTRODUCTION

At the present stage of society development, in the conditions of industrial production volumes and the world's vehicle-amount growth, the depletion of fossil fuels is proceeding at high rates. The burning of traditional motor fuel by cars causes significant damage to the environment [1]–[3]. The reduction of consumption of scare fuels of petroleum origin by road transport is one of the priority problems of society's sustainable development.

Considerable attention is paid to research on the possibility of improving environmental sustainability and the energy efficiency of vehicles [4]–[8].

Liquefied petroleum gas (LPG) [9]–[12], compressed natural gas (CNG) [4], [6]–[10], [13]–[17] and liquefied natural gas (LNG) [2], [11], [12] are environmentally friendly alternatives to traditional liquid motor fuels, and it is advisable to expand their use.

Meanwhile, there is currently no systematic approach to reducing the consumption of traditional motor fuels by expanding the use of alternative fuels by the fleet. This leads to the irrational over-consumption of petroleum origin fuels, inhibits the development of the alternative motor fuel market, and worsens the environmental situation.

Thus, it is necessary to develop and implement mechanisms for the optimal use of oil origin fuels, to stimulate the development of alternative motor fuels in order to create sustainable, ecologically balanced economic development that takes into account the interests of the future generations.

2 THE THREAT OF THE FUEL AND ENERGY RESOURCES CONSUMPTION GROWTH

The modern level of civilization development is characterized by a rapid growth of the Earth's population (Table 1) and its ever growing need for exhaustible fuel and energy resources. An extensive scenario for the world energy development leads to the aggravation of the socio-economic and environmental problems, which requires the effective actions to shift to rational energy consumption.

WIT Transactions on Ecology and the Environment, Vol 236,
www.witpress.com, ISSN 1743-3541 (on-line)
doi:10.2495/AIR190191

Table 1: The Earth's population dynamics, mln. people*.

Region	1950	1970	1990	2010	2015	2030	2050	2100
World including:	**2,525**	**3,682**	**5,310**	**6,930**	**7,349**	**8,501**	**9,725**	**11,213**
Africa	229	366	632	1,044	1,186	1,679	2,478	4,387
Asia	1,394	2,120	3,202	4,170	4,393	4,923	5,267	4,889
Europe	549	657	721	735	738	734	707	646
L. America	169	288	447	600	634	721	784	721
North America	172	231	281	344	358	396	433	500
Australia, etc.	13	20	27	36	39	47	57	71

*Based on the data from the United Nations [18], [19].

In 2017, the primary energy consumption in the world was 13.5 billion TOE. The main energy resources consumers are China – 3.1 billion TOE (23.2% of total energy consumption), the USA – 2.2 billion TOE (16.5%), and India – 0.75 billion TOE (5.5%). Russia occupies the fourth place, consuming 698.3 million TOE of the primary energy, which is equal to 5.2% of the total global consumption [20], [21].

Over the past 50 years, the primary energy consumption has increased by 3.7 times (Fig. 1). High rates of growth in fuel and energy resources consumption require the adoption of effective solutions in the field of their rational use and alternative energy development.

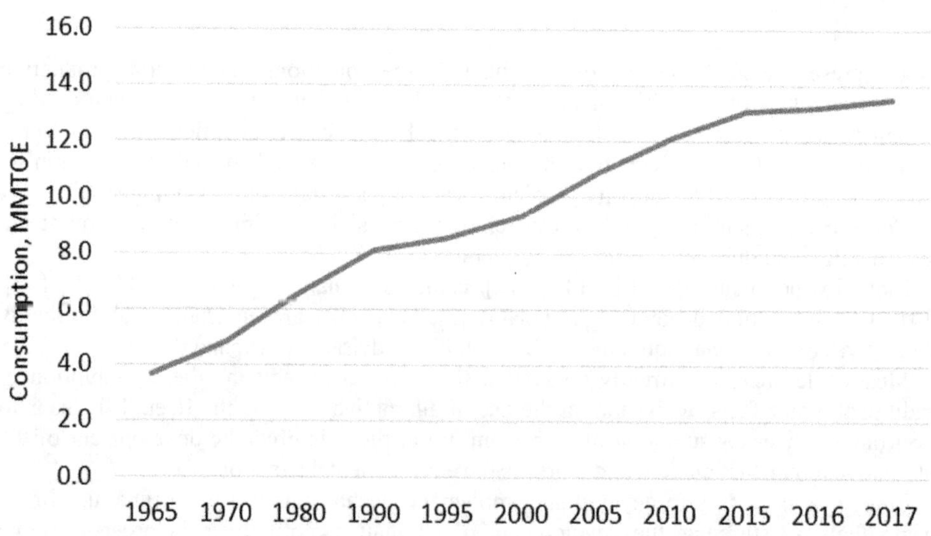

Figure 1: World primary energy consumption dynamics (based on data from BP [20], [21]).

3 THE ROLE OF THE ROAD TRANSPORT IN THE REDUCTION OF FUEL AND ENERGY RESOURCES CONSUMPTION

During the last 100 years, oil has been playing a central role in the world balance of energy consumption. In 2017, the share of oil in the total energy consumption was 34.2%, coal – 27.6%, natural gas – 23.4%, while the total contribution of nuclear energy, hydropower and renewable energy sources was only 14.8% (Table 2) [20], [21].

Table 2: Energy resources consumption in 2017, MMTOE.

Region	Energy resources						Total consumption
	Oil	Natural gas	Coal	Nuclear energy	Hydro-power	Renewable energy	
China	608	207	1,893	56	262	107	**3,132**
USA	913	636	332	192	67	95	**2,235**
India	222	47	424	8	31	22	**754**
RF	153	365	92	46	41	0.3	**698**
World total	**4,622**	**3,156**	**3,731**	**596**	**919**	**487**	**13,511**

The main consumers of petroleum are transport, industry, energy production, and other sectors. Within that, the priority role is taken by transport (Fig. 2). The main role of transport in oil consumption in the long term is confirmed by the medium- and long-term forecasts made by the BP and the International Energy Agency (IEA).

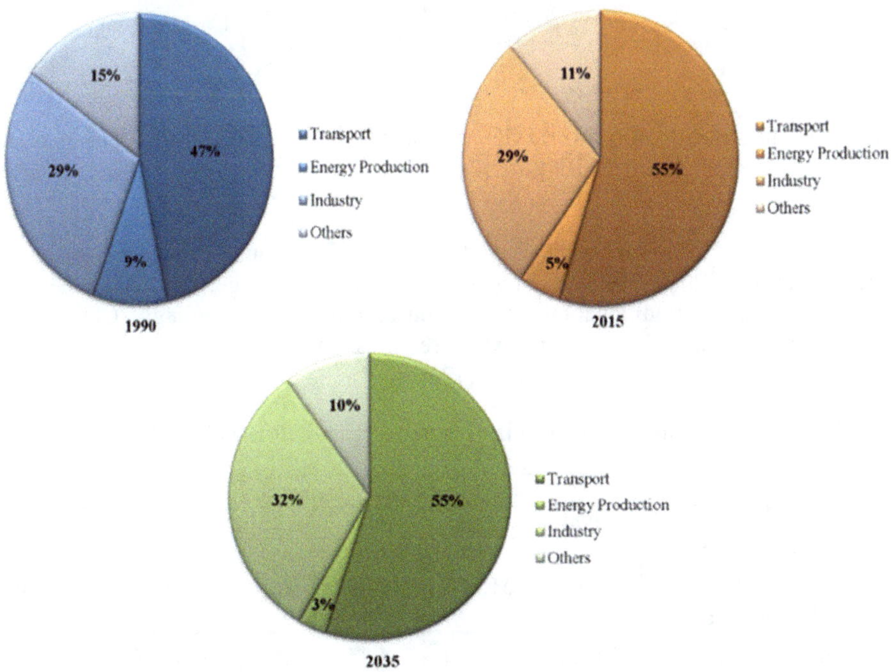

Figure 2: Oil consumption by sectors (based on data from BP [21], [22]).

The fuel combustion by transport causes significant damage to the environment, both from pollution by toxic substances and from CO_2 emissions [6], [7].

Carbon dioxide is the main component of technogenic emissions entering the atmosphere when burning organic fuels, thus being an indicator of the scarce hydrocarbon fuels consumption and the level of civilization development as a whole. The increase of CO_2 emission contributes to global warming and indicates the depletion of fuel and energy resources.

In the structure of CO_2 emission from the transport sector, the main contribution is made by cars, light trucks, and passenger cars weighing up to 3.5 tons (43.3%), as well as by cargo vehicles (22.2%) (Fig. 3).

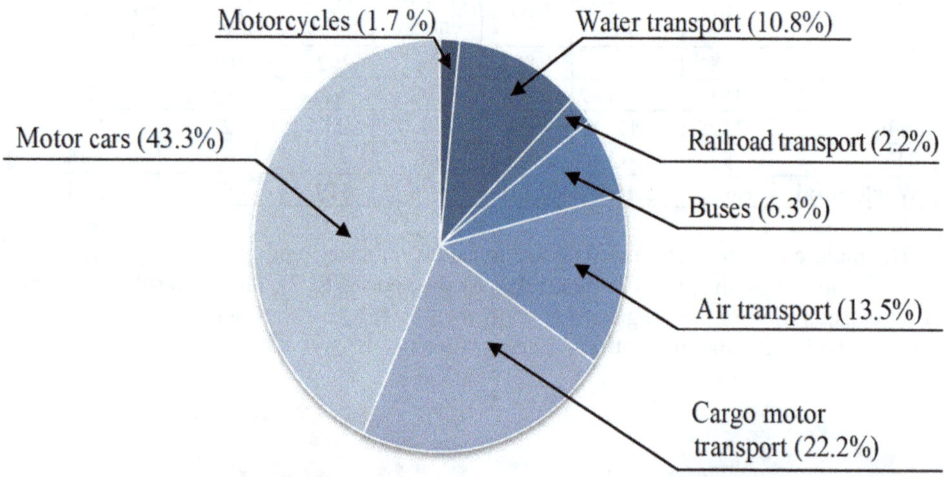

Figure 3: The structure of the world CO_2 emission from the vehicle sector of the economy (based on data from BP [20], [21]).

The data in Table 3 show more than a twofold increase in the number of cars in the country during the period of 2000–2017, which was mainly due to passenger cars. Thus, the search for ways to reduce the traditional motor fuels consumption and, accordingly, CO_2 emission by cars, is a priority task, the solution to which requires urgent action considering the current high rate of the population motorization [7], [23].

Table 3: The dynamics of the amount of cars in Russia [24], [25].

№	Name	Years					
		2000	2005	2010	2012	2015	2017
1	**Amount of vehicles, thousands, including:**	**25,394**	**31,210**	**40,662**	**45,471**	**51,290**	**50,600**
1.1	Passenger cars	20,353	25,570	34,354	38,792	44,200	46,500
1.2	Trucks	4,401	4,848	5,414	5,751	6,200	3,700
1.3	Buses	640	792	894	928	890	400

4 THE WAYS OF REDUCING THE PETROLEUM FUELS CONSUMPTION BY ROAD TRANSPORT

The conducted research allowed us to offer two main ways to reduce the deficit petroleum-derived fuel consumption by vehicles: 1) the increase of the fuel economy of cars with gasoline and diesel engines and 2) the use of alternative fuel and energy (Fig. 4) [23], [26], [27].

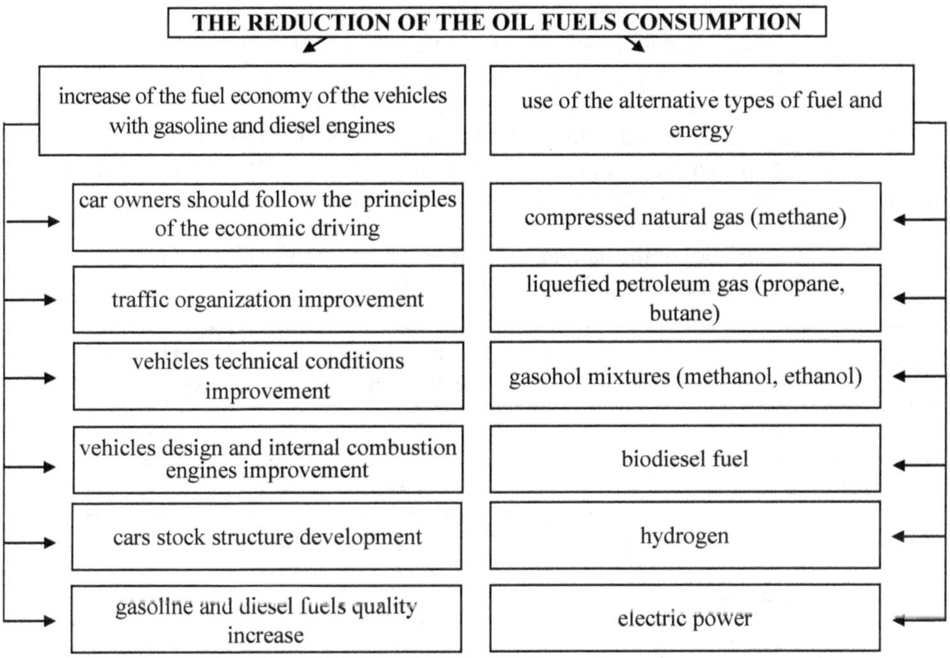

Figure 4: The main ways to reduce petroleum fuels consumption by road transport.

It should be noted that for Russia in the short and medium term, the emphasis should be placed on the quality improvement of traditional motor fuels as one of the most effective ways of increasing vehicle fuel efficiency. One of the possible ways to solve this problem is the effective use of the economic incentive mechanisms of oil products producers, aimed at the improvement of the produced gasoline and diesel fuel environmental characteristics [23], [26]–[29].

The second direction of reducing petroleum fuel consumption by vehicles is the transition to the alternative fuels and energy. It should be underlined that the development of alternative energy in Russia is hampered by a number of objective factors. Russia is rich in traditional fuel and energy resources; as a consequence, the development of alternative energy was not at the forefront.

The raw orientation of the Russian economy extensively exploits traditional energy and hinders the development of alternatives, which requires significant financial resources. The high monopolization of the energy market by the vertically integrated companies is also a serious deterrent to the development of alternative energy in Russia. One cannot ignore the other problems of introducing each type of alternative energy in Russia: financial, geographic (especially climatic), technical, etc.

The change of energy carriers is a painful and long historical period, in which the changes in production technologies are reflected. Meanwhile, the most developed countries at the present stage actively introduce the latest developments in the field of alternative energy. Accordingly, to get the positive results in the matters of alternative energy in the long term, it is necessary to act today.

Despite the obstacles hampering the development of alternative energy in Russia, there are indisputable prerequisites for an increase in the amount of vehicles using gas-cylinder fuel and other alternative types of motor fuel in the country's total automobile fleet.

The main share of the fuel consumption by the Russian vehicle fleet belongs to gasoline, while this country ranks second in the world (after Iran with their reserves of 34 trillion m^3) in the field of the prospected proven reserves of natural gas (Table 4) [30], that creates a huge potential for the transition to a highly environmentally friendly gas-cylinder fuel.

Table 4: Characteristics of the natural gas proven reserves and production at the beginning of 2017 (based on data from BP [20], [21]).

Region	**Gas resources, trillion m^3**	**The share in the world resources, %**
North America	10.8	5.6
Latin America	8.2	4.2
Europe and Eurasia, including:	62.2	32.1
Russian Federation	**35.0**	**18.1**
The Near and Middle East	79.1	40.9
Australia and the rest of Asia	19.3	10.0
Africa	13.8	7.1
World, total	**193.5**	**100.0**

As for the electric vehicles and hybrid electric vehicles [31], they are not so relevant for Russia, primarily due to the climatic conditions: the harsh climate in most parts of the country significantly limits their mass use. Also, today, the Russian consumer is not ready to significantly overpay for a car with an environmental performance improvement. The same applies to the vehicles running on biofuel [32]. Under conditions of the large natural gas reserves in Russia, the development of road transport consumption of the other types of alternative energy is only possible in the long term.

5 THE MAIN PROBLEMS OF THE ALTERNATIVE TYPES OF MOTOR FUEL MARKET DEVELOPMENT AND THEIR SOLUTION

The main alternative to the petroleum-derived fuels used by Russian transport in the medium term is compressed natural gas (methane) and liquefied petroleum gas (propane-butane mixture) [9], [13]–[17].

Methane is a compressed natural gas supplied to the specialized automobile gas-filling compressor stations (AGSCS) via pipelines. Propane-butane mixture is obtained from oil and condensed petroleum gases. Considering the fact that, firstly, the widespread use of propane-butane as an automotive fuel does not completely abolish dependence on oil; and secondly, our country ranks second in the world in proven prospected reserves of natural gas (Table 4), the use of compressed natural gas is the main alternative to traditional motor fuels.

Despite the fact that Russia ranks second in the world in proven prospected reserves of the natural gas (Table 4), today it is necessary to solve a number of problems for the widespread introduction of the compressed natural gas as a motor fuel. The main ambitions of the Government in the issue of transport large-scale gasification are registered in the Russian Government decree of May 13th, 2013, № 767-r [33], according to which a set of measures should be developed and submitted so that by 2020 in the subjects of the Russian Federation, the level of natural gas use as a motor fuel by public transport and by municipal services transport would reach 50% of the total number of the equipment units (in the cities with a population of more than 1,000,000 people). However, today it is difficult to distinguish the active measures that should be performed by the Government and all the motor fuel market participants to achieve these parameters.

The conducted researches have allowed to reveal the basic problems on a way of the road transport mass gasification in Russia, and also to offer their possible solutions (Figs 5 and 6).

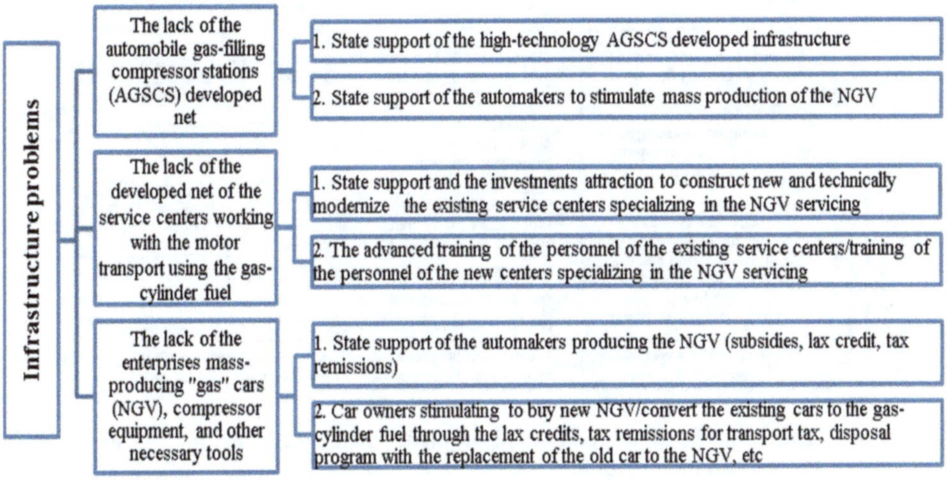

Figure 5: Infrastructure problems of the motor vehicles mass gasification in Russia and ways to solve them.

To date, the infrastructure problems are the major for the mass gasification of transport (Fig. 5). There is not a sufficient number of automobile gas-filling compressor stations (AGSCS) for the large-scale gasification of the car stock. This problem resembles a vicious circle: on the one hand, the development of the gas engine fuel market is hampered by the poorly developed infrastructure, and on the other hand, the infrastructure is absent due to the shortage of natural gas vehicles.

The large-scale gasification of vehicles is a complicated task, requiring a system approach to solve it. In addition to the basic infrastructure problems of this issue, the studies carried out have revealed a number of additional directions for solving the problems of vehicle gasification (Fig. 6).

It is important to note that the transfer of public transport (city buses, fixed-route taxis) to gas, subject to all safety requirements, will lead to a much greater environmental and economic effect than the transfer of the personal vehicles, due to the higher energy efficiency of the trip per passenger.

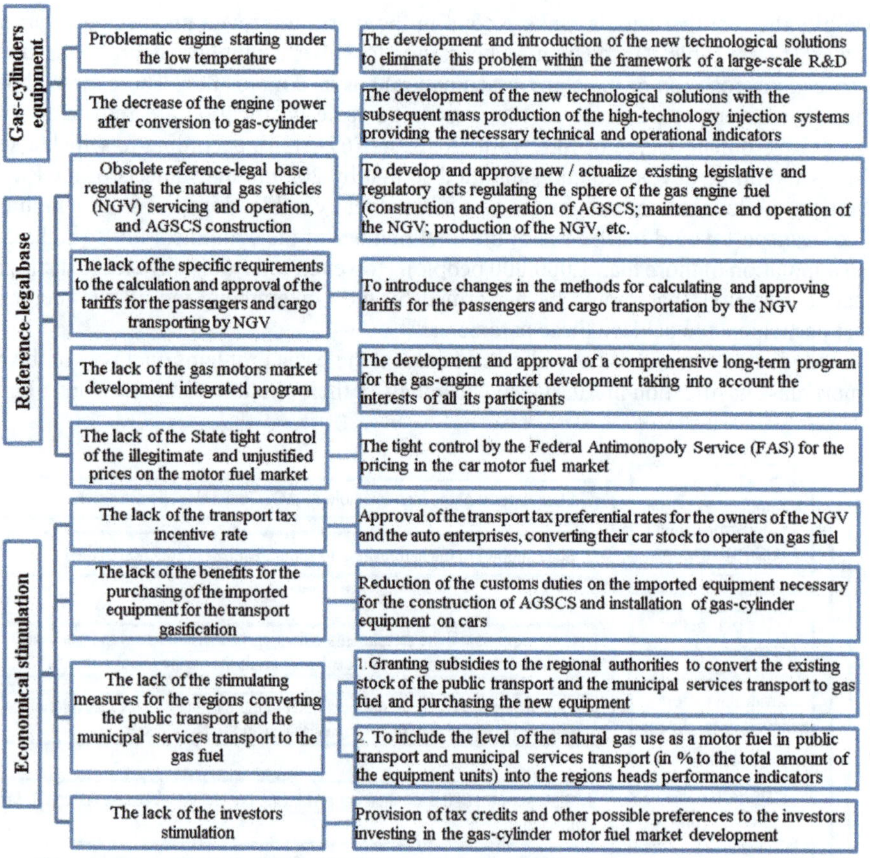

Figure 6: The complex of the additional problems of the vehicles mass gasification in Russia and their solution.

To implement the “modal shift”, it is necessary to stimulate the population to abandon a personal car in favour of public transport. To solve this problem, a well-developed infrastructure of public transport is needed, employing the increase in the traffic frequency and the number of vehicles, the increase in the routes number, the improvement of the transport comfort, and its speed increase.

The consumption of gas fuel, taking into account the significant disposable reserves of the natural gas, the potential of the associated petroleum gas processing, and the complex environmental situation in Russia, should increase. The development of the gas-cylinder fuel market in the country is possible only with substantial state support. The significant dependence of vehicle gasification level on State support is evidenced by the accumulated experience of the countries actively developing in this direction.

Considering the high complexity of solving the problem of the significant share of replacing traditional non-renewable energy resources with new sources of energy in the motor transport sector of Russia, it is necessary to integrate the efforts of all the participants in this matter: the scientists, the government, the society, the automakers, and the fuel producers to resolve it.

6 CONCLUSION

In Russia, under the existing conditions of the country's high rate of vehicle-amount growth and the reduction of oil reserves, it is necessary to implement a set of measures to improve the fuel efficiency of vehicles with gasoline and diesel engines, as well as to develop alternative motor fuels. Taking into account the high level of natural gas reserves in the Russian Federation, there is a huge potential for switching to the highly ecological gas-cylinder fuel. An integrated solution of the main problems of vehicle gasification by the selected directions will allow a reduction of scarce fuel and energy resources consumption, reducing air pollution and improving the ecological situation in the country.

ACKNOWLEDGEMENT

This research was supported by Act 211 Government of the Russian Federation, contract № 02.A03.21.0006.

REFERENCES

[1] Schiavon, M. et al., Assessing the air quality impact of nitrogen oxides and benzene from road traffic and domestic heating and the associated cancer risk in an urban area of Verona (Italy). *Atmospheric Environment*, **120**, pp. 234–243, 2015.

[2] Golubeva, A., Magaril, E., Magaril, R., Panepinto, D & Viggiano, F., Managing the environmental adaptation of vehicles operation. *WIT Transactions on Ecology and the Environment*, vol. 223, WIT Press: Southampton and Boston, pp. 161–169, 2017.

[3] Yambyshev, F.D. & Shigabutdinov, R.M., Investigating the pollution of the atmosphere by motor transport. *International Journal of Engineering and Technology*, **7**, pp. 231–234, 2017.

[4] Magaril, E., Increasing the efficiency and environmental safety of vehicle operation through improvement of fuel quality. *International Journal of Sustainable Development and Planning*, **10**(6), pp. 880–893, 2015.

[5] Magaril, E., Improvement of the environmental and operational characteristics of vehicles through decreasing the motor fuel density. *Environmental Science and Pollution Research*, **23**(7), pp. 6793–6802, 2016.

[6] Anisimov, I., Ivanov, A., Chikishev, E., Chainikov, D. & Reznik, L., Assessment of gas cylinder vehicles adaptability for operation at low ambient temperature conditions. *WIT Transactions on Ecology and the Environment*, **190**(1), pp. 685–695, 2014.

[7] Trofimenko, Y.V., Grigoreva, T.Y. & Evgenev, G.I., Energy-saving problems of road facilities in Russia. *WIT Transactions on Ecology and the Environment*, vol. 190, WIT Press: Southampton and Boston, pp. 535–542, 2014.

[8] Magaril, E., Magaril, R., Al-Kayiem, H.H., Skvortsova, E., Anisimov, I. & Rada, E.C., Investigation on the possibility of increasing the environmental safety and fuel efficiency of vehicles by means of gasoline nano-additive. *Sustainability*, **11**(7), 2019.

[9] Raslavičius, L., Keršysa, A., Mockus, S., Keršienė, N. & Starevičius, M., Liquefied petroleum gas (LPG) as a medium-term option in the transition to sustainable fuels and transport. *Renewable and Sustainable Energy Reviews*, **32**, pp. 513–525, 2014.

[10] Johnson, E., LPG: a secure, cleaner transport fuel? A policy recommendation for Europe. *Energy Policy*, **31**, pp. 1573–1577, 2003.

[11] Masi, M., Experimental analysis on a spark ignition petrol engine fuelled with LPG (liquefied petroleum gas). *Energy*, **41**, pp. 252–260, 2012.

[12] Streimikiene, D., Baležentis, T. & Baležentienė, L., Comparative assessment of road transport technologies. *Renewable and Sustainable Energy Reviews*, **20**, pp. 611–618, 2013.

[13] MacLean, H.L. & Lave, L.B., Evaluating automobile fuel/propulsion system technologies. *Progress in Energy and Combustion Science*, **29**(2003), pp. 1–69, 2003.
[14] Frick, M., Axhausen, K.W., Carle, G. & Wokaun, A., Optimization of the distribution of compressed natural gas (CNG) refueling stations: Swiss case studies. *Transportation Research Part D: Transport and Environment*, **12**, pp. 10–22, 2007.
[15] Hekkert, M.P., Hendriks, F.H.J.F., Faaij, A.P.C. & Neelis, M.L., Natural gas as an alternative to crude oil in automotive fuel chains well-to-wheel analysis and transition strategy development. *Energy Policy*, **33**, pp. 579–594, 2005.
[16] Aslam, M.U., Masjuki, H.H., Kalam, M.A., Abdesselam, H., Mahlia, T.M.I. & Amalina, M.A., An experimental investigation of CNG as an alternative fuel for a retrofitted gasoline vehicle. *Fuel*, **85**, pp. 717–724, 2006.
[17] De Carvalho Jr., A.V., Natural gas and other alternative fuels for transportation purposes. *Energy*, **10**, pp. 187–215, 1985.
[18] United Nations (UN), World population prospects: The 2015 revision, key findings and advance tables. UN report, p. 1, 2015.
[19] UN, Department of Economic and Social Affairs website. www.un.org/development/desa/en/.
[20] British Petroleum (BP), Investor tools. http://tools.bp.com/energy-charting-tool.aspx#/st/carbon/dt/emissions/unit/MTCO2/region/NOA/SCA/EU/MIE/AFR/AP/view/map/.
[21] BP, Statistical review of world energy. www.bp.com/en/global/corporate/energy-economics/statistical-review-of-world-energy/downloads.html.
[22] BP, Energy outlook 2035. www.bp.com/en/global/corporate/energy-economics/energy-outlook-2035.html.
[23] Golubeva, A. & Magaril, E., Improved economic stimulation mechanism to reduce vehicle CO_2 emissions. *WIT Transactions on the Built Environment*, vol. 130, WIT Press: Southampton and Boston, pp. 485–489, 2013.
[24] Federal State Statistics Service, www.gks.ru.
[25] RIA Novosti, The number of cars in Russia exceeded 56 million, 2016. http://ria.ru/society/20160220/1377940767.html.
[26] Golubeva, A. & Magaril, E., Economic stimulation to decrease the CO_2 emission by the motor transport, *UrFU Bulletin. Economy and Management Series*, **15**(3) pp. 359–381, 2016. (In Russian.)
[27] Golubeva, A. & Magaril, E., Environmental tax as an instrument of economic stimulation to improve the quality of motor fuels. *WIT Transactions on Ecology and the Environment*, vol. 192, WIT Press: Southampton and Boston, pp. 149–159, 2015.
[28] Mayburov, I. & Leontyeva, Y., Transport tax in Russia as a promising tool for the reduction of airborne emissions and the development of the road network. *WIT Transactions on Ecology and the Environment*, vol. 198, WIT Press: Southampton and Boston, pp. 391–401, 2015.
[29] Mayburov, I. & Leontyeva, Y., Reducing the negative impact of motor transport on the environment: Prospects for the use of fiscal instruments in Russia. *WIT Transactions on Ecology and the Environment*, vol. 186, WIT Press: Southampton and Boston, pp. 863–874, 2014.
[30] Magaril, E.R., Abrzhina, L.L. & Golubeva, A.S., Ecological and economic problems and prospects of using the fuel and energy resources. *UrFU Bulletin. Series: Economy and management*, vol. 5, pp. 114–130. (In Russian.)
[31] Huang, Y., Surawski, N.C., Organ, B., Zhou, J.L., Tang, O.H.H. & Chan, E.F.C., Fuel consumption and emissions performance under real driving: Comparison between

hybrid and conventional vehicles. *Science of the Total Environment 2019*, **659**, pp. 275–282.

[32] Al-Kayiem, H.H., Wahhab, H.A.A., Magaril, E. & Aziz, A.R.A., Performance and emissions investigation of a single cylinder diesel engine using enhanced blend biodiesel by nanoparticles. *AIP Conference Proceedings 2018*, **2035**, 2018.

[33] Garant, Russian Government decree of May 13th, 2013, № 767-r on the regulation of the relations in the sphere of the gas motor fuel use, including natural gas, as a motor fuel. http://base.garant.ru/70376264/#ixzz4bNyL5R1x. Accessed on: 22 Jun. 2018.

CHILDREN'S EXPOSURE TO INDOOR AIR IN SCHOOLS: IMPACT ON WHEEZING

JULIANA P. SÁ, PEDRO T. B. S. BRANCO, MARIA C. M. ALVIM-FERRAZ, FERNANDO G. MARTINS & SOFIA I. V. SOUSA
Laboratory for Process Engineering, Environment, Biotechnology and Energy (LEPABE), Faculty of Engineering, University of Porto, Portugal

ABSTRACT

Wheezing is a common symptom in childhood and has been associated with air pollution. Children spend a large part of their time in school, this being the most important indoor environment apart from home. However, studies on the impact of children's indoor air pollution exposure at schools on respiratory health are scarce. Thus, this study aimed to assess the impact of children's exposure to indoor air pollution in a total of five urban nursery and primary schools on active wheezing. Multivariate logistic regression models were used to estimate the associations, adjusted for sex, age group (pre-school/primary school) and parental history of asthma. A microenvironmental modelling approach was used to estimate indoor air pollution exposure to each of the pollutants exceeding legislation limit values (CO_2, formaldehyde and $PM_{2.5}$), as the sum of the product of time spent by the child in different indoor school microenvironments and the time-averaged concentration measured in each microenvironment. Measurements were performed in 11 classrooms, two bedrooms and two canteens in Porto, Portugal. A total of 164 completed parent-reported questionnaires derived from the International Study of Asthma and Allergies in Childhood allowed to identify active wheezing (at least one wheeze episode in the previous 12 months) in 16.5% of the studied children. Although not statistically significant, the studied children's exposure to indoor air pollution in nursery and primary schools seemed to be associated with an increase in the odds of having active wheezing especially for $PM_{2.5}$ (OR = 1.57, p-value = 0.675). These results highlight the importance of applying indoor air pollution mitigation measures in nursery and primary schools. The impacts of those measures, on both indoor air quality and children's respiratory health, should be evaluated in future studies.

Keywords: indoor air pollution, nursery and primary schools, children, wheezing.

1 INTRODUCTION

Children constitute a sensitive population to environmental contaminants' exposure, including indoor air pollutants, since their lungs and immune system are still under development and they have a relatively higher amount of air inhalation. Effectively, evidence has been made that indoor air pollution (IAP) impacts children's respiratory health [1].

Wheezing is a very common health respiratory symptom on childhood and one of the most common causes of morbidity and hospitalization among infants and young children, and respiratory infections [2]. Approximately 25–30% of infants have at least one wheezing episode, and nearly half of the children have a history of wheezing by six years of age [3]. Wheezing is often considered the expression of an acute infection, being also frequently associated with other respiratory diseases [4]. In fact, wheezing could be an indicator of asthma and chronic obstructive pulmonary disease, being these the most common causes of wheezing [5]. Besides, asthma and wheezing are among the most frequent reasons for children's visits to paediatricians [6].

Apart from home, school is the most important indoor environment for children, in where they spend a great part of their day. However, indoor air quality (IAQ) in schools has been less studied than in other buildings, and consequently the attention given to the adverse children's health effects in these environments is scarce [1]. Nevertheless, in Portugal, indoor air quality limit values are defined for new and existent commercial and services buildings,

WIT Transactions on Ecology and the Environment, Vol 236,
www.witpress.com, ISSN 1743-3541 (on-line)
doi:10.2495/AIR190201

including schools, for pollutants such as particulate matter ($PM_{2.5}$ and PM_{10}), carbon dioxide, carbon monoxide, volatile organic compounds, formaldehyde and radon [7].

Existing literature focused on the association between children's exposure to indoor air pollutants and respiratory health is also limited, and mostly focusing on asthma, thus neglecting childhood wheezing although it is more common and easier to diagnose [8]–[13].

Moreover, information on the prevalence of childhood wheezing covering Portuguese children remains scarce. Furthermore, the impact of exposure to IAP in nursery and primary schools on childhood wheezing has not been extensively studied. Thus, this study mainly aimed to assess the impact of children's exposure to IAP in urban nursery and primary schools on active wheezing.

2 METHODOLOGY

The study population consisted of pre-schoolers (3–5 years old) and primary school children (6–10 years old) attending three nurseries and two primary schools involved in the INAIRCHILD project [14]. These schools were located in urban context in Porto district, in the north of Portugal (41°N, 8°W).

According to Branco et al. [15], a microenvironmental modelling approach seemed to be the best methodology to assess individual children's exposure to air pollution, whereby it was adopted for this study. Based on the concepts of calculating exposure introduced by Fugas [16], Duan [17], [18] and Ott [19], the present work resorted to eqn (1) to estimate individual children's exposure

$$E_i = \sum_{j=1}^{J} C_{ij} t_{ij}, \tag{1}$$

where E_i is the exposure of the ith individual, C_{ij} is the concentration of the pollutant measured in the jth microenvironment of the ith individual, t_{ij} is the time spent by the ith individual in the jth microenvironment, and J is the number of different microenvironments (ME) considered.

The adopted approach was based on both continuous monitoring of indoor air pollutants in the distinct indoor ME and data of time–location patterns for a day (daily exposure). Therefore, data from daily time–location patterns, i.e. the time spent by each individual (child) in each indoor ME (classrooms, bedrooms and canteens) was obtained from the school timetable and validated by the educator/teacher of the class. In turn, indoor continuous sampling of CO_2, formaldehyde and $PM_{2.5}$ were carried out during occupancy periods in 11 classrooms, two bedrooms and two canteens, and hourly mean values were calculated. Detailed characterization of sampling and IAQ of these ME was previously described [20] and conclusions were that CO_2, formaldehyde and $PM_{2.5}$ have exceeded Portuguese legislation limit values [7], thus they were selected to integrate the IAP exposure evaluation in the present study. After being calculated, IAP exposure was dichotomized in "under" or "above" limit values by considering the limit values in the Portuguese legislation for IAQ as the cut-off, namely: (i) 2250 mg/m^3 for CO_2 (plus 30% of margin of tolerance as no mechanical ventilation system was working in the room); (ii) 100 µg/m^3 for formaldehyde; and (iii) 25 µg/m^3 for $PM_{2.5}$ (plus 100% of margin of tolerance as no mechanical ventilation system was working in the room), for the association with active wheezing.

A total of 164 ISAAC-derived questionnaires (International Study of Asthma and Allergies in Childhood derived questionnaires) were completed by parents or guardians of children attending the studied ME, accepted to participate in the study and signed an informed consent according to the Helsinki Declaration developed by the World Medical Association, allowing collecting individual information of sex, age, parental history of asthma, and respiratory symptoms, namely wheezing and dyspnoea. Active wheezing (reporting at least one wheeze episode in the previous 12 months) was considered the main health outcome in this study.

Moreover, this study was approved by both the Ethics Commission of Universidade do Porto and the Ethics Commission for Health of Centro Hospitalar Universitário de São João, Porto.

Descriptive statistics were used to express the characteristics of individuals, IAP exposure and health outcomes. Multivariate logistic regression models were used to estimate the odds ratios (OR) of the associations between IAP exposure and active wheezing, adjusted for age group (pre-schoolers or primary school children), sex and parental history of asthma. Statistical analyses were performed with R software version 3.5.3 [21]. The level of statistical significance was set at 0.05.

3 RESULTS AND DISCUSSION

3.1 Study site and population

Children's daily indoor exposures to CO_2, formaldehyde and $PM_{2.5}$ were summarized (minimum, median, mean, maximum and interquartile range (IQR)) in Table 1.

Table 1: Summary of children daily exposures to CO_2, formaldehyde and $PM_{2.5}$ in the studied microenvironments.

	Minimum	Median	Mean	Maximum	IQR
CO_2 (mg/m^3)	1706	2641	2579	3877	487.4
Formaldehyde (µg/m^3)	19.96	64.58	71.67	277.7	73.38
$PM_{2.5}$ (µg/m^3)	20.52	32.56	41.64	67.88	38.10

IQR = interquartile range.

When comparing to the limit values of the Portuguese legislation for IAQ, the majority of the individuals were exposed to indoor air pollutants above that threshold. In fact, the worst case was $PM_{2.5}$ (92.4% of individuals exposed to concentrations above the threshold), followed by CO_2 (61.0%) and finally formaldehyde (26.8%).

Characteristics of the studied population (mean age 5.4 years old), including respiratory symptoms, were summarized in Table 2.

The study population was gender balanced, including exactly the same number of male and female children. The number of children not born in Portugal was residual. Important to consider is the number of the studied children having at least one asthmatic parent (23.9%). The prevalence of ever wheezing and dyspnoea (at least once in lifetime) was 28.7% and 8.5%, respectively, while the prevalence of active wheezing and dyspnoea (at least one episode in the previous year) was 16.4% and 5.5%, respectively. Effectively, in both ever and previous year, the prevalence of wheezing was higher than the prevalence of dyspnoea, highlighting the importance of that respiratory outcome. From those reporting active wheezing (16.4%), one to three diurnal attacks in the previous year were frequently reported (15.2%), while only 1.2% reported more than 12 diurnal attacks in the same period. On the other hand, nocturnal episodes of wheeze were not so common as diurnal, as the majority of those reporting active wheezing did not present any nocturnal attack (9.1%), while 3.0% and 4.3% reported having less than one and at least one nocturnal attack of wheezing per week, respectively. Exercise induced and speech-limiting wheeze attacks were not so common, being presented in only 2.4% and 1.8% of the study population, respectively.

Table 2: Characterization of study population, including respiratory heath symptoms (with respective 95% confidence intervals).

	%	95% Confidence interval
Age group		
Pre-schooler	61.6	54.1–69.0
Primary school children	38.4	31.0–45.9
Sex		
Male	50.0	42.3–57.7
Female	50.0	42.3–57.7
Born in Portugal		
Yes	99.4	98.2–100.0
No	0.6	0.0–1.8
Asthmatic parent		
No	76.1	69.5–82.6
Yes	23.9	17.4–30.5
Ever symptoms (lifetime prevalence)		
Wheeze	28.7	21.7–35.6
Dyspnoea	8.5	4.3–12.8
Active symptoms (last year)		
Wheeze	16.4	10.8–22.1
Dyspnoea	5.5	2.0–9.0
Prevalence of wheeze in the last year		
Attacks number		
None	0.0	0.0–0.0
1 to 3	15.2	9.7–20.7
4 to 12	0.0	0.0–0.0
>12	1.2	0.0–2.9
Nocturnal attacks number		
None	9.1	4.7–13.6
< 1 night/week	3.0	0.4–5.7
≥ 1 night/week	4.3	1.2–7.4
Exercise induced	2.4	0.1–4.8
Speech-limiting attacks	1.8	0.0–3.9

Active wheezing prevalence in the present study could be compared with the prevalence reported in previous studies as they used a similar methodology (data reported in ISAAC-derived questionnaires). In primary school children from Viseu, Portugal, Martins et al. [22] reported a slightly lower active wheezing prevalence (11.7%) when compared with that of the present study. Also for primary school children (aged 7–9 years old) from two municipalities in Western Cape Province, South Africa, Olaniyan et al. [8] reported a slightly lower prevalence of active wheezing (12.9%) than in the present study. In another study in Portugal, Branco et al. [12] reported an active wheezing prevalence of 17.1% for children with less than 5 years old attending nursery schools in Porto and Bragança districts, which was very similar to the prevalence calculated in the present study. On the other hand, Norbäck et al. [23] reported an active wheezing prevalence of 19.7% in pre-schoolers (aged 3–6 years old) from day care centres in seven cities of China, which was slightly higher than that calculated in the present study.

3.2 Modelling associations between IAP exposure and active wheezing

Table 3 summarizes the results from the multivariate logistic regression models representing the association between exposure to CO_2, formaldehyde and $PM_{2.5}$, and active wheezing, namely odds ratios (OR), correspondent 95% confidence intervals (95% CI) and *p*-values. A separate model was built for each indoor air pollutant studied. Crude (unadjusted) and adjusted models were obtained.

Table 3: Crude and adjusted odds ratios (OR and aOR), and respective 95% confidence intervals (95% CI) and significance (*p*-values) of the associations between exposure to CO_2, formaldehyde and $PM_{2.5}$, and active wheezing.

	crude OR (95% CI)	**crude *p*-value**	**aOR (95% CI)**	**adjusted *p*-value**
Exposed to CO_2	0.83 (0.36–1.95)	0.676	0.71 (0.29–1.73)	0.449
Age group: Pre-schooler	3.08 (1.10–8.66)	**0.033***	2.79 (0.96–8.07)	**0.044***
Sex: Male	2.64 (1.08–6.49)	**0.034***	2.27 (0.90–5.70)	0.074
Asthmatic parent: Yes	1.89 (0.76–4.66)	0.168	1.85 (0.72–4.73)	0.207
Exposed to formaldehyde	1.30 (0.52–3.24)	0.580	1.09 (0.41–2.90)	0.861
Age group: Pre-schooler	3.08 (1.10–8.66)	**0.033***	2.61 (0.89–7.63)	0.064
Sex: Male	2.64 (1.08–6.49)	**0.034***	2.25 (0.90–5.64)	0.076
Asthmatic parent: Yes	1.89 (0.76–4.66)	0.168	1.82 (0.71–4.70)	0.223
Exposed to $PM_{2.5}$	1.66 (0.20–13.9)	0.640	1.57 (0.17–14.4)	0.675
Age group: Pre-schooler	3.22 (1.10–9.47)	**0.033***	2.49 (0.81–7.67)	0.097
Sex: Male	3.59 (1.22–10.6)	**0.020***	3.07 (1.01–9.31)	**0.037***
Asthmatic parent: Yes	1.23 (0.43–3.49)	0.702	1.07 (0.36–3.20)	0.906

95% CI = 95% of confidence interval; OR = odd ratio; aOR = adjusted odd ratio; *Significant at p-value < 0.05 calculated by likelihood ratio (LR) test.

Although not statistically significant, children's exposure to indoor $PM_{2.5}$ in nursery and primary schools seemed to be associated with an increase in the odds of having active wheezing in childhood (aOR = 1.57, *p*-value = 0.675). Although less significant, similar results were found for formaldehyde exposure (aOR = 1.09, *p*-value = 0.861). On the other hand, although also not statistically significant CO_2 presented an opposite behaviour (aOR = 0.71, *p*-value = 0.449), which was expected as CO_2 is not an air pollutant per se, although it is usually used as a global indicator of air change rate (ventilation).

Previous studies in the literature reported that PM has a negative impact on human respiratory health, namely in children [24], [25]. Moreover, short-term $PM_{2.5}$ exposures in schools were linked to increased hospital admissions and emergency department visits for respiratory effects, particularly wheezing [10]. In all models, being a male child, a pre-schooler or having at least one asthmatic parent increased the odds of having active wheezing.

These results were in accordance with some previous studies in the literature, namely with a study in primary school children (aged 9–11 years old) from Malta that also reported a direct association between wheezing and high exposure to indoor formaldehyde and $PM_{2.5}$ [26]. Rawi et al. [10] stated significant associations between wheezing and indoor $PM_{2.5}$ concentrations in preschools (aged 5–6 years old) from Selangor, Malaysia and no significant association between indoor CO_2 concentration and respiratory symptoms. Another study performed both in primary schools and homes of schoolchildren living in Malaysia (Kuala Lumpur and Negeri Sembilan), reported no significant associations between exposure to indoor $PM_{2.5}$ and current wheezing [9]. A Portuguese study in public primary schools also located in Porto concluded that higher exposure to $PM_{2.5}$ increased the odds of wheeze among children, and that indoor exposure to formaldehyde was related with wheeze in the past year [11]. Another Portuguese study performed in day care centres from Porto and Lisbon (mean age of participants: 3.1 years old) described a significant relation between CO_2 concentration and reported wheezing in the previous 12 months [27]. However, these comparisons with other previous studies should be interpreted with care, mainly because some of those considered IAP concentrations rather than IAP exposures, and others were carried out also in other microenvironments rather than only nursery and/or primary schools (e.g. home).

The main strength of the present study was the use of a ME modelling approach to estimate children's individual exposure to IAP in nursery and primary schools. On the other hand, a limited number of nursery and primary schools, and consequently a limited number of participants (children) were the main limitations of this study.

4 CONCLUSIONS

As far as the author's knowledge goes, this was the first study assessing the impact of children's exposure to IAP (CO_2, formaldehyde and $PM_{2.5}$) in urban nursery and primary schools on active wheezing, using a ME modelling approach to estimate IAP exposure.

This study concluded that children was very often exposed to levels of IAP higher than legislated inside nursery and primary schools, namely $PM_{2.5}$, CO_2 and formaldehyde. The prevalence of childhood active wheezing was 16.4%, in accordance with previous studies in the literature for similar study populations. Although not statistically significant, the studied children's exposure to IAP in nursery and primary schools, especially $PM_{2.5}$ (OR = 1.57, p-value = 0.675), seemed to be associated with an increase in the odds of having active wheezing. These results highlighted the importance of applying IAP mitigation measures in nursery and primary schools to reduce its levels, which is expected to mitigate the negative effects on children's respiratory health.

Future developments include the enrolment of more schools and participants (children), as well as to extend the analysis to physician diagnosed respiratory health outcomes, namely asthma. Moreover, the impact of mitigation measures should also be evaluated in future studies.

ACKNOWLEDGEMENTS

This work was financially supported by: project UID/EQU/00511/2019 – Laboratory for Process Engineering. Environment. Biotechnology and Energy (LEPABE) funded by national funds through FCT/MCTES (PIDDAC); project PTDC/SAU-SAP/121827/2010 funded by FCT, COMPETE, QREN and EU; and project POCI-01-0145-FEDER-032391, funded by FEDER funds through COMPETE2020 – Programa Operacional Competitividade e Internacionalização (POCI) and by national funds (PIDDAC) through FCT/MCTES.

REFERENCES

[1] Annesi-Maesano, I., Baiz, N., Banerjee, S., Rudnai, P., Rive, S. & Group, S., Indoor air quality and sources in schools and related health effects. *Journal of Toxicology and Environmental Health, Part B*, **16**(8), pp. 491–550, 2013.

[2] Emenius, G., Svartengren, M., Korsgaard, J., Nordvall, L., Pershagen, G. & Wickman, M., Indoor exposures and recurrent wheezing in infants: A study in the BAMSE cohort. *Acta Paediatrica*, **93**(7), pp. 899–905, 2004.

[3] Weiss, L.N., The diagnosis of wheezing in children. *American Family Physician*, **77**(8), pp. 1109–1114, 2008.

[4] Bousquet, J. & Kaltaev, N., *Global Surveillance, Prevention and Control of Chronic Respiratory Diseases: A Comprehensive Approach*, World Health Organization, 2007.

[5] Ng, M.C.W. & How, C.H., Recurrent wheeze and cough in young children: Is it asthma? *Singapore Medical Journal*, **55**(5), pp. 236–241, 2014.

[6] Hu, F.B., Persky, V., Flay, B.R., Zelli, A., Cooksey, J. & Richardson, J., Prevalence of asthma and wheezing in public schoolchildren: Association with maternal smoking during pregnancy. *Annals of Allergy, Asthma and Immunology*, **79**(1), pp. 80–84, 1997.

[7] Portaria number 353-A/2013, Legislation from Ministérios do Ambiente, Odenamento do Território e Energia, da Saúde e da Solidariedade, Emprego e Segurança Social – Diário da República – 1ª Série, N°. **253**(9), pp. 6644(2)–6644, 2013.

[8] Olaniyan, T. et al., Asthma-related outcomes associated with indoor air pollutants among schoolchildren from four informal settlements in two municipalities in the Western Capc Province of South Africa. *Indoor Air*, **29**(1), pp. 89–100, 2019.

[9] Abidin, Z.E., Semple, S., Rasdi, I., Ismail, S.N.S. & Ayres, J.G., The relationship between air pollution and asthma in Malaysian schoolchildren. *Air Quality, Atmosphere and Health*, 7(4), pp. 421–432, 2014.

[10] Rawi, N.A.M.N., Jalaludin, J. & Chua, P.C., Indoor air quality and respiratory health among Malay preschool children in Selangor. *BioMed Research International*, 2015.

[11] Madureira, J. et al., Indoor air quality in schools and its relationship with children's respiratory symptoms. *Atmospheric Environment*, **118**, pp. 145–156, 2015.

[12] Branco, P.T.B.S. et al., Asthma prevalence and risk factors in early childhood at Northern Portugal. *Revista Portuguesa de Pneumologia*, **22**(3), pp. 146–150, 2016.

[13] Raaschou-Nielsen, O. et al., Long-term exposure to indoor air pollution and wheezing symptoms in infants. *Indoor Air*, **20**(2), pp. 159–167, 2010.

[14] Sousa, S.I.V., Ferraz, C., Alvim-Ferraz, M.C.M., Vaz, L.G., Marques, A.J. & Martins, F.G., Indoor air pollution on nurseries and primary schools: impact on childhood asthma: Study protocol. *BMC Public Health*, **12**(1), p. 435, 2012.

[15] Branco, P.T.B.S., Alvim-Ferraz, M.C.M., Martins, F.G. & Sousa, S.I.V., The microenvironmental modelling approach to assess children's exposure to air pollution: A review. *Environmental Research*, **135**, pp. 317–332, 2014.

[16] Fugas, M., Assessment of total exposure to an air pollutant. *Proceeding of the International Conference on Environmental Monitoring*, Las Vegas Nevada Institute of Electrical and Electronic Engineers Inc., New York, 1975.

[17] Duan, N., Micro-environment types: A model for human exposure to air pollution. Stanford University, Department of Statistics, 1981.

[18] Duan, N., Models for human exposure to air pollution. *Environment International*, **8**(1–6), pp. 305–309, 1982.

[19] Ott, W.R., Concepts of human exposure to air pollution. *Environment International*, 7(3), pp. 179–196, 1982.

[20] Sá, J., Branco, P., Alvim-Ferraz, M., Martins, F. & Sousa, S., Evaluation of low-cost mitigation measures implemented to improve air quality in nursery and primary schools. *International Journal of Environmental Research and Public Health*, **14**(6), p. 585, 2017.

[21] R Core Team, R: A language and environment for statistical computing. R Foundation for Statistical Computing, Vienna, Austria. www.R-project.org/. Accessed on: Apr. 2019.

[22] Martins, P. et al., Airways changes related to air pollution exposure in wheezing children. *European Respiratory Journal*, **39**(2), pp. 246–253, 2012.

[23] Norbäck, D. et al., Sources of indoor particulate matter (PM) and outdoor air pollution in China in relation to asthma, wheeze, rhinitis and eczema among pre-school children: Synergistic effects between antibiotics use and PM_{10} and second hand smoke. *Environment International*, **125**, pp. 252–260, 2019.

[24] Jones, A.P., Indoor air quality and health. *Atmospheric Environment*, **33**(28), pp. 4535–4564, 1999.

[25] Nandasena, S., Wickremasinghe, A.R. & Sathiakumar, N., Indoor air pollution and respiratory health of children in the developing world. *World Journal of Clinical Pediatrics*, **2**(2), pp. 6–15, 2013.

[26] Fsadni, P., Bezzina, F., Fsadni, C. & Montefort, S., Impact of school air quality on children's respiratory health. *Indian Journal of Occupational and Environmental Medicine*, **22**(3), pp. 156–162, 2018.

[27] Carreiro-Martins, P. et al., CO_2 concentration in day care centres is related to wheezing in attending children. *European Journal of Pediatrics*, **173**(8), pp. 1041–1049, 2014.

HEALTH AND ECONOMIC IMPACTS OF OZONE SHIP-RELATED AIR POLLUTION IN PORTUGAL

RAFAEL A. O. NUNES[1], MARIA C. M. ALVIM-FERRAZ[1], FERNANDO G. MARTINS[1], JUKKA-PEKKA JALKANEN[2], HANNA HANNUNIEMI[2] & SOFIA I. V. SOUSA[1]
[1]LEPABE – Laboratory for Process, Environment, Biotechnology and Energy, Faculty of Engineering, University of Porto (FEUP), Portugal
[2]Finnish Meteorological Institute, Finland

ABSTRACT

Air pollution is the leading cause of the global burden of disease from the environment, entailing substantial economic consequences. International shipping is a significant source of NO_x, SO_2, CO and PM, which contributes to the increase in O_3 levels that can cause known negative health impacts. Thus, this study aimed to estimate the health impacts of ozone ship-related air pollution in Portugal in 2015. To determine the health-related effects of this sector, shipping emissions were obtained from an Automatic Identification System based emission inventory using the Ship Traffic Emission Assessment Model (STEAM). Their contributions to SOMO35 (sum of ozone daily 8-h maximum means over 35 ppb in the calendar year, expressed in ppb per day) levels in Portugal were modelled using the EMEP/MSC-W chemistry transport model (simulations with and without shipping emissions). Log-linear functions based on WHO-HRAPIE relative risks for each health endpoint (all-cause and respiratory mortality, and cardiovascular and respiratory hospital admissions) were used to estimate the attributable fractions. Then, the excess burden of disease was calculated by multiplying the attributable fractions with the baseline incidence of each health endpoint and the population at LAU2 level. Costs associated with the health impacts were estimated as the product of the excess burden of disease and its unit health cost value. Shipping emissions contributed to an increase of 21% in the number of deaths for all-cause mortality and respiratory diseases, as well as hospital admissions for cardiovascular and respiratory diseases. A total external cost of around 180 (86–266) M€ a yr^{-1} was estimated for these health endpoints in 2015. These results show that O_3-related air pollution from ships is a considerable problem affecting the Portuguese population.

Keywords: shipping emissions, modelling, air pollution, impact pathway approach, health effects.

1 INTRODUCTION

International shipping has been recognized as a significant source of pollution namely concerning nitrogen oxides (NO_x), sulphur oxides (SO_x), CO and particulate matter (PM), also contributing to the increase of O_3 levels. Nevertheless, the associated effects on premature mortality and morbidity are not yet well documented [1]–[4]. Once 70% of emissions from ships in international routes occur until 400 km from the coast and can be easily transferred hundreds of kilometres towards the mainland, ships have the potential to contribute significantly to air quality degradation and consequent effects on human health in coastal areas, as well as in inner areas [2], [5]. Shipping emissions are regulated by the International Maritime Organization (IMO) according to the Annex VI "Prevention of Air Pollution from Ships" under the International Convention for the Prevention of Pollution from Ships (MARPOL) [6]. For Portugal, air quality standards for ambient air are defined in *Decreto-Lei nº 102/2010*, which imposes limits for human health protection [7].

In the last 25 years, emissions from land-based sources (traffic, heating, industrial production, power generation, etc.) have decreased substantially in Europe due to new and more effective air pollution and climate policies. Emissions of NO_x, SO_x and fine particulate matter have been reduced by about 52%, 83% and 28%, respectively [8], [9]. An opposite behaviour was verified for ship emissions which have been neglected for a long time, making

WIT Transactions on Ecology and the Environment, Vol 236,
www.witpress.com, ISSN 1743-3541 (on-line)
doi:10.2495/AIR190211

shipping one of the least and last regulated anthropogenic emission sources [8], [10], [11]. Therefore, significant progress has been made to study air quality impact of ship emissions all over the world, however only a few also studied their negative effects on human health [1], [3], [12]–[14]. Since studies on the impact of ship emissions on human health are still scarce and Portugal has a very important geo strategic position in international maritime transport, it is relevant to study the impacts of ship emissions on health in this region. Moreover, as far as known there are no studies concerning health impacts of ozone ship-related air pollution in Portugal. Thus, this study aimed to reduce this gap by estimating the health impacts of ozone ship-related air pollution in Portugal for 2015.

2 MATERIALS AND METHODS

2.1 Ship emissions inventory

The shipping emissions inventory of Portugal for 2015 was obtained based on the full bottom-up Ship Traffic Emission Assessment Model (STEAM). STEAM allows to assess emissions from each individual ship combining the Automatic Identification System (AIS) shipping activity information (both terrestrial and satellite-based, provided by Orbcomm) with the detailed technical characteristics of each ship and engine type (collected from various sources and archived for over 90,000 ships) and emission factors of a particular ship type, size, machinery type and operative mode. Based on this information, the model can predict the power consumptions and loads of main and auxiliary engines and the fuel consumption of the ship with reasonableness based on resistance that each ship requires to overcome at a specified speed [15].

2.2 Atmospheric simulations

To evaluate the contributions of shipping emissions on O_3 concentrations over Portugal, atmospheric simulations have been performed with the chemical transport model EMEP/MSC-W. The model was configured for two scenarios, with (w/ship case) and without shipping emissions (wt/ship case) for 2015 with a horizontal resolution of 0.1° x 0.1° (long–lat), vertical structure of the model with 34 layers and a data output time step of 1 h. Grid emissions in the same resolution from other sources divided in 11 SNAP (Selected Nomenclature for Air Pollution) sectors like public power, industry, other stationary combustion, road transport among others were taken from official European emission inventories provided by the Centre on Emission Inventories and Projection (CEIP) that are reported under the Convention on Long-range Transboundary Air Pollution (CLRTAP) and National Emission Ceilings (NEC) Directive [15]. Emissions from shipping sector considered in the CEIP inventory were excluded to perform the simulations. Moreover, contribution to emissions of the dust from Sahara, emissions of NO_x from lightning and emissions from forest and vegetation fires from the "Fire INventory from NCAR version 1.5" [16] were also considered. More details on the model can be found in Simpson et al. [17] and NMI [18]. The meteorological data for 2015 was generated by the state-of-the-art Integrated Forecast System model (IFS) of the European Centre for Medium-Range Weather forecasts (ECMWF).

2.3 Health impact assessment

To assess the excess of health burden attributable to ozone ship-related air pollution SOMO35 indicator (sum of ozone daily 8-h maximum means over 35 ppb in the calendar year, expressed in ppb per day) was used. SOMO35 was applied because it is considered the most robust indicator for O_3 health impact assessments in global model calculations [19]. Log-linear functions (based on WHO-HRAPIE relative risks) for each health endpoint (derived from available epidemiologic studies) were used to estimate the relative risks (RR), as recommended by WHO [20] and other authors [21]–[25]. RR for linear-log functions were estimated using the following equation:

$$RR_{log-linear} = e^{\beta(C-C_0)}, \tag{1}$$

where C is the SOMO35 level (converted to μg m^{-3}), C_0 cut-off value (considered zero in calculations) and β-coefficient relates the change in the RR to a unit change in air pollutant concentration (an increase of the risk per 10 μg m^{-3}).

After RR calculations, the attributable fractions (AF) were calculated following the attributable risk as:

$$AF = (RR - 1)/RR. \tag{2}$$

To estimate the excess burden of disease (EBD) the increment in the number of deaths and additional cases due to ozone ship-related air pollution over 2015 was estimated, using eqn (3):

$$\Delta EBDs = BI \times AF \times Pop, \tag{3}$$

where BI is the baseline incidence of the selected health endpoint for a given population and Pop is the population within the age group of interest.

Data about population by single year of age of Portugal at LAU2 level (civil parish) was obtained from the Eurostat 2011 Census database hub [26].

All-cause mortality baseline rate was obtained from the National Statistical System of Portugal for 2015 (INE, 2018). Respiratory mortality baseline rate was obtained from the European Health for All database (HFA-DB). Data on hospital admissions for respiratory and cardiovascular causes was taken from the OECD Stat database [27]. Health endpoints, baseline incidences, and RR used in this study are listed in Table 1.

2.4 Assessment of socio-economic external costs of the burden of disease

To estimate the economic cost, the burden of disease unit health costs (cost per case of illness) were used. The exposure cost for a particular health endpoint was calculated as the product of the Exposure Response Function (ERF) and its unit health cost value (eqn (4)).

Table 1: Health endpoints, baseline incidences, and RR used in this study.

Health endpoints	Age group	BI per 10^5	RR (95% CI)
All-cause mortality	Adults (age ≥ 30 years)	1,048	1.0029 (1.0014–1.0043)
Respiratory mortality	30 to 44 years	3.0	1.014 (1.005–1.024)
	45 to 59 years	16.4	
	60 to 74 years	76.3	
	75+ years	976.2	
Cardiovascular hospital admissions	>64 years	1,112.4	1.0089 (1.0050–1.0127)
Respiratory hospital admissions	>64 years	919.1	1.0044 (1.0007–1.0083)

BI = baseline incidence; RR = relative risk; CI = confidence interval.

$$Exposure\ cost = (ERF) \times (Cost\ per\ case\ of\ illness\ or\ death). \quad (4)$$

Deaths were valued using the value of statistical life (VSL), which is how much society is willing to pay to avoid an anonymous death [28].

Following the benefit transfer approach that takes into account differences in income levels between two places, country-specific VSL for Portugal in 2015 was estimated using the formula recommended in OECD [29], based on an extensive meta-study performed by OECD [30]. VSL was adjusted according to the eqn (5).

$$VSL\ C\ 2015 = VSL\ EU\ 2005 \times \left(\frac{YC}{YEU}\right)^{\beta} \times (1 + \Delta P + \Delta Y)^{\beta}, \quad (5)$$

where: $VSL\ EU$ 2005 is the VSL of the average of EU27 countries (USD 3.6 million in 2005); YC is the Gross Domestic Product (GDP) per capita at the purchasing power parity (PPP) in 2015; YEU is the average GDP per capita of EU27 countries at PPP in 2015; β is the income elasticity of VSL, which measures the percentage increase in VSL for a percentage increase in income (the value of 0.8 was established by OECD); PPP is the purchasing power parity-adjusted exchange rate in 2005; ΔP is the percentage increase in consumer price from year 2005 to 2015 (measured by consumer price index (CPI) that reflects the inflation or changes in the cost to the average consumer of acquiring a basket of goods and services); and ΔY is the percentage change in real GDP per capita growth from the reference year to 2015 (derived from real GDP per capita annual growth). All values previously described were obtained for the reference years and 2015 from the OECD data portal [31]. Table 2 summarizes the unit values used in this study.

Table 2: Unit cost values for VSL and respiratory and cardiovascular hospital admissions used in this study.

Parameter	Unit value	Units
Value of Statistical Life (VSL)	2.60	M€/death
Respiratory hospital admissions	2 414	€/hospital admission
Cardiovascular hospital admissions	2 414	€/hospital admission

3 RESULTS AND DISCUSSION

Table 3 presents the estimated number of premature deaths and additional cases associated with exposure to ozone ship-related air pollution for 2015 in Portugal, as well as the external costs. Fig. 1(a)–(d) shows the spatial distribution of premature deaths and additional cases.

Additional 69 (95% CI: 33, 102) deaths due to all causes, 34 (95% CI: 12, 57) deaths due to respiratory diseases, as well as 46 (95% CI: 26, 66) and 23 (95% CI: 4, 43) hospital admissions due to cardiovascular and respiratory diseases, respectively, were estimated due to ozone ship-related air pollution for 2015 in Portugal. Overall, shipping emissions contributed to increase in 21% the number of deaths and additional cases of all health endpoints considered. As can be seen from Fig. 1, most mortality and additional cases occurred close to Portugal coastline. These results seemed to be related with a greater susceptibility of these zones to shipping emissions and because those are densely populated areas. It is also important to emphasise that results showed additional cases of hospital admissions due to respiratory diseases in some inner parishes of Portugal. The number of deaths due to respiratory diseases and the number of hospital admissions due to cardiovascular and respiratory diseases were compared with the total number of deaths and

Table 3: Total number of additional cases and external costs in Portugal associated with ozone ship-related air pollution in 2015.

Air pollutant	Health endpoint	Additional deaths/cases (95% CI)	External costs (95% CI) (values in M€)
O_3	All-cause mortality	69 (33–102)	179 (85.7–265)
	Respiratory mortality	34 (12–57)	88.4 (31.2–148)
	Cardiovascular hospital admissions	46 (26–66)	0.111 (0.0628–0.159)
	Respiratory hospital admissions	23 (4–43)	0.0555 (0.096.6–0.104)

CI = confidence interval.

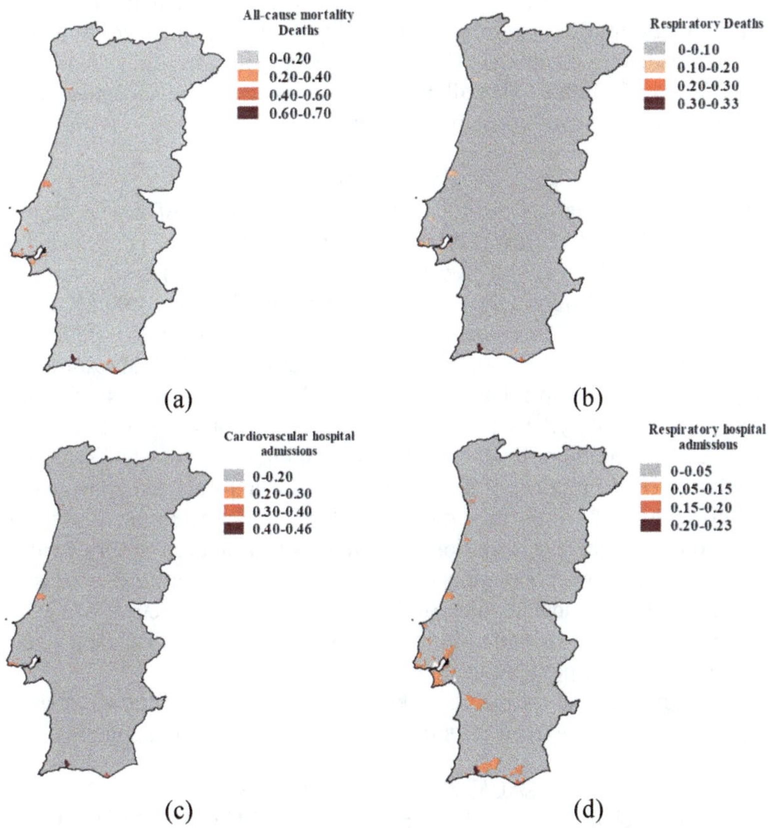

Figure 1: Spatial distribution (at LAU2 level) of: (a) Deaths due to all causes; (b) Deaths due to diseases of respiratory system; (c) Hospital admissions due to diseases of cardiovascular system; and (d) Hospital admissions due to diseases of respiratory system.

additional cases reported for Portugal during 2015. According to the results, the number of deaths due to respiratory diseases and the additional cases due to diseases of cardiovascular and respiratory systems attributable to ozone ship-related exposure represented 0.3%, 0.2% and 0.1% of the total, respectively. Moreover, and more specifically, the number of deaths calculated from all-cause mortality was compared with the number of premature deaths attributable to O_3 exposure during 2015 for Portugal reported on the Air quality in Europe – 2018 report performed by the European Environmental Agency (EEA). Results showed that the number of deaths due to O_3 ship-related exposure calculated in this study represented 23% of the total number of deaths reported by the EEA. Results confirm that O_3 related air pollution from ships seemed to be a considerable problem affecting the Portuguese population, mainly in coastal areas.

Regarding exposure costs, it was estimated 79 M€ (95% CI: 85.7–265 M€) related with the number of additional deaths from all causes, 88.4 M€ (95% CI: 31.2–148 M€) from deaths due to respiratory diseases as well as 0.111 M€ (95% CI: 0.0628–0.159 M€) and 0.0555 M€ (95% CI: 0.096.6 €–0.104 M€) related with additional hospital admissions due to cardiovascular and respiratory diseases, respectively. A total exposure cost of around 180 M€ yr^{-1} (95% CI: 86–266 M€ yr^{-1}) resulting of O_3 related air pollution from ships was estimated in Portugal for 2015 (respiratory mortality could not be added to the overall costs estimate to avoid double counting). The results above described show that human health external costs due to O_3 related air pollution from ships cannot be neglected when the total external costs of shipping emissions are considered.

4 CONCLUSIONS

This study allowed estimating health and economic impacts of ozone ship-related air pollution in Portugal for 2015. It was estimated 69 (95% CI: 33–102) deaths and costs of 79 M€ (95% CI: 85.7–265 M€) due to all causes, 34 (95% CI: 12–57) deaths and costs of 88.4 M€ (95% CI: 31.2–148 M€) from deaths due to respiratory diseases, 46 (95% CI: 2,666) and 23 (95% CI: 4.43) hospital admissions and costs of 0.111 M€ (95% CI: 0.0628–0.159 M€) and 0.0555 M€ (95% CI: 0.096.6 €–0.104 M€) due to cardiovascular and respiratory diseases, respectively, due to ozone ship-related air pollution. A total exposure cost of around 180 M€ yr^{-1} (95% CI: 86 –266 M€ yr^{-1}) was estimated in Portugal for 2015. This study shows that ozone ship-related air pollution increased 21% the number of deaths and cases of all health endpoints. Moreover, the number of deaths due to ozone ship-related exposure calculated in this study represented 23% of the total number of deaths reported by the EEA for Portugal in 2015, which confirms that ship emissions may have a substantial impact on human health, mainly, in areas with high population density and near the Portuguese coastline. These findings contribute to the understanding of the comprehensive impacts of ozone ship-related air pollution, supporting the view that shipping emissions are a significant source air pollution that affects the Portuguese population.

Future developments should include the study of other air pollutants and of different health endpoints, as well as extend the study to the Iberian Peninsula domain.

ACKNOWLEDGEMENTS

This work was financially supported by: project UID/EQU/00511/2019 – Laboratory for Process Engineering, Environment, Biotechnology and Energy – LEPABE funded by national funds through FCT/MCTES (PIDDAC); project "LEPABE-2-ECO-INNOVATION" – NORTE-01-0145-FEDER-000005, funded by Norte Portugal Regional Operational Programme (NORTE 2020), under PORTUGAL 2020 Partnership Agreement, through the European Regional Development Fund (ERDF) and project EMISSHIP (POCI – 01 – 0145 – FEDER – 032201), funded by FEDER funds through COMPETE2020 – Programa Operacional Competitividade e Internacionalização (POCI) and by national funds (PIDDAC) through FCT/MCTES.

REFERENCES

[1] Brandt, J. et al., Assessment of past, present and future health-cost externalities of air pollution in Europe and the contribution from international ship traffic using the EVA model system. *Atmospheric Chemistry Physics*, **13**, pp. 7747–7764, 2013.

[2] Corbett, J.J., Winebrake, J.J., Green, E.H., Kasibhatla, P., Eyring, V. & Lauer, A., Mortality from ship emissions: A global assessment. *Environmental Science & Technology*, **41**, pp. 8512–8518, 2007.

[3] Sofiev, M. et al., Cleaner fuels for ships provide public health benefits with climate tradeoffs. *Nature Communications*, **9**, 2018.

[4] Nunes, R.A.O., Alvim-Ferraz, M.C.M., Martins, F.G. & Sousa, S.I.V., Assessment of shipping emissions on four ports of Portugal. *Environmental Pollution*, **231**(2), pp. 1370–1379, 2017.

[5] Eyring, V. et al., Transport impacts on atmosphere and climate: Shipping. *Atmospheric Environment*, **44**, pp. 4735–4771, 2009.

[6] International Maritime Organization (IMO), Prevention of Air Pollution from Ships. www.imo.org/en/OurWork/Environment/PollutionPrevention/AirPollution/Pages/Air-Pollution.aspx. Accessed on: 11 Apr. 2018.

[7] *Decreto-Lei nº 102/2010*. Diário da República 1.ª série – Nº 186. Ministério do Ambiente e do Ordenamento do Território.

[8] Aulinger, A., Matthias, V., Zeretzke, M., Bieser, J., Quante, M. & Backes, A., The impact of shipping emissions on air pollution in the greater North Sea region – Part 1: Current emissions and concentrations. *Atmospheric Chemistry Physics*, **16**, pp. 739–758, 2016.

[9] European Environment Agency (EEA), *Emissions of the Main Air Pollutants in Europe*, Copenhagen, 2017.

[10] Aksoyoglu, S., Baltensperger, U. & Prévôt, A.S.H., Contribution of ship emissions to the concentration and deposition of air pollutants in Europe. *Atmospheric Chemistry Physics*, **16**, pp. 1895–1906, 2016.

[11] Chen, D. et al., Contribution of ship emissions to the concentration of $PM_{2.5}$: A comprehensive study using AIS data and WRF/Chem model in Bohai Rim Region, China. *Science Total Environment*, **610–611**, pp. 1476–1486, 2018.

[12] Jonson, J.E., Jalkanen, J.-P., Johansson, L., Gauss, M. & Van Der Gon, H.A.C.D., Model calculations of the effects of present and future emissions of air pollutants from shipping in the Baltic Sea and the North Sea. *Atmospheric Chemistry Physics*, **15**, pp. 783–798, 2015.

[13] Antturi, J. et al., Costs and benefits of low-sulphur fuel standard for Baltic Sea shipping. *Journal of Environmental Management*, **184**, pp. 431–440, 2016.

[14] Liu, H. et al., Health and climate impacts of ocean-going vessels in East Asia. *Nature Climate Change*, **6**, pp. 1037–1041, 2016.

[15] EMEP/CEIP, Emissions as used in EMEP models. www.ceip.at/ms/ceip_home1/ceip_home/webdab_emepdatabase/emissions_emepmodels/. Accessed on: 10 Feb. 2019.

[16] Wiedinmyer, C. et al., The Fire INventory from NCAR (FINN): a high resolution global model to estimate the emissions from open burning. *Geoscientific Model Development*, **4**, pp. 625–641, 2011.

[17] Simpson, D. et al., The EMEP MSC-W chemical transport model – Technical description. *Atmospheric Chemistry Physics*, **12**, pp. 7825–7865, 2012.

[18] Norwegian Meteorological Institute (NMI), *EMEP/MSC-W Model Unofficial User's Guide*, 2017.
[19] Faridi, S. et al., Long-term trends and health impact of $PM_{2.5}$ and O_3 in Tehran, Iran, 2006–2015. *Environmental International*, **114**, pp. 37–49, 2018.
[20] World Health Organization (WHO), AirQ+: Software Tool for Health Risk Assessment of Air Pollution. www.euro.who.int/en/health-topics/environment-and-health/air-quality/activities/airq-software-tool-for-health-risk-assessment-of-air-pollution. Accessed on: 25 Feb. 2019.
[21] Li, Y., Henze, D.K., Jack, D. & Kinney, P.L., The influence of air quality model resolution on health impact assessment for fine particulate matter and its components. *Air Quality, Atmosphere & Health*, **9**, pp. 51–68, 2016.
[22] Cárdaba Arranz, M., Muñoz Moreno, M.F., Armentia Medina, A., Alonso Capitán, M., Carreras Vaquer, F. & Almaraz Gómez, A., Health impact assessment of air pollution in Valladolid, Spain. *BMJ Open*, **4**, p. e005999, 2014.
[23] Natalie, M. et al., Urban and transport planning related exposures and mortality: A health impact assessment for cities. *Environmental Health Perspectives*, **125**, pp. 89–96, 2017.
[24] Anenberg, S.C. et al., Survey of ambient air pollution health risk assessment tools. *Risk Analysis*, **36**, pp. 1718–1736, 2016.
[25] Yorifuji, T. et al., Health impact assessment of PM_{10} and $PM_{2.5}$ in 27 southeast and east Asian cities. *Journal of Occupational and Environmental Medicine*, **57**, 2015.
[26] European Statistical System (ESS), 2011 Census Hub, 2011 Census database. https://ec.europa.eu/CensusHub2/query.do?step=selectHyperCube&qhc=false. Accessed on: 25 Feb. 2019.
[27] Organisation for Economic Co-operation and Development (OECD), Health Status. https://stats.oecd.org/index.aspx?queryid=30123#. Accessed on: 25 Feb. 2019.
[28] Organisation for Economic Co-operation and Development (OECD), *Valuing Mortality Risk Reductions in Regulatory Analysis of Environmental, Health and Transport Policies: Policy Implications*, 2011.
[29] Organisation for Economic Co-operation and Development (OECD), *The Cost of Air Pollution Health Impacts of Road Transport*, 2014.
[30] Organisation for Economic Co-operation and Development (OECD), *Mortality Risk Valuation in Environment, Health and Transport Policies*, 2012.
[31] Organisation for Economic Co-operation and Development (OECD), OECD Data. https://data.oecd.org/. Accessed on: 25 Feb. 2019.

REAL-LIFE EMISSION FACTOR ASSESSMENT FOR BIOMASS HEATING APPLIANCES AT A FIELD MEASUREMENT CAMPAIGN IN STYRIA, AUSTRIA

RITA STURMLECHNER[1], CHRISTOPH SCHMIDL[1], ELISA CARLON[1], GABRIEL REICHERT[1], HARALD STRESSLER[1], FRANZISKA KLAUSER[1], JOACHIM KELZ[1], MANUEL SCHWABL[1], BERNADETTE KIRCHSTEIGER[2], ANNE KASPER-GIEBL[2], ERNST HÖFTBERGER[1] & WALTER HASLINGER[1]
[1]BIOENERGY 2020+ GmbH, Fixed Bed Conversion Systems, Austria
[2]Institute of Chemical Technologies and Analytics, Vienna University of Technology, Austria

ABSTRACT

Biomass combustion is a major contributor to ambient air pollution. Thus, knowing the real-life emissions of biomass heating systems is crucial. Within the project Clean Air by biomass a field measurement campaign was conducted. 15 biomass heating appliances were tested in households at the end user according to their usual operation. Emission factors for gaseous and particulate emissions, as well as for the genotoxic and carcinogenic substance benzo(a)pyrene, were evaluated and compared to current proposed European and Austrian emission factors used for emission inventories. Moreover, the shares of particles and benzo(a)pyrene in hot and cooled flue gas were determined. Results showed a high variability of emissions in the field. Highest values and ranges occurred for room heaters (TSPtotal: 226 mg/MJ). Biomass boilers showed clearly lower emission factors (TSPtotal: 184 mg/MJ) in the field than room heaters and also than the proposed European and Austrian emission factors, in many cases. Emission factors for tiled stoves showed a similar trend (TSPtotal: 67 mg/MJ). The share of condensable particles in the flue gas was remarkable. Especially benzo(a)pyrene was found mostly in the condensable fraction of the particles.

Keywords: biomass combustion, field measurement campaign, emission factors, benzo(a)pyrene, condensable particles.

1 INTRODUCTION

Domestic biomass heating appliances are very common in Austria and all over Europe. Thus, knowing their emissions in real life is important, since they contribute to air quality issues. Especially PM emissions from biomass combustion contributes significantly to PM pollution in Europe [1]–[3]. Moreover, focus on the polycyclic aromatic hydrocarbon benzo(a)pyrene (BaP) is given, due to its genotoxic and carcinogenic impact on human health [3], [4]. Real-life emission factors (EF) are used for emission inventories, air quality modelling or the prediction of air pollution impact on human health [5]–[7]. EF can be evaluated in two ways, by either close to real-life lab testing or by field measurements. One example of lab testing is the *beReal* test protocol for firewood room heaters [8]. During field measurements it was shown that this lab test method can reflect real-life situations [9]. Other lab tests showed a high variability of emission results at different close to real-life testing methods [10]–[13].

Close to real-life testing methods include transient conditions, like ignition, preheating or load changes of the heating appliances in order to reflect real-life conditions. Nevertheless, the broad variety of framework conditions (e.g. user influence, chimney design, heat output dimensioning etc.) cannot be considered in harmonized lab tests. Thus, field measurements are required in order to evaluate the broad range of emissions in real life.

In Austria, a comprehensive field measurement campaign was conducted by Spitzer et al. [14] in 1998. 173 biomass heating appliances were tested and average emission factors were evaluated. Within the project *BioMaxEff* [15], 16 newly installed biomass boilers were tested

WIT Transactions on Ecology and the Environment, Vol 236, © 2019 WIT Press
www.witpress.com, ISSN 1743-3541 (on-line)
doi:10.2495/AIR190221

in the field at nominal load and varying load conditions close to real-life operation. EF need to reflect real-life conditions as close as possible. Thus, they have to be updated regularly, since the stock of biomass heating appliances is continuously changing.

In the project *Clean Air by biomass* a field measurement campaign was conducted measuring 15 different biomass heating appliances in the field at the end user. This study presents an overview of gaseous, particulate and BaP emission results and a comparison to current emission factors for Europe and Austria, respectively. Moreover, particle emissions and their chemical composition (BaP) in the hot and undiluted flue gas and in the cooled and diluted flue gas were compared.

2 MATERIAL AND METHODS

Within the *Clean Air by biomass* project a field measurement campaign was conducted, measuring 15 biomass heating appliances in total. These are:

- 6 room heaters (RH)
- 6 biomass boilers (BB)
- 3 tiled stoves (TS)

Table 1 gives an overview of the tested appliances. Their year of construction, nominal heat output, classification according to the respective EN standard and the evaluated emission parameters in the field are given. The appliances were installed in single family or farmhouses. Whereas boilers represented the main heating source, room heaters were used as additional heating source. Tiled stoves were used as both.

2.1 Testing procedure

During field measurements, all tested appliances were operated by the end user according to their usual operation habits. Fuel was also provided by the end user in order to reflect real-life conditions. Measurement of gaseous emissions, temperature and draught conditions of the flue gas was done continuously. Particulate emissions were measured discontinuously in each batch (RH and TS) or test phase (BB). Measurements lasted until the end of a heating cycle. For room heaters and tiled stove this was determined by the end user or at a maximum of three continuous batches. For boilers, three or four test phases were measured to evaluate different combustion phases – at least ignition and full load operation.

2.2 Measurement equipment and set up

For the field measurements two different equipment sets were used. Depending on the local situation in the field (mainly space demand and accessibility), the more extensive or the standard set was chosen (Table 2).

Measurements were done with a logging interval of 1 s for flue gas draught (p), temperature (T) and gaseous emissions, i.e. oxygen (O2), carbon dioxide (CO2), carbon monoxide (CO) and organic gaseous compounds (OGC). Particulate emissions (TSP and TSP40) were either measured according to the standard VDI 2066-1 (Standard set) or according to a new in-house developed TSP sampling procedure (Extensive set). Based on the method of Klauser et al. [16], it was adapted to enable simultaneous measurements of hot and cooled particles with only one suction device (it was necessary to have only one sampling point in the flue duct at the end user in the field). Thereby, a suction tube with a planefilter directly after the sampling nozzle and an additional planefilter after a diluter were used. The

Table 1: Overview of tested appliances.

	Appliance	Year of construc-tion	Nominal heat output	EN standard	Evaluated parameters	Tech-nology
1	Firewood room heater	2013	8 kW	EN 13240	CO, OGC, TSP, TSP40, BaP	RH (n=6)
2	Firewood room heater	2017	8 kW	EN 13240	CO, OGC, TSP, TSP40, BaP	
3	Firewood room heater	~1997	8 kW	ÖNORM M 7520	CO, OGC, TSP, TSP40, BaP	
4	Firewood room heater	2013	7.3 kW	EN 13240	CO, OGC, TSP, TSP40, BaP	
5	Firewood insert	~2010	n.a.	EN 13229	CO, TSP	
6	Firewood cooker	2016	20 kW	EN 12815	CO, OGC, TSP, TSP40, BaP	
7	Wood chip boiler	2001	40 kW	EN 303-5	CO, OGC, TSP, TSP40, BaP	BB (n=6)
8	Wood chip boiler	2001	40 kW	EN 303-5	CO, TSP	
9	Wood chip boiler	~1985	~60 kW	n.a.	CO, TSP	
10	Boiler using saw dust	1962	250 kW	n.a.	CO, OGC, TSP, TSP40, BaP	
11	Boiler using saw dust	2018	350 kW	EN 303-5	CO, OGC, TSP, TSP40	
12	Firewood boiler	2010	35.6 kW	EN 303-5	CO, TSP	
13	Firewood tiled stove	1985	n.a.	n.a.	CO, TSP	TS (n=3)
14	Firewood tiled stove	~1970	n.a.	n.a.	CO, OGC, TSP, TSP40	
15	Firewood slow heat release appliance	2016	n.a.	EN 15250	CO, OGC, TSP, TSP40	

n.a.: not available

Table 2: Overview of measured parameters with the two measurement equipment sets.

Measurement	Measured parameters	Number of measured
Extensive measurement set	CO, CO_2, O_2, OGC, T, p, TSP, TSP40, BaP	BB: n = 3 (4); TS: n = 2; RH: n = 5
Standard measurement set	CO, CO_2, O_2, T, p, TSP,	BB: n = 3 (2); TS: n = 1; RH: n = 1

first one is for the determination of particles in the hot flue gas (TSP) and was heated up continuously to 130°C. The second filter is for the determination of particles in cooled flue gas (<40°C), which includes condensable organic compounds (TSP40). The sum of both gives the total sampled particles (TSPtotal). Fig. 1 shows a scheme of the used suction device. After the measurements the filters were sent to an analyzing lab for the determination of benzo(a)pyrene (BaP) concentrations (Section 2.4).

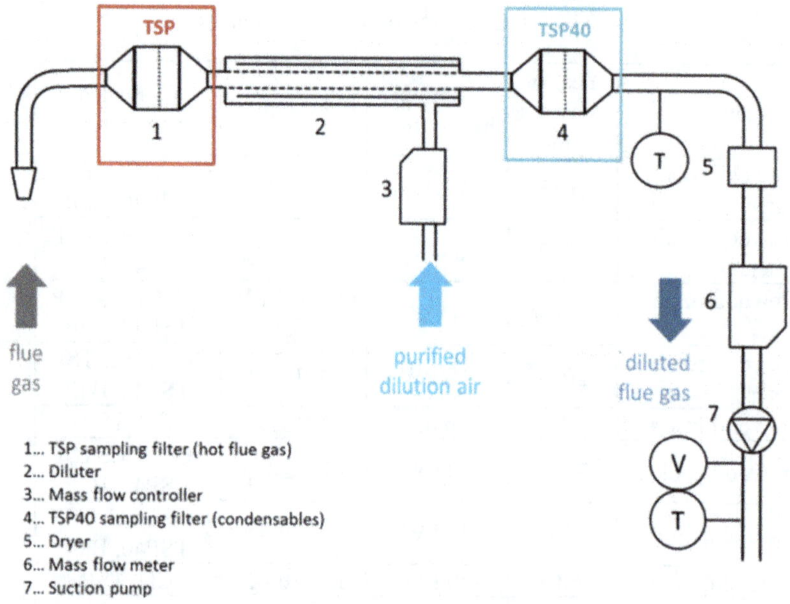

Figure 1: Scheme of measurement equipment (suction device) for particle sampling (TSP and TSP40).

2.3 Data evaluation

Data evaluation was done for each appliance for a whole heating cycle (Section 2.1). Thereby, for gaseous emissions (CO and OGC) a time weighted mean was calculated over the whole measurement period. Results are given in mg/m³ at standard temperature and pressure (STP), referred to 13% O_2 in dry flue gas. For particulate emissions (TSP and TSP40) results are firstly calculated for each measured batch/phase according to Klauser et al. [17]. The sum of TSP and TSP40 specifies the result for TSP_{total}. These results are given as well in mg/m³ at standard temperature and pressure (STP), referred to 13% O_2 in dry flue gas. Afterwards a time-weighted mean was calculated for the result of the whole heating cycle.

The proposed European and Austrian EF are given in mg/MJ. Thus, emission concentration results of the field measurements were transferred to mg/MJ based on combustion calculation [18] and according to the fuel composition. For the final evaluation results were aggregated according to the type of appliances. Means, medians, maxima and minima were calculated for the presentation of the results.

2.4 Chemical characterization of sampled particles

The TSP and TSP40 filters, taken during the field measurements, were characterized regarding BaP concentrations. This procedure was done according to the protocol of DIN EN 15549 using a solvent mixture of dichloromethane and cyclohexane for extraction. The chemical analysis itself was done with a GC-MS. Results are given in µg/m³. They are transferred to µg/MJ according to the fuel composition (same calculations as for other emissions – Section 2.3).

2.5 Fuel

Fuel for the field measurements was provided by the end user. To evaluate fuel properties, representative samples of the used fuel were taken and analyzed for carbon (C), hydrogen (H), nitrogen (N), water (w) and ash (a) content. The analyses were done according to the standards ISO 16948 (C, H, N), ISO 18134-2 (w) and ISO 18122 (a). The mean results are given in Table 3. All chemical parameters (C, H, N) of all tested fuels are in a narrow range, so the calorific values of the fuels are on a comparable level. The water content is in a range between 6 and 16 wt%. This means, that firewood as well as wood chips and saw dust are already dried. Especially for wood chips this indicates, that the fuel is provided by the user itself and stored in an appropriate way. The ash content is very low for all fuels.

Table 3: Overview of chemical analysis of fuels, used in the field measurements.

	C (wt%)	H (wt%)	N (wt%)	w (wt%)	a (wt%)
Mean	49.5	6.1	0.1	10.7	0.7

3 RESULTS AND DISCUSSION

3.1 Emission results

Fig. 2 gives an overview of the emission results of the tested appliances, according to the type of technology. On average, room heaters have the highest emissions. This technology shows mean values of 4029 mg/MJ for CO, 712 mg/MJ for OGC, 138 mg/MJ for TSP and 226 mg/MJ for TSP_{total}. The lowest gaseous emissions occur at boilers (CO: 914 mg/MJ, OGC: 56 mg/MJ). Particulate emissions were lowest at tiled stoves (TSP: 58 mg/MJ, TSP_{total}: 67 mg/MJ). Medians (red cross) indicate that for room heaters and partly for boilers, a few higher values increase the mean values. In case of room heaters these values occurred with very bad user operation of the appliance (e.g. restricted air supply or overload batches). At boilers mainly the old boiler using saw dust as fuel had higher emissions.

Moreover, emissions have a high variability (which is common for emissions measured in the field), indicated by the whiskers (maximum and minimum) in Fig. 2. Especially room heaters have a wide range in emissions due to prevalent transient conditions during firewood burning and due to more possible influencing factors of the end user.

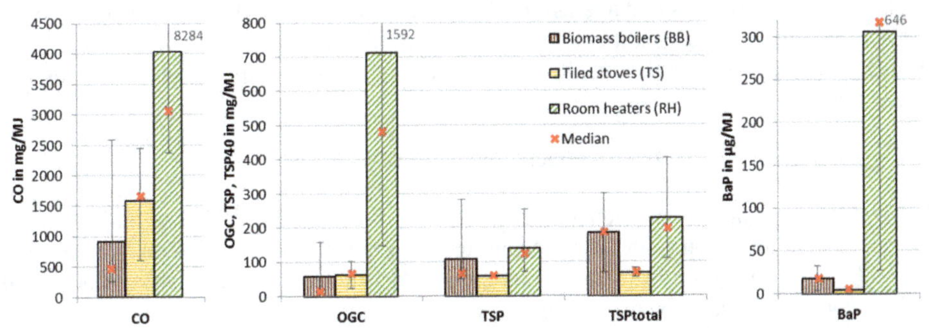

Figure 2: Overview of emission results (mean values) of the field measurement campaign. The whiskers indicate maximum and minimum values.

The study of Spitzer et al. [14] showed similar values for room heaters, but much higher values for boilers for CO (RH: 4463 mg/MJ, BB: 4303 mg/MJ) and OGC (RH: 664 mg/MJ, BB: 448 mg/MJ). TSP results are on a comparable level for both technologies (RH: 148 mg/MJ, BB: 90 mg/MJ). This indicates, that the stock of boilers changed to more advanced technologies with lower gaseous emissions.

The current field measurement results are comparable to the emissions of the study of Ozgen et al. [11]. There, lab tests with three consecutive batches at a firewood stove including the ignition and preheating batch were conducted. The measured emissions showed values of 3196 mg/MJ for CO and 110 mg/MJ for TSP.

Another study from Pettersson et al. [12] presents results from lab tests with two full load batches at a firewood stove. This study also takes maloperation into account, i.e. restricted air supply and moist wood. Emissions are 3600 mg/MJ for CO, 820 mg/MJ for OGC and 140 mg/MJ for TSP_{total}. OGC and TSP_{total} are higher than the presented values of the field measurements and of the study of Ozgen et al. [11]. This reveal that unfavorable user habits can increase those emissions.

The results of the field measurements in the *BioMaxEff* [15] project showed lower emissions for boilers on average (CO: 243 mg/MJ, OGC: 5 mg/MJ, TSP: 15 mg/MJ). However, these results only include measurements at new installed pellet boilers, which were not tested in the current study.

3.1.1 BaP emissions

As illustrated in Fig. 2 it is even more obvious, that BaP emissions are highest for room heaters in the field. This technology shows an average value of 305 µg/MJ compared to 18 µg/MJ for boilers and 4 µg/MJ for tiled stoves. Also, the range of BaP emissions in the field is very high. Results indicate that the user could be a major influencing factor on BaP formation. However, further investigations are needed to evaluate most important influencing factors of the formation of BaP.

The study of Klauser et al. [10] showed clearly lower values for BaP with a maximum of 86 µg/MJ for firewood appliances. However, only new and advanced technologies were tested. Moreover, the operation of the appliance at *beReal* testing follows the manual of the manufacturer, so maloperation is omitted.

Ozgen et al. [11] found BaP emissions of 204 µg/MJ which is on average lower, but in the range of the current study. Pettersson et al. [12] found higher values (BaP: 610 µg/MJ) at restricted air supply and the use of moist wood for testing. Compared to the current study this is at the level of the maximum value of the measurements (646 µg/MJ).

Klauser et al. [16] measured average BaP emissions at a state-of-the-art wood chip boiler of 3.5 µg/MJ at starting conditions. Since, these conditions are normally more likely to formulate BaP emissions, the result is clearly lower compared to the emissions of the current study.

3.2 Comparison to current emission factors for Europe and Austria

Fig. 3 shows the single emission results for OGC, BaP, TSP and TSP_{total} of the tested appliances in comparison to currently suggested emission factors for Europe (EMEP) and Austria (AEF). In Table 4, the different emission factors for Europe and Austria, which were used for comparison, are given.

The comparison of the EF of the field measurements (Fig. 2 and Fig. 3) and the suggested EF (Table 4 and Fig. 3), shows that EF for boilers measured in this study are much lower than the proposed European EF, when using EMEP_d (conventional boilers), except for 1

TSP result. This outlier is the result of the old boiler fired with saw dust. The 95% confidence interval given for European EF, cover this higher value. The lowest results of the field measurement campaign are well reflected by EMEP_e (pellet boilers), although other technologies were tested, except for particulates which are higher by trend. EMEP_c (advanced/eco-labelled boilers) show higher EF for gaseous emissions (CO, OGC) and represents particulates and BaP, compared to the field measurements. Tiled stoves are compared with EMEP_b (high-efficiency stove) and EMEP_c (advanced/eco-labelled stove). EMEP_b shows higher EF than the emissions of tiled stoves of this study. EMEP_c would represent the emissions of tiled stoves, except for OGC the proposed EF is higher. EMEP_a (conventional stove), EMEP_b (high efficiency stove) and EMEP_c (advanced/eco-labelled stove) are used for comparison with room heaters. The range of those EF comprises the values of the field emissions well, with exception of some cases. For BaP 3 clearly higher results are measured. Nevertheless, they are within the 95% confidence interval of EMEP_a and EMEP_b.

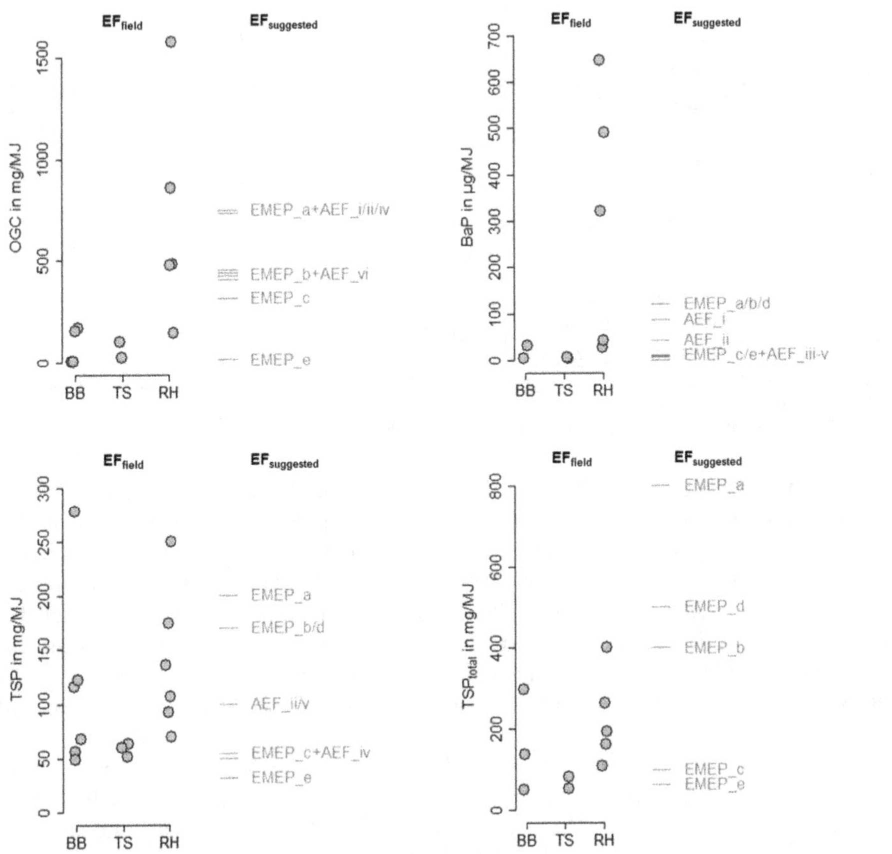

Figure 3: Emission factor results for OGC, BaP, TSP and TSP_{total} of the field measurements (EF_{field}) in comparison to current suggested European (EMEP) and Austrian (AEF) emission factors (EF_{sugg}), differentiated by technology (BB: boilers, TS: tiled stoves, RH: room heaters).

Table 4: Overview of emission factors for Europe [5] and Austria [6] in mg/MJ.

Abbreviation	Description	CO	OGC	TSP	TSP_{total}	BaP
EMEP_a	Conventional stoves	4000	750*	200	800	121
EMEP_b	High-efficiency stoves	4000	437.5*	170	400	121
EMEP_c	Advanced/eco-labelled stoves and boilers	2000	312.5*	54	100	10
EMEP_d	Conventional boilers <50kWth	4000	437.5*	170	500	121
EMEP_e	Pellet stoves and boilers	300	12.5*	32	62	10
AEF_i	Wood stoves and cooking stoves	3955.9	736.35*	148		86.25**
AEF_ii	Tiled wood stoves and masonry heaters	2345.3	422.5*	100		42.5**
AEF_iii	Natural-draft wood boilers	3483	437.5*	75		8.75**
AEF_iv	Forced-draft wood boilers	3234.2	406.25*	50		0.5**
AEF_v	Wood chips boilers with conventional technology	2400	540.5*	100		6**
AEF_vi	New wood stoves and cooking stoves	2345.3	453.61*			

*EF are given for NMVOC; based on Anderl et al. 2017 for biomass combustion a methane share of 25% is in OGC [6]; $EF_{OGC} = EF_{NMVOC} + 0.25\ EF_{NMVOC}$.
**Austrian EF are only given for the sum of 4 PAHs [6]; the share of BaP is given as 25%; $EF_{BaP} = 0.25\ EF_{PAHs}$.

Austrian EF for boilers are in general too high for gaseous emissions (CO and OGC) in comparison to the study results. Particulates are well represented by AEF_iii (natural-draft wood boiler), AEF_iv (forced-draft wood boiler) and AEF_v (wood chips boiler with conventional technology). There is one EF for tiled stoves (AEF_ii…tiled wood stoves and masonry heaters). This EF overestimates emissions in comparison to the conducted field measurements of this study. For all room heaters, AEF_i (wood stoves and cooking stoves) represents a reliable mean value. Some emissions are clearly higher, others are clearly lower than this EF. AEF_vi (new wood stoves and cooking stoves) represents the lower values of this study.

Especially for room heaters it is very hard to define appropriate EF even when distinguishing between different technologies, since the variability is very high. Moreover, the operation of room heaters is influenced by many factors regarding the framework, foremost user operation habits. The availability of a 95% confidence interval like it is available for European EF, is a good measure for modeling best or worst-case scenarios of air quality in a region. Nevertheless, this wide range of emissions in the field makes it hard to predict air quality or conduct emission inventories.

3.3 Differences of particulate measurements

For the field measurement, tested with the extensive equipment, TSP and TSP40 samples are compared (Fig. 4). Thus, an evaluation of the share of condensable particles is possible. Moreover, the fraction of BaP in the condensable particles is determined. Results show that a high share of TSP_{total} is in the hot flue gas (TSP). However, for room heaters 33% can be found in the condensable fraction (TSP40). Moreover, emissions of BaP are mainly found in the condensable fraction. For air quality issues the sum of both fractions is fundamental, since a fraction of volatile organic compounds condensates at ambient temperatures. Thus, measuring this sum of particles is necessary to get reliable emission factors for TSP_{total}. Otherwise, emissions could be underestimated. This is even more important regarding BaP and the impact of air quality on human health.

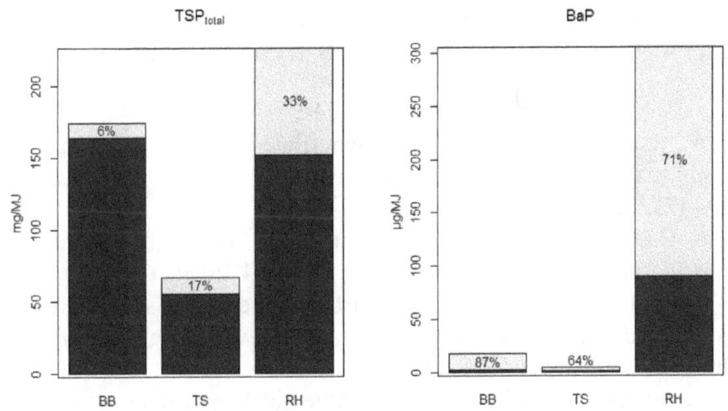

Figure 4: Comparison of particle measurement and the chemical characterisation in hot (dark) and cooled (bright) flue gas.

4 CONCLUSION

In this study results of a field measurement campaign, including 15 biomass heating appliances are presented. The following conclusions can be drawn:

The results of this study in comparison to the study of Spitzer et al. indicate that for boilers a technological development towards lower gaseous emissions had happen. For room heaters this was not found, even though the results of official type tests assume such a development. However, room heaters are more influenced by the end user. End user training and also focus on the optimization of real-life conditions during heating (including ignition, preheating and load changes, which is not tested at official type tests) could have a high potential for future improvement.

The proposed European and Austrian EF show higher values by trend for boilers and tiled stoves. For room heaters EF are in the range of emissions of this study. Combined with the comparison to the results of Spitzer et al. of 1998, this reveals that EF, especially for boilers and tiled stoves, need to be updated regularly. Moreover, due to the high variability of the field measurement results it is very hard to define EF. This high variability is mainly caused by framework conditions (user influence, installation in the house, dimensioning, chimney…).

Close to real-life lab tests can be a good opportunity to reflect the technological development of heating appliances at close to real-life conditions. Nevertheless, but they cannot reflect the whole framework. Since EF are used for emission inventories, they have an impact on policy decision makers and regulations. Thus, field measurements should be fostered in order to get a good overview of the emissions in real life and even more reliable EF.

The comparison of particles in the hot and cooled flue gas reveal that the amount of condensable particles is unneglectable. Moreover, BaP is predominantly found in the condensable fraction of particles. Hence, EF should consider condensable particles. Otherwise TSP_{total} and BaP can be underestimated.

ACKNOWLEDGEMENTS

The project was supported by the project partners Styrian Energy Agency, Technical University of Vienna, Impulszentrum Vorau and chimney sweeper Plesar and received funding from the Austrian Research Promotion Agency under the COMET program.

REFERENCES

[1] Bari, M.A., Baumbach, G., Kuch, B. & Scheffknecht, G., Wood smoke as a source of particle-phase organic compounds in residential areas. *Atmospheric Environment*, **43**, pp. 4722–4732, 2009.

[2] Guerreiro, C.B.B., Foltescu, V. & de Leeuw, F., Air quality status and trends in Europe. *Atmospheric Environment*, **98**, pp. 376–384, 2014.

[3] Denier van der Gon, H.A.C. et al., Particulate emissions from residential wood combustion in Europe – revised estimates and an evaluation. *Atmospheric Chemistry and Physics*, **15**, pp. 6503–6519, 2015.

[4] IARC, Polynuclear aromatic compounds: Chemical, environmental and experimental data. IARC monographs on the evaluation of the carcinogenic risk of chemicals to humans. International Agency for Research on Cancer: Lyon, 1983.

[5] EEA, *EMEP/EEA Air Pollutant Emission Inventory Guidebook 2016: Technical Guidance to Prepare National Emission Inventories*. European Environment Agency, 2016.

[6] Anderl, M. et al., Austria's informative inventory report (IIR). Report, no. REP-0609, Umweltbundesamt GmbH: Vienna, 2017.

[7] Stabile, L. et al., A novel approach to evaluate the lung cancer risk of airborne particles emitted in a city. *Science of the Total Environment*, **656**, pp. 1032–1042, 2019.

[8] Reichert, G. et al., beReal – development of a new test method for firewood roomheaters reflecting real life operation. *ETA-Florence Renewable Energies*, 2016.

[9] Rönnback, M., Persson, H., Jespersen, M.G. & Jensen, J.H., Deliverable 7.1 *Documentation and Evaluation of Field Data Demonstration*. Danish Technological Institute, 2016.

[10] Klauser, F. et al., Emission characterization of modern wood stoves under real-life oriented operating conditions. *Atmospheric Environment*, **192**, pp. 257–266, 2018.

[11] Ozgen, S. et al., Emission factors from small scale appliances burning wood and pellets. *Atmospheric Environment*, **94**, pp. 144–153, 2014.

[12] Pettersson, E., Boman, C., Westerholm, R., Boström, D. & Nordin, A., Stove performance and emission characteristics in residential wood log and pellet combustion, Part 2: Wood stove. *Energy Fuels*, **25**, 315–323, 2011.

[13] Tissari, J., Hytönen, K., Lyyränen, J. & Jokiniemi, J., A novel field measurement method for determining fine particle and gas emissions from residential wood combustion. *Atmospheric Environment*, **41**, 8330–8344, 2007.

[14] Spitzer J., Enzinger, P., Fankhauser, G., Fritz, W., Golja, F. & Stiglbrunner, R., Emissionsfaktoren für feste Brennstoffe. Joanneum research report, Graz, 1998.

[15] Verma, V.K. et al., BioMaxEff. Final report, Wieselburg, 2014.

[16] Klauser, F. et al., Development of a compact technique to measure benzo(a)pyrene emissions from residential wood combustion, and subsequent testing in six modern wood boilers. *Biomass and Bioenergy*, **111**, pp. 288–300, 2018.

[17] Klauser, F. et al., Effect of oxidizing honeycomb catalysts integrated in a firewood room heater on gaseous and particulate emissions, including polycyclic aromatic hydrocarbons (PAHs). *Energy & Fuels*, **32**, pp. 11876–11886, 2018.

[18] Kaltschmitt, M., Hartmann, H. & Hofbauer, H., *Energie aus Biomasse*, Springer-Verlag: Berlin and Heidelberg, 2009.

ESTIMATION OF FUEL LOSS AND SPATIAL–TEMPORAL DISPERSION OF VEHICULAR POLLUTANTS AT A SIGNALIZED INTERSECTION IN DELHI CITY, INDIA

RAJNI DHYANI, NIRAJ SHARMA & MUKTI ADVANI
Transport Planning and Environment Division, CSIR-Central Road Research Institute, India

ABSTRACT

Vehicular traffic is considered one of the major contributors to urban air pollution in a city like Delhi. Signalized traffic intersections are considered as urban hotspots due to their high air pollution levels which generally exceed the air quality standards specified by the regulatory agencies. The high air pollution levels mainly result from the idling of a large number of motor vehicles at these signalized intersections resulting in fuel losses and corresponding emissions leading to deteriorating air quality. There are ~950 signalized traffic intersections in Delhi city. A large number of the population works and lives close to these traffic intersections and are exposed to these air pollutants. Therefore, it becomes imperative to understand the spatial–temporal distribution of vehicular emissions resulting because of the idling of motor vehicles at these intersections. In the present study, the fuel loss estimation due to the idling of vehicles was carried out at Lodhi Road intersection using idling fuel consumption data carried out on various test vehicles representing the Delhi traffic. The fuel loss was converted into emissions using fuel-based Intergovernmental Panel on Climate Change (IPCC) emission factors. The emissions have been estimated in terms of greenhouse gas (CO_2, CH_4, and N_2O) emissions and indirect greenhouse gas (CO, NO_x, and NMVOC). The results indicated that daily there was ~230 litre of petrol, ~100 litre of diesel and ~170 kg of compressed natural gas (CNG) loss due to idling at Lodhi Road intersection. The resultant emissions load estimated was ~1300 kg of CO_2, 13 kg of NO_x, 90 kg of CO per day. The spatial extent of vehicular pollutant has been assessed with the help of the CAL3QHC model.

Keywords: emissions, fuel loss, idling, signalized intersection, CAL3QHC model.

1 INTRODUCTION

The traffic intersections are considered as air quality hotspot because of generally poor air quality at these places which often exceeds the stipulated air quality standards. The main reasons for this is high vehicular traffic activities, vehicular idling due to traffic delay, red time signal etc. A vehicle run on a city road trip consists of idling, acceleration, cruise, and deceleration modes [1]. The share of these different stages depends upon driver behaviour, level of traffic congestion at traffic intersections etc. Further, the fuel consumption of vehicles depends upon factors viz. type and category of vehicles, engine technology, inspection and maintenance (I & M) practices etc along with the driving cycle of the city [2], [3]. Large variations in traffic flows, increased queuing time and constricted road geometries makes signalized road intersections as pollution hotspots [4], [5]. The delays at traffic intersections result in idling fuel losses and corresponding emissions which deteriorate the air quality of surrounding areas. A study carried out by Sharma et al. [6] estimated idling fuel losses at 950 signalized traffic intersections in Delhi to be ~262,703 kL petrol, ~145,284 kL diesel, 248 kt CNG and 10,202 kL LPG per year. Further, if these losses converted into monetary terms they estimated to be around 5.9 billion USD per year (1 US$ = 70 Indian Rs.).

The people working, commuting or spending time at these signalized traffic intersections gets exposed to high level of air pollutants. Complexity in dispersion of air pollution at urban intersection makes it difficult to manage air quality at these hotspots. Thus, to manage the air

WIT Transactions on Ecology and the Environment, Vol 236, © 2019 WIT Press
www.witpress.com, ISSN 1743-3541 (on-line)
doi:10.2495/AIR190231

quality at these urban hotspots, it becomes more important to understand the extent of spread and dispersion behaviour of pollutant at traffic intersections. The dispersion of air pollutants at traffic intersections get influenced by various factors viz. intersection geometry, surface roughness, wind speed and wind direction etc.. Vehicular pollution dispersion models have been used to predict present and future air quality along the roads/highways. They have been used and have helped in improving the understanding of the pollution dispersion at traffic intersections. There are many line source models (AERMOD, CALINE4, and ADMS) which have been used regularly to predict air quality along road/highway corridors modelling. In the present study, Gaussian equation based CAL3QHC model have been used predict the air quality at a selected traffic intersection. CAL3QHC is an exclusive traffic intersection/ hotspot model. The model has been used in various studies to predict air quality at traffic intersections, however, very few studies have been carried out in India at traffic intersections as well as using this model.

Thus, in the present study, authors have attempted an indicative exercise to estimate emissions due to the idling of vehicles at a signalized intersection in Delhi city. Further, spatial–temporal extent of the vehicular pollution (under mixed traffic conditions) at Lodhi Road intersections has been assessed with help of Gaussian based CAL3QHC intersection model.

2 METHODOLOGY

Delhi city has been experiencing exponential growth in the motorized vehicle population from ~4.3 million in the year 2005 to nearly 10 million vehicles in 2016 [7]. Delhi registered vehicles constitutes nearly 4.22% of the vehicular population of India [8]. Road transport is the main mode of transportation in the city. There are ~950 signalized traffic intersections in Delhi which could be divided into high (>0.2 million vehicles/day), medium (0.1–0.2 million vehicles/day) and low (<0.1 million vehicles/day) traffic intersections according to their traffic volume. The private mode of transportation viz., car, and two-wheelers constitutes ~85% of all traffic volume of Delhi city. Delhi has the highest traffic density, therefore regular traffic congestion and delays at various traffic intersections are not unfamiliar in the city. The selected signalized traffic intersection is located at Lodhi Road (28°35'28.06"N, 77°13'44.24"E) at an elevation of 707 feet (Fig. 1). Lodhi road intersection is a low capacity intersection (<0.1 million vehicles/day) with an average daily traffic volume of ~92,000 vehicles. The Lodhi road intersection is located in a mix of government offices, schools, colleges and residential complexes. However, the surroundings of selected site are relatively greener, spacious and have planned development as compared to most of the Delhi city.

2.1 Estimation of idling fuel losses and corresponding emissions

The methodology for estimation of fuel loss and corresponding emissions at Lodhi Road signalized intersection consists of two parts: (i) carrying out fuel consumption studies at idling on various test vehicles (representatives of vehicles fleet plying on selected city/state). These tests simulate and help in estimating the fuel consumed during idling of vehicles at various signalized intersections; and (ii) converting the fuel losses as estimated in step (i) into emissions by employing appropriate emission factors and various other input parameters collected from primary surveys or secondary sources.

For the estimation of idling fuel consumption, different categories and vintage of (two-wheelers, bike scooters, cars, etc.) petrol-powered and diesel-powered vehicles were tested [6], [9]. The vehicles tested were broadly categorized in the following categories: (i) 2W–4S

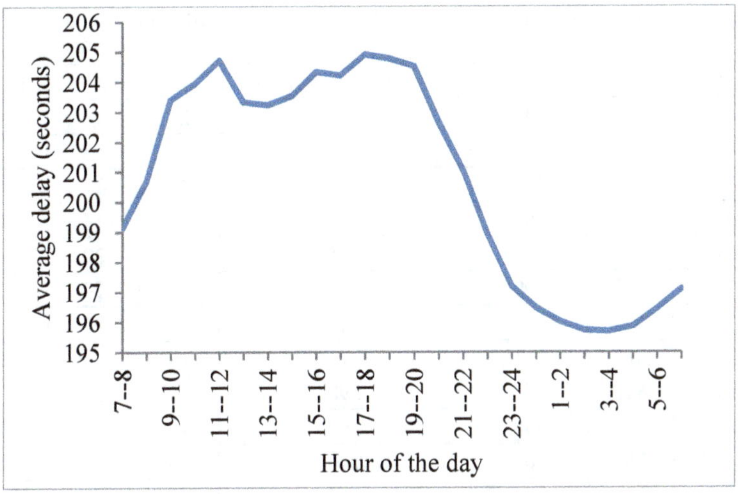

Figure 1: Average delay at Lodhi Road signalized traffic intersection.

(two wheelers–4 stroke-motorcycle); (ii) car–petrol; (iii) car–diesel; (iv) LCV (light commercial vehicles)–diesel; and (v) bus–diesel, HCV (heavy commercial vehicles), 3W (three wheelers)–auto, which form the fleet composition in Delhi.

The total number of vehicles idling at selected intersection due to delay has been estimated by cumulative input–output flow polygon approach [10] are shown in Table 1. From Table 1, maximum idling traffic could be observed during evening hours (17.00–18.00 and 18.00–19.00 hours).

Further, the maximum waiting time could be observed during morning and evening traffic peak hours. During late night hours, traffic delay was observed to be less as compared to day time hours due to low traffic count at intersection (Fig. 1).

Idling fuel consumption was estimated in terms of mass units (kg) then converted in terms of equivalent energy units by applying the fuel specific net calorific values (NCV) for emission estimations (TJ/t). Idling emissions (kg/hour) were estimated using IPCC [11], [12] default pollutant-specific (CO_2, CH_4, CO, N_2O, NO_x and NMVOC) emission factors (kg/TJ) for each fuel type used [13]. The detailed methodology for estimation of idling fuel consumption and corresponding emissions can be referred from [6].

2.2 CAL3QHC model description

CAL3QHC is a Gaussian based mathematical model to predict vehicular pollutants viz., Carbon monoxide (CO) or other inert pollutant from vehicular at traffic intersections. CAL3QHC is a consolidation of the CALINE3 line source dispersion model and an algorithm that estimates the length of the queues formed by idling vehicles at signalized intersections [14]. The model predicts the air pollution concentrations from both moving and idling vehicles. CAL3QHC requires all input parameters necessary to run CALINE4, on addition to these input idling emission rates, number of moving lanes, approach traffic volume at each link, idling emission factors signal cycle length, red time signal, clearance lost time, Saturation flow rate, signal type, and arrival type etc. at the intersection is required. Table 2 presents the input parameters collected/estimated for air quality prediction by CAL3QHC model at Lodhi Road intersection.

Table 1: Hourly idling traffic at Lodhi Road signalized intersection.

Time	Car	2W–4S	Auto rickshaw (3W)	Bus	LCV	HCV
7–8	274	136	94	24	18	0
8–9	394	203	148	21	13	1
9–10	555	338	234	42	17	0
10–11	628	367	255	31	14	0
11–12	654	369	325	25	32	1
12–13	602	299	255	23	22	1
13–14	596	304	246	22	22	2
14–15	612	311	275	14	20	0
15–16	642	373	298	15	23	1
16–17	664	358	269	17	24	0
17–18	758	365	260	23	20	1
18–19	756	366	256	21	12	0
19–20	671	394	262	12	27	0
20–21	564	291	205	18	16	1
21–22	455	238	149	11	14	0
22–23	338	109	73	10	12	4
23–24	171	50	49	5	9	3
24–1	99	29	39	0	6	4
1–2	61	16	21	1	4	3
2–3	30	16	10	0	3	3
3–4	27	10	12	0	5	3
4–5	31	13	28	1	5	5
5–6	84	29	39	9	5	7
6–7	135	53	54	14	6	1
Total	9800	5035	3855	357	348	42
% share	50.4	25.9	19.8	1.8	1.8	0.2

The CAL3QHC model has been used to predict CO concentration under prevailing traffic and meteorological conditions during morning (10–11 am) and evening (6–7 pm) traffic peak hours. The CO being the indicator pollutant for vehicular activities was chosen for the present study.

2.3 Estimation of emission factors

In the present study, hourly emission factors were estimated for twelve links. Out of twelve links, eight were free flow links and four were idling links. Following methodologies were used to estimate the emission factors for both the cases.

2.3.1 Free flow traffic emission factors

The weighted emission factor used in free flow traffic conditions was estimated for the eight links which have free flow conditions. Due to absence of vehicle speed based emission factors, the weighted emission factor (WEF) were estimated on the basis of vehicle kilometre

Table 2: Input parameters for CAL3QHC model.

S. no.	Parameters	Unit/value	Source
(i)	CO measurement (hourly)	mg/m^3 or µg/m^3	Secondary [15]
Measurement of fuel consumption and free flow/idling emission factors			
(i)	Traffic volume	Vehicles/hour	Primary
(ii)	Categorization of vehicles based on fuel type (petrol, diesel, CNG, LPG) technology type (2 strokes, 4 strokes)	No. of vehicles	Primary and secondary
(iii)	Age profile/vintage of vehicle (fuel station survey)	Number or %	Primary and secondary
(iv)	Fuel consumption by different categories of vehicles during idling (petrol, diesel, CNG and LPG vehicles)	ml/10 min	Primary and secondary [9]
(v)	Average hourly time delay at intersection for each category of vehicle	Seconds (s)	Primary
(vi)	Net calorific value (NCV)	TJ/10^{3}tonne	Secondary [11], [12]
(vii)	Emission factor (EF) (idling)	tonne/TJ	Secondary [11], [12]
(viii)	Emission factor (EF) (free flow traffic)	g/km	Secondary [16]
Meteorological data			
(i)	Wind speed	m/s	Secondary [15]
(ii)	Stability class	Pasquill–Gifford (P–G)	Secondary
(iv)	Mixing height	m	Secondary [17]
(v)	Background concentration	ppm	Primary
(vi)	Multiple wind direction	degree (°)	Primary
(vii)	Wind direction	degree (°)	Secondary [15]
(viii)	Wind direction increment angle	degree (°)	Primary
(ix)	Start angle	degree (°)	Primary
(x)	End angle	degree (°)	Primary
Road geometry			
(i)	Road length	meter (m)	Primary
(ii)	Link height	meter (m)	Primary
(iii)	Mixing zone width	meter (m)	Primary
Queue parameters			
(i)	Average signal cycle length	seconds (s)	Estimated
(ii)	Average red time length	seconds (s)	Estimated
(iii)	Clearance lost time	seconds (s)	Estimated
(iv)	Approach traffic volume	vehicles per hour (vph)	Primary
(v)	Idle emission factor	gram/vehicle-hour (g/v-hr)	Primary
(vi)	Saturation flow rate	vehicle/hour/lane (v/hr/lane)	Estimated
(vii)	Signal type	Pre-timed/ actuated/semi-actuated	Primary
(viii)	Arrival type	Worst/below average/average/ above average/ best progressing	Primary (observed)

travelled based (VKT) emission factors provided by CPCB [16]. The WEF is a function of vehicle emission factor (vehicle category, type, fuel type, age profile, vintage etc.) and vehicle activity (traffic volume). The equation for calculation of WEF is as follows:

$$WEF = [\Sigma(j)\,\Sigma(ky)\ \mathrm{N\ (j,\ ky).\ EF\ (i,\ j,ky)}]/\mathrm{Total\ no.\ of\ vehicles}, \tag{1}$$

where:

- *WEF* is weighted emission factor (g/km);
- *N (j, ky)* is number of vehicles of a particular type j and age k in year y;
- *EF(i,j,ky)* is emission factor for component i for the vehicle type j and age k in year y (g/km);
- *i* is pollutant component (viz. CO);
- *j* is type of vehicle (i.e. 2W, 3W (auto rickshaw), cars, bus, truck etc.);
- *k* is age of the vehicle in year y.

2.3.2 Idling traffic emission factors

Idling emission factor for CO was estimated and expressed in terms of g/vehicle-hour. The following equation has been used for estimation of–emissions during idling by various categories of vehicles (i)

$$E_i = \frac{[\Sigma V_{j\,(ky)}.FC_{(j,f,ky)}.D_f.T_{(Jj,f,ky)}.\,NCV_f.EF_{(f,i)}]}{3.6} * 10^6, \tag{2}$$

where:

- E_i = Total emission of pollutants type *i* (kg/h);
- $V_{j\,(ky)}$ = Vehicle type *j* of vintage *ky* (no. of vehicles);
- $FC_{(j,\,f,\,ky)}$ = Fuel consumption during idling by vehicle type *j* using fuel type *f* of vintage *ky* (l/h);
- D_f= Density of fuel *f* (kg/l);
- $T_{(j,\,f,\,ky)}$ (s)= Time delay at traffic intersections by vehicle type *j* of fuel type *f* of vintage *ky*;
- NCV_f=Net calorific value for fuel type *f* (Tera-joule/tonne (TJ/kt));
- $EF_{(f,\,i)}$= Emission factor for fuel type *f* of pollutant type *i* (tonnes/Tera-joule (t/TJ)).

2.4 Meteorological parameters

The micro-meteorological data such as wind speed, wind direction was taken from the IMD air quality monitoring station at Lodhi Road [15]. The hourly mixing height values were obtained from the Indian Meteorological Department (IMD) [17]. The worst case wind angle was considered in the present study with 0 degree as start wind angle and 360 as end angle with 5 degree wind increment angle.

Tables 3 and 4 presents the link-wise input data collected/estimated for the prediction of air quality at the selected traffic intersection. Table 3 presents the input used for free flow traffic link and Table 4 presents the input required for the idling traffic/queue links.

Table 3: Traffic input parameters for free flow traffic links in CAL3QHC model.

Link	No. of vehicles (emission factor, g/mile)		Wind speed (m/s)		Mixing height (m)		Mixing zone width (m)
	10–11 am	6–7 pm	10–11 am	6–7 pm	10–11 am	6–7 pm	
Link 1	1557 (2)	1563 (1.9)	1.1	1.3	1110	1800	14
Link 3	2112 (2)	2130 (1.9)	1.1	1.3	1110	1800	14
Link 4	1729 (2)	1738 (2)	1.1	1.3	1110	1800	14
Link 6	1307 (2)	1549 (2)	1.1	1.3	1110	1800	14
Link 7	1870 (2)	1644 (2)	1.1	1.3	1110	1800	15
Link 8	1388 (2.1)	1817 (2)	1.1	1.3	1110	1800	15
Link 10	1388 (2.1)	1603 (2)	1.1	1.3	1110	1800	15
Link 11	1388 (2)	1603 (1.9)	1.1	1.3	1110	1800	15

Table 4: Traffic input parameters for idling traffic/queue links in CAL3QHC model.

Input parameters	Link 2		Link 5		Link 9		Link12	
	10–11 am	6–7 pm	10–11 am	6–7 pm	10–11 am	6–7 pm	10–11 am	6–7 pm
Average signal cycle length (s)	281	290	281	290	281	290	281	290
Red time cycle (s)	105	107	105	107	105	107	105	107
Clearance lost time (s)	3	3	3	3	3	3	3	3
Approach traffic volume(v/hr)	1201	1581	1386	1381	1186	1308	1373	1360
Idling emission factor (g/hr-v)	2.4	2.3	3.4	3	3.1	4	3.6	4
Saturation flow rate (v/hr/lane)	1800	1800	1800	1800	1800	1800	1800	1800

3 RESULTS AND DISCUSSIONS

3.1 Idling fuel loss and corresponding emissions

The fuel consumption during idling at Lodhi road intersection was estimated for different type and categories of vehicles for 24-hour period. Petrol driven vehicles (especially, private vehicles viz. cars and two-wheelers) are predominantly more in Delhi as compared to public and commercial transport vehicles. Therefore, it was observed that among all fuel type, losses of petrol was highest followed by CNG, diesel and LPG (Table 5).

The diurnal variation of idling emissions viz. NO_x, CO, N_2O, CH_4, NMVOC, and CO_2 has been shown in Fig. 2. Thus, there were idling emission losses to the tune of ~1300 kg of CO_2, 13 kg of NO_x, 90 kg of CO per day. Further, peak idling emissions could be observed during morning and evening peak traffic hours. During night time to early morning hours, when there is less idling traffic at intersection the estimated idling emissions were found to be comparatively very low.

Table 5: Idling fuel loss at Lodhi Road intersection.

Idling fuel loss	Quantity (per day)
Petrol	232.14L
Diesel	97.8L
CNG	169.8kg
LPG	10.27L

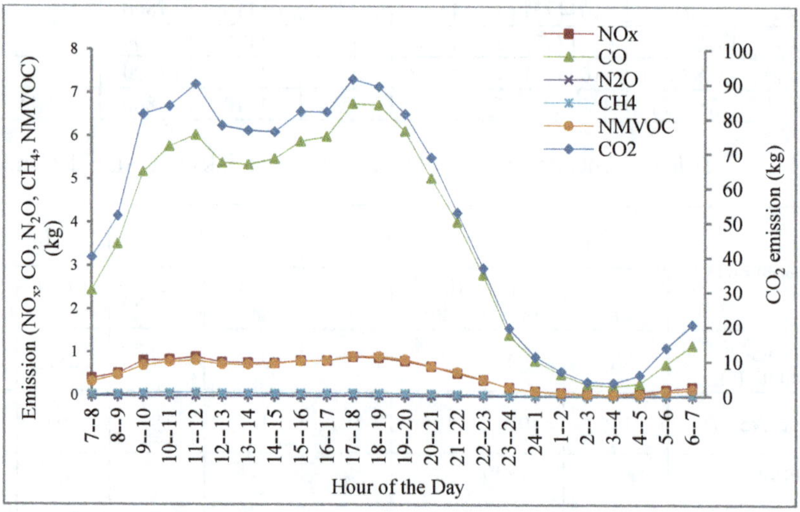

Figure 2: Diurnal variation in idling emissions at Lodhi Road signalized traffic intersection.

3.2 Spatial–temporal extent of emissions at signalized intersection

The spatial extent of vehicular pollutant has been assessed with the help of the CAL3QHC model [14]. Like every typical four arm signalized traffic intersection in India, each arm of Lodhi road intersection has three traffic movements – free left, straight and right turn. Out of these three movements, straight and right movement traffic has to idle at intersection (Fig. 3). In the present study, the spatial temporal extent of pollutants has been observed for morning and evening peak hours only.

The CO concentrations were estimated at pre-identified receptors locations at 2 m, 5 m and 10 m from mixing zone width at its four arms and at centre (middle of intersection). Worst case scenario (in terms of wind angle) selected for the present study and 60 minute average concentration of CO was predicted. The background concentration (input in model) was kept zero to observe only the vehicular emissions contribution to the surroundings. During morning peak hour (10–11 am) the highest concentration was 343.5 $\mu g/m^3$ (0.34 mg/m^3) observed at 2 m away from mixing zone width, at distance of 10 m the concentration was zero, indicating absence of contribution from vehicular emission. During the evening peak hour (6–7 pm), the highest CO concentration was observed at centre of intersection (1030.5 $\mu g/m^3$ or 1.03 mg/m^3) followed by 2 m (687 $\mu g/m^3$), 10 m (572.5 $\mu g/m^3$) and 20 m (343.5 $\mu g/m^3$) away from mixing zone width, thus indicating reduction in vehicular pollution as the distance from the intersection (or zone of influence) increases (Fig. 4).

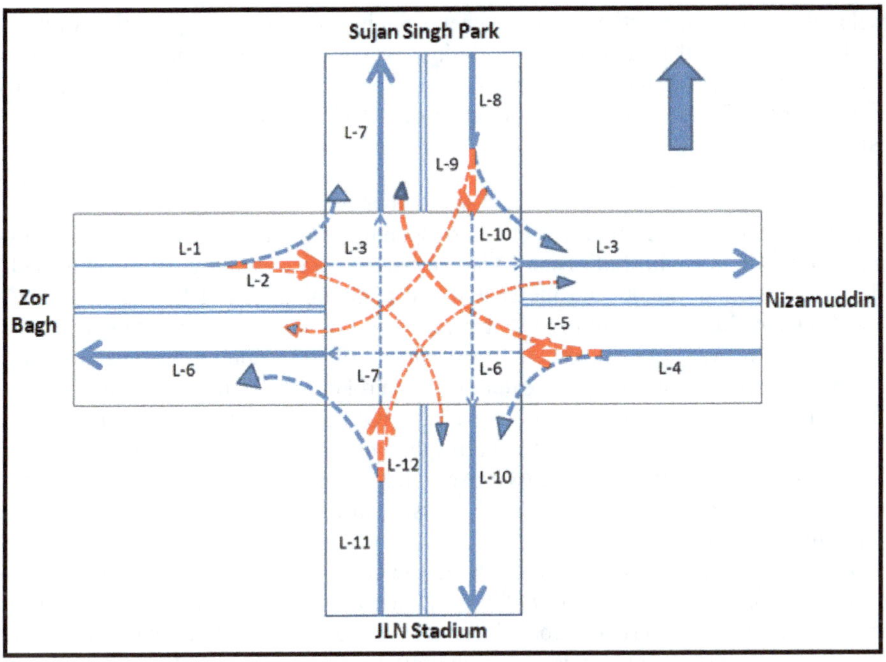

Figure 3: Traffic movement and selected links at Lodhi Road signalized traffic intersection (L = link).

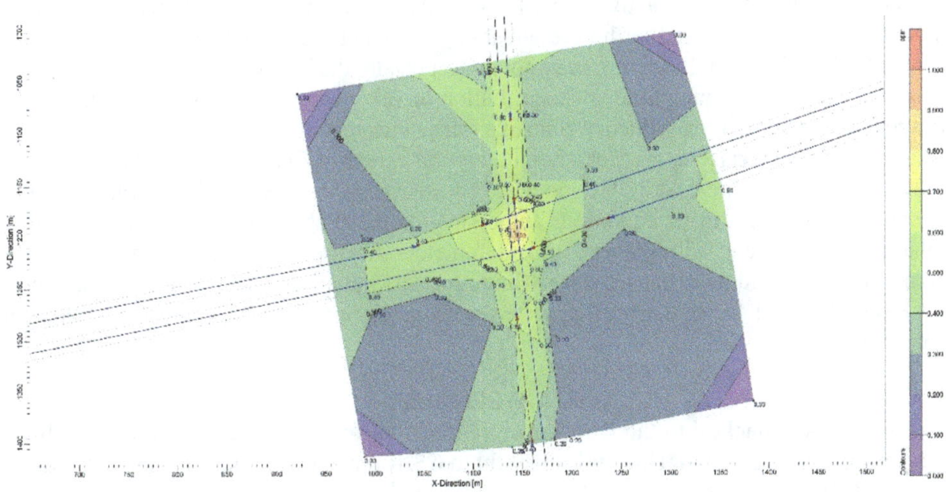

Figure 4: Contour map of CO concentration along Lodhi Road intersection (1800–1900 hours).

During night time (e.g. 2–3 am), due to low traffic volume (~70 vehicles/link/hour), the average signal cycle length (72 seconds) and red time signal (20 seconds) also get reduced which results in smaller queue length and less idling, thus lower CO contribution from vehicles at intersection. Further, ~100 m away from intersection during any time of the day, the vehicular pollutant influence was negligible. When compared to the average observed CO concentration (~1.5 mg/m^3) during evening hour (6–7 pm) from ambient air quality station at Lodhi Road [15], the performance of the CAL3QHC model was found to be satisfactory.

4 CONCLUSIONS

In the present study, idling fuel loss and corresponding emissions were estimated at Lodhi Road intersection in Delhi (India) and an indicative exercise was carried out with the help of CAL3QHC model to understand the spatial and temporal extent of CO dispersion at the selected intersection. The Lodhi road intersection is low traffic capacity intersection with traffic volume of ~92,000 vehicles/day; out of these ~19,000 vehicles idle on an average for ~200 seconds per hour in a day. Large number of petrol driven vehicles (cars, two-wheelers) results in highest loss of petrol fuel followed by CNG (three-wheelers, buses), Diesel (LCV, HCV) and LPG. The fuel loss results corresponding emissions, with highest being CO_2 followed by NO_x, CO, etc. The same exercise could be applied at any signalized intersection to get the estimate of fuel loss and related emission throughout the city in order to find out the enormity of the problem.

Apart from estimation of idling fuel losses and corresponding emissions the spatial and temporal distribution of pollutants holds an important aspect in air quality management studies at these intersections. The spatial and temporal dispersion of pollutants delineate the zone of influence, which could help further in drafting the air quality and traffic management plans. Further, it is necessary to understand the dispersion at intersections because of the exposure level and impact on health of people (due to idling emission and resultants poor air quality) who resides, work or pay short visit at intersections/hotspots due to variety of reasons (viz. commuting, business activity, etc.). The CAL3QHC is Gaussian based intersection model which has been used in the present study to predict the CO concentration. The high concentration observed during the peak hours, (especially during evening peak hours), gives clear indication of the magnitude of contribution of vehicles in deteriorating the air quality in surrounding areas or its influence zone of a traffic intersection. However, the present study is an indicative exercise to present spatial temporal extent of CO in an urban hotspot using CAL3QHC model. The CAL3QHC model has been used in very few studies in India that too under heterogeneous traffic conditions and owing to unavailability or lack of understanding related to traffic and idling/queue link parameters requirement. Studies that are more comprehensive need to be carried out in future for performance evaluation and validation of CAL3QHC model for Indian traffic (heterogeneous or mixed traffic) and meteorological conditions.

ACKNOWLEDGEMENTS

The authors are thankful to the Director of CSIR-CRRI for kindly permitting us to publish the present paper. Dr Rajni Dhyani is thankful to CSIR for providing her financial assistance through CSIR-Research Associate fellowship.

REFERENCES

[1] Goel, R. & Guttikunda, S.K., Evolution of on-road vehicle exhaust emission in Delhi. *Atmospheric Environment,* **105**, pp. 78–90, 2015.

[2] Ishaque, M. & Noland, R., Simulated pedestrian travel and exposure to vehicle emissions. *Transportation Research Part D*, **13**(1), pp. 27–46, 2008.

[3] Pandian, S., Gokhale, S. & Goshal, A.K., Evaluating effects of traffic and vehicle characteristics on vehicular emissions near traffic intersections. *Transportation Research Part D*, **14**, pp. 180–196, 2009.

[4] Goel, A. & Kumar, P., Zone of influence for particle number concentrations at signalised traffic intersections. *Atmospheric Environment*, **123**, pp. 25–38, 2015.

[5] Soulhac, L., Garbero, V., Salizzoni, P., Mejean, P. & Perkins, R., Flow and dispersion in street intersections. *Atmospheric Environment,* **43**, pp. 2981–2996, 2009.

[6] Sharma, N., Kumar, P.V.P., Dhyani, R., Sekhar, C.R. & Ravinder, K., Idling fuel consumption and of emission of air pollutants at selected signalized intersections in Delhi. *Journal of Cleaner Production*, **212**, pp. 8–12, 2019.

[7] Transport Department, Government of National Capital Territory of Delhi (GNCTD), Total vehicles registered zonewise up to 31 March 2016. http://delhi.gov.in/wps/wcm/connect/doit_transport/Transport/Home/. Accessed on: 15 Apr. 2017.

[8] Ministry of Road Transport and Highways, Government of India (MoRTH), *Road Transport Yearbook (2015–2016)*. http://morth.nic.in/showfile.asp?lid=3141. Accessed on: 23 May 2017.

[9] Central Road Research Institute (India) (CRRI), Evaluation of economic loss due to idling of vehicles at signalized intersection and mitigation measures (ELSIM), funded under 12th five year plan, 2017.

[10] Kyte, M. & Tribelhorn, M., Operation, analysis and design of signalized intersection (final report): A module for the introductory course in transportation engineering. Department of Transportation, University Transportation Centers Program, 2014.

[11] Intergovernmental Panel on Climate Change (IPCC), *Revised Guidelines for National Greenhouse Gas Inventories, Volume 2 Energy*, 1996.

[12] Intergovernmental Panel on Climate Change (IPCC), *Guidelines for National Greenhouse Gas Inventories, Work Book Volume 2*, 2006.

[13] Singh, A., Gangopadhyay, S., Nanda, P.K., Bhattacharya, C., Sharma, C. & Bhan, C., Trends of greenhouse gas emission from the road transport sector in India. *Science of Total the Environment,* **390**(1), pp. 124–131, 2008.

[14] User's guide to CAL3QHC version 2.0: A modelling methodology for predicting pollutant concentrations near roadway intersections (revised). EPA-454/R-92-006 (revised). USEPA Office of Air Quality Planning and Standards, Research Triangle Park, NC 27711, September 1995.

[15] Central Pollution Control Board (CPCB), Central control room for air quality management: All India. Average report criteria. https://app.cpcbccr.com/ccr/#/caaqm-dashboard-all/caaqm-landing/data. Accessed on: 10 Mar. 2018.

[16] Central Pollution Control Board (CPCB), Status of pollution generated from road transport in six mega cities. Ministry of Environment Forest and Climate Change, Government of India. http://cpcb.nic.in/cpcbold/upload/NewItems/NewItem_215_Report. pp. 127–137, 2015.

[17] Attri, S.D., Singh, S., Mukhopadhyay, B. & Bhatnagar, A.K., Atlas of hourly mixing height and assimilative capacity of atmosphere in India. Met. Monograph No. Environment Meteorology-01/2008, Indian Meteorological Department, Government of India: New Delhi, 2008.

SPECIAL SESSION: TRANSPORT AND AIR QUALITY IN PORTUGAL

(JOINT SESSION WITH URBAN TRANSPORT 2019)

USING AIR QUALITY MODELLING AND EMISSION PROJECTIONS AS A SUPPORT TO THE FIRST AIR POLLUTION CONTROL PROGRAM UNDER NEC DIRECTIVE TARGETS FOR 2030

JOANA FERREIRA[1], DIOGO LOPES[1], SÍLVIA COELHO[1], ALEXANDRA MONTEIRO[1], MYRIAM LOPES[1], DÍLIA JARDIM[2], FILIPA MARQUES[2], FILOMENA BOAVIDA[2] & ANA I. MIRANDA[1]
[1]Department of Environment and Planning & CESAM, University of Aveiro, Portugal
[2]Portuguese Environment Agency, Portugal

ABSTRACT
The new National Emission Ceilings (NEC) Directive was transposed into Portuguese law in 2018 creating the obligation to comply with the established NEC for Portugal and to develop a National Air Pollution Control Programme (NAPCP). An analysis of emission trends and projections, and potential air quality and environmental impacts, was performed as a support to the NAPCP. Notwithstanding the emissions reduction trend, NEC noncompliance cases are still foreseen for SO_x, NMVOC and $PM_{2.5}$ under a 2030 current level legislation scenario. In terms of air quality impacts, a likely compliance with air quality objectives for health protection is expected for NO_2 and $PM_{2.5}$, with an increase of O_3 levels. Based on the obtained results, it is possible to identify the need for further emission reduction measures and for the air quality simulation of other scenarios.
Keywords: NEC directive, emission projections, air quality impacts, modelling.

1 INTRODUCTION

The latest report by the European Environment Agency raises awareness for the high population in Europe still exposed to dangerous levels of atmospheric pollution that can lead to premature mortalities. Beyond the impact on human health, air pollutants also have adverse effects on the environment, namely causing crop losses and acidification of soils and surface waters [1]. To ensure the reduction of these negative impacts, the revised directive on National Emissions Ceilings (NEC), published in 2016, sets stricter targets, expressed as a percentage reduction of 2005 emissions, for the air pollutants included in the previous regulation (SO_2, NO_X, NMVOC and NH_3), and an additional ceiling for $PM_{2.5}$ [2]. The new reduction commitments for 2030 are designed to implement the reduction of air quality impacts by 2030 established in the Communication on a Clean Air Programme for Europe [3]. Compliance with these commitments is also expected to contribute to the European Union's long-term objective of achieving levels of air quality in line with the guidelines of the World Health Organization. Full compliance with European air quality standards can be achieved in the short to medium term by focusing on the implementation of existing policy plus Member State action, and thus substantially tighter emission reduction commitments are needed.

The NEC Directive fine tunes the provisions and requires each Member State to draw-up, adopt and to regularly update a national air pollution control programme (NAPCP), taking into account the wider air quality challenges of the Member State with a view to ensuring that the 2020, 2025 and potentially also the 2030 NECs are met in time, contributing effectively to the achievement of the European air quality objectives and without significantly impacting air quality in neighbouring Member States.

The background information and the policies and measures selected for implementation, to be included in the NAPCP, shall not only be determined through the evaluation of the

WIT Transactions on Ecology and the Environment, Vol 236, © 2019 WIT Press
www.witpress.com, ISSN 1743-3541 (on-line)
doi:10.2495/AIR190241

extent of the emission reductions to be achieved, but also on the obligation to prevent or reduce the transboundary air pollution impact, to contribute to the achievement of European air quality objectives and to ensure coherence with any other relevant national plans or programmes. To this effect, Member States should take into account the need to reduce emissions in zones and agglomerations affected by excessive air pollutant concentrations and/or in those that contribute significantly to air pollution in other zones and agglomerations. NAPCPs should, therefore, contribute to the successful implementation of air quality plans enacted under the European Air Quality Directive.

Portugal is still facing atmospheric pollution problems, especially in urban areas, and had to develop and implement air quality plans to further reduce emissions aiming to improve air quality [4]–[7]. In this sense, it is important to assess and control national emissions, considering their spatial variability, in an integrated way, with the future perspective, to comply with the new NEC Directive commitments and, at the same time, fulfil the air quality objectives/standards.

To support the NAPCP development for Portugal, an analysis of emission trends and projections, and an evaluation of impacts on air quality and on the environment should be performed. In this context, this work aims to present the outcomes of the ongoing project FUTURAR – Air quality in Portugal in 2030 – a policy support (http://futurar.web.ua.pt/en/project), in what concerns the:

1. Analysis of the national emission projections submitted under the Convention on Long-range Transboundary Air Pollution (LRTAP), against the NEC established for Portugal;
2. Assessment of the impacts on air quality and ecosystems of these projections;
3. Evaluation of potential NEC and EU air quality targets compliance, and identification of possible non-compliant areas in the country.

2 METHODOLOGY

This study focuses on the evaluation of emission projections and their impacts on air quality and environment under the NEC Directive. Firstly, an analysis of emission trends and latest submitted projections for 2020, 2025 and 2030 was carried out aiming to verify the compliance of NEC considering the current legislation until 2030 (CLE scenario, without additional measures). Then, an estimation of air quality impacts and an evaluation of the legislation fulfilment were performed.

The Air Quality Directive defines annual limit values of 40 and 25 $\mu g.m^{-3}$ for NO_2 and $PM_{2.5}$ concentrations respectively, and establishes a maximum daily 8-hour mean of 120 $\mu g.m^{-3}$ for O_3, that could not be exceeded in more than 25 days per year. To evaluate the air pollution impacts and identify potential areas of non-compliance of the air quality targets defined in the EU Air Quality Directive (2008/50/EC), the air quality modelling system WRF-CAMx was applied over Portugal with a spatial resolution of 0.05°, driven by a coarser domain over Europe at 0.25° by a two-way nesting approach.

The air quality modelling system includes the Weather Research & Forecasting (WRF) (Version 3.7.1) model [8], and the Comprehensive Air Quality Model with Extensions (CAMx) (Version 6.4) [9]. Two simulations were performed: base case (as reference) and future scenario. The reference scenario considered the latest gridded emissions available (EMEP 2015) [10], disaggregated by sector and pollutant, for the simulation domain resolution of 0.05°. For the future scenario, the 2030 national projections were disaggregated assuming the same spatial distribution of emissions by activity sector and pollutant. All other input data (meteorology, initial and boundary conditions of the European domain) were kept the same in both simulations.

To assess the impacts of air pollutants on vegetation and natural ecosystems, the Air Quality Directive defines critical levels (for SO_2 and NO_X) and the Accumulated Ozone exposure over a Threshold of 40 ppb (= 80 $\mu g.m^{-3}$) (AOT40). The long-term critical level for SO_2 is 20 $\mu g.m^{-3}$ as an annual mean and this value is applied during the winter period (from 1 October to 31 March). While for the NO_X, the annual air quality guideline is 30 $\mu g.m^{-3}$. Regarding the AOT40, this parameter is the sum of the difference between hourly concentrations greater than 80 $\mu g.m^{-3}$ during the summer period (May to July) using only the one-hour values between 8 am and 8 pm of each day. This indicator is designed for the protection of crops and forests and the target value is 18 000 (($\mu g.m^{-3}$).h).

In this work, the critical levels and AOT40 were estimated for the outputs from the air quality modelling system and using the Portuguese land use dataset at high spatial resolution [11], to provide the air pollution risks only on the grid cells with more than 50% of vegetation.

3 EMISSION TRENDS AND PROJECTIONS

Under the United Nations Economic Commission for Europe (UNECE) Convention on Long-Range Transboundary Air Pollution of 1979 (CLRTAP) and related Protocols, Member States report annual emission inventory information from 1990 until the current year minus 2 years. Ferreira et al. [12] analysed the evolution of the Portuguese national emissions and the compliance of old NEC Directive in 2010. The emission inventories reported to the EC show that emissions in Portugal presented a consistent decreasing trend for all pollutants until 2014, and slight decreases and increases from 2015 onwards. The 2010 targets were achieved with measures implemented at the time, which allowed Portugal to even surpass the defined goals.

As an additional requirement, emission projections for SO_2, NO_X, NH_3, NMVOCs, PM and, if available, black carbon (BC) must be reported. The projections shall cover the years 2020, 2025, 2030 and, where available, 2040 and 2050. Projected emissions are used to assess whether or not Member States are on track towards meeting their reduction commitments for 2020 and 2030 [13].

Considering that NEC targets are based on emission reduction percentages of the 2005 emissions and based on the latest projections reported by Portugal, Fig. 1 presents the national emissions for 2005 and 2015 (considered as the reference year for the present study) and the Portuguese projections for 2030, by activity sector.

Comparing 2005 and 2015 emissions, it can be concluded that Portugal is performing well in reducing its atmospheric emissions. However, efforts are still required to further reduce the 2015 levels. The analysis of emission projections indicates that national commitments based on current legislation are not enough to achieve NEC in 2030 and additional measures are needed, namely, to control SOx, NMVOC and $PM_{2.5}$ emissions. Moreover, actions are mainly required for the industrial sector as the main contributor to the referred pollutants.

4 AIR QUALITY AND ENVIRONMENTAL IMPACTS

4.1 Air quality impacts

The outputs of the WRF-CAMx system for 2015 and 2030 were processed to get annual averages and differences between the reference and the future scenario over the study region as represented in Fig. 2.

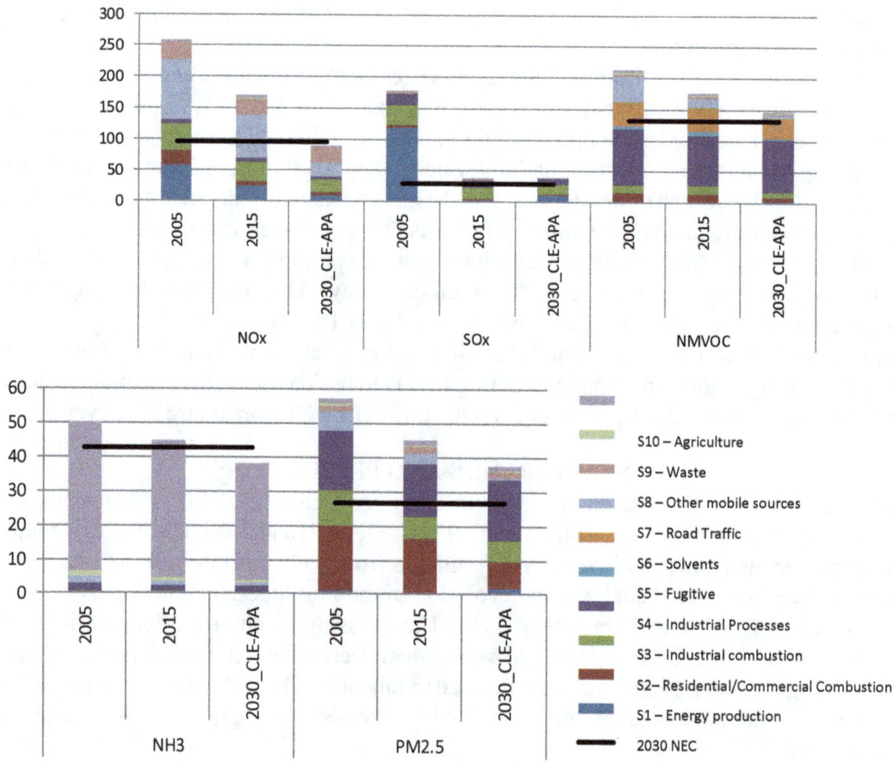

Figure 1: National emissions (in kton) by SNAP activity sector for 2005 and 2015 (reported in 2017) and emission projections reported in 2017 for CLE scenario by the Portuguese Environment Agency.

According to the air quality results, a decrease of NO_2 annual average concentrations is expected for 2030 mainly in the urban areas of Porto and Lisbon, with a maximum decrease of 15 $\mu g.m^{-3}$. Regarding $PM_{2.5}$, the model simulates increases and decreases for the annual average. A maximum decrease of 10 $\mu g.m^{-3}$ was obtained and increases up to 15 $\mu g.m^{-3}$ were achieved in particular areas, where large industrial point sources are located. For O_3, the results indicate averaged increases of 5 to 20 $\mu g.m^{-3}$.

To evaluate the potential compliance of the air quality objectives in 2030 based on the national emission projections, the legislated parameters were computed for the future scenario air quality results: NO_2 and $PM_{2.5}$ annual average concentrations and the number of exceedances of the O_3 target value. The maps presented in Fig. 3 exhibit the likely, uncertain and unlikely compliance of the objectives based on the targets set by the Air Quality Directive.

Compared to a reference scenario based on 2015 emissions spatial distribution in Portugal, the impacts of CLE projections for 2030 considering the same share in the spatial distribution of emissions by pollutant and sector reveal a likely compliance of air quality objectives for NO_2 and $PM_{2.5}$ objectives in the whole country – Continental Portugal. For O_3, the estimated impact indicates a higher area with non-compliance in 2030. This can be justified by the fact

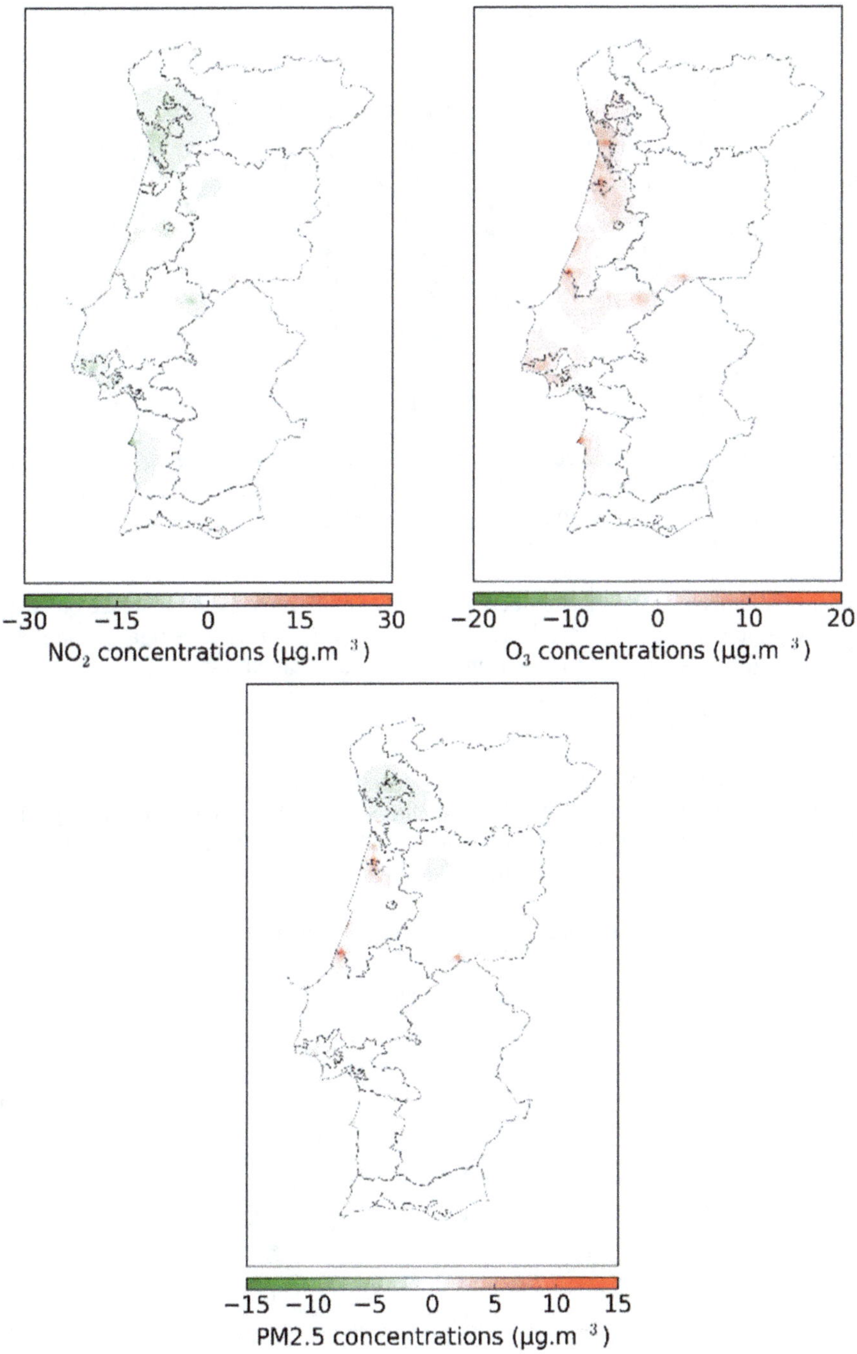

Figure 2: Differences between the annual average concentrations modelled by WRF-CAMx for reference and the future scenario, for NO_2, O_3 and $PM_{2.5}$ pollutants.

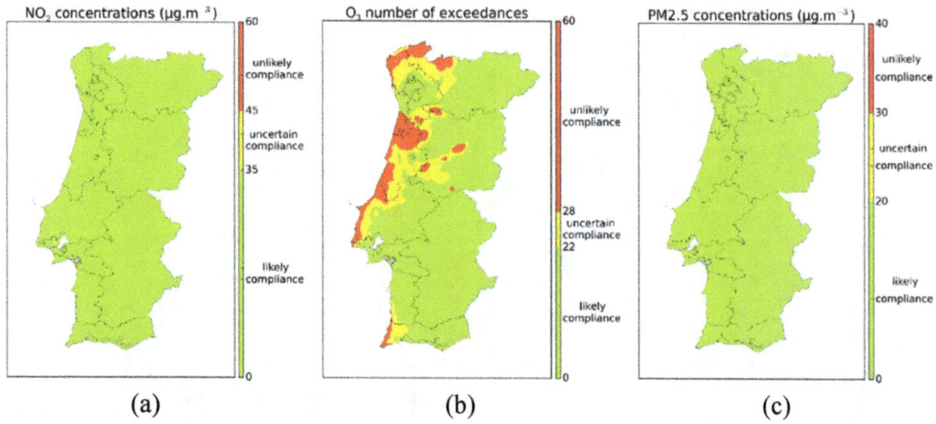

Figure 3: Compliance of air quality objectives based on Portuguese emission projections for 2030. (a) NO_2 annual limit value; (b) Maximum number of exceedances of O_3 target value; (c) $PM_{2.5}$ annual limit value.

that projected emissions estimate a decrease of NO_X emissions and an increase on NMVOC emissions, which is particularly reflected in rural areas. A further investigation on the NO_X/VOC ratio regimes is required to better understand the obtained results.

4.2 Environmental impacts

The vegetation and ecosystem potential impacts of emission projections for 2030 were estimated by the calculation of critical levels (for SO_2 and NO_X) and AOT40. Figs. 4 and 5 display the AOT40 and SO_2 and NO_X obtained results, respectively, for crop and forest areas in the country according to the Portuguese land use database.

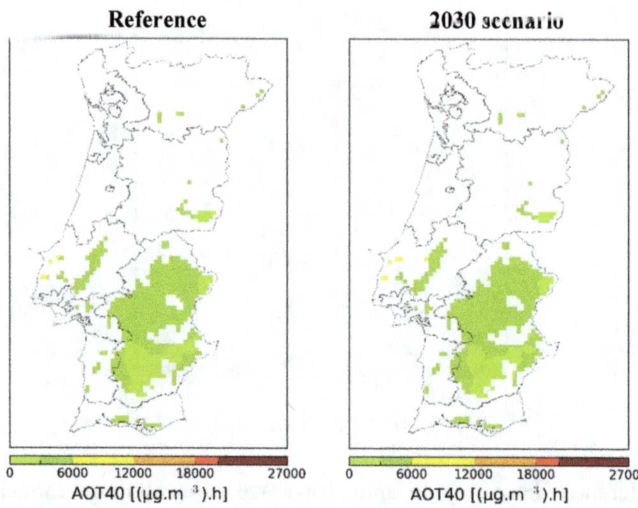

Figure 4: AOT40 levels in crop and forested areas obtained for the reference case and 2030 scenario based on O_3 modelled concentrations.

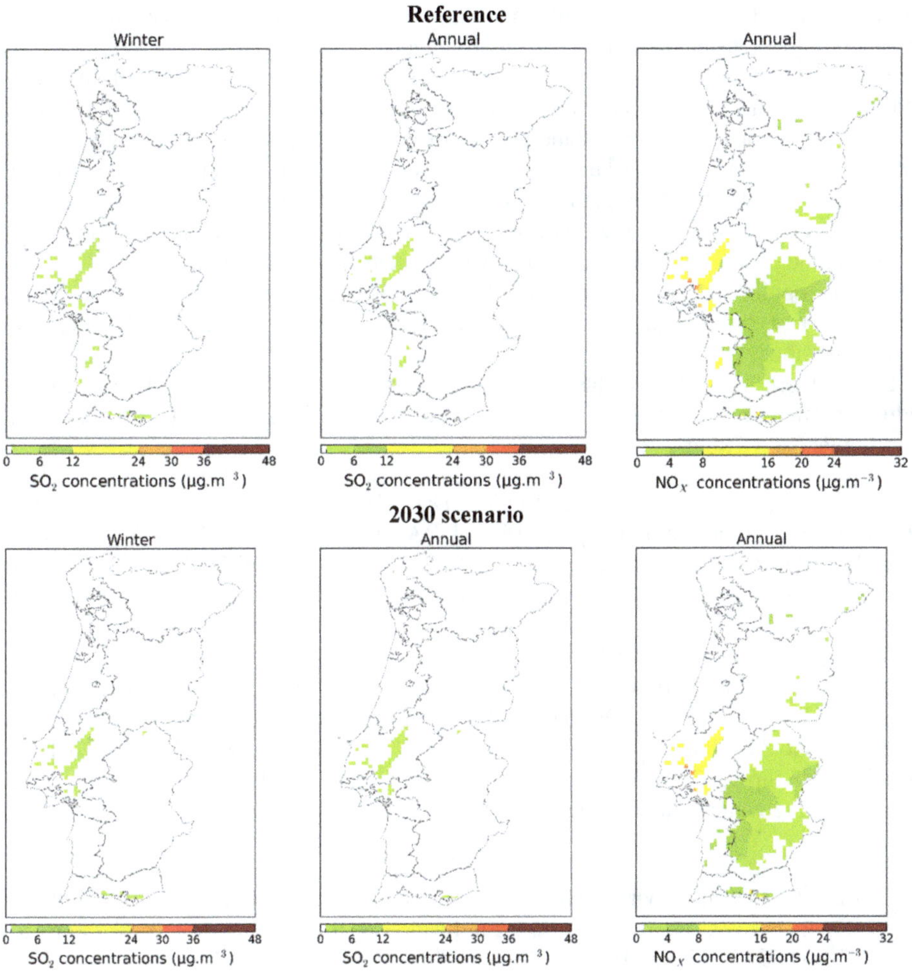

Figure 5: Critical levels for SO_2 and NO_X obtained for the reference and the future scenarios based on modelled concentrations.

AOT40 ozone values are below the limit value, and with similar patterns and levels, for both reference and future scenarios, indicating a likely compliance of the legislation (Fig. 4). Concerning the estimated SO_2 critical levels, both for winter and entire year, values are below the legal target for the reference case, and even a slight decrease is expected in the future, especially in the south of the country. For NO_X, critical levels calculated for reference and future scenarios reveal an uncertain compliance of the target value in the agricultural area northeast of Lisbon, along the Tagus river basin.

5 FINAL REMARKS

In this work an analysis of atmospheric emission trends and projections was performed in the scope of the National Emission Ceilings Directive that should be accomplished from 2030 onwards, aiming to achieve the established ceilings in 2030. Moreover, as a support to the development of the National Air Pollution Control Program (NAPCP), required by the NEC

Directive, the WRF-CAMx air quality modelling system was used to evaluate the impacts of emission projections on air quality and environment.

The emission projections considering the current legislation until 2030 reveal that additional measures for further emission reductions are required to attain the NEC targets established for Portugal. Notwithstanding, in terms of air quality impacts, a likely compliance with air quality objectives for health protection is expected for NO_2 and $PM_{2.5}$. However, a degradation of air quality regarding O_3 is potentially foreseen. Regarding impacts on vegetation and ecosystems, there are no special concerns, except for NO_X critical levels, although modelling results show a tendency of reducing risk in the future.

It must be highlighted that, in this study, land use change was not considered for the 2030 simulation nor the climate forcing, since the objective was to evaluate the impacts of future emissions.

Future work will include the design and assessment of the most efficient emission reduction measures to comply with the NEC and minimize impacts on air quality, environment and health.

ACKNOWLEDGEMENTS

This work was supported by the project FUTURAR (PTDC/AAG-MAA/2569/2014-POCI-01-0145-FEDER-016752) funded by FEDER, through COMPETE2020 – Programa Operacional Competitividade e Internacionalização (POCI), and by national funds (OE), through FCT/MCTES. J. Ferreira is funded by national funds (OE), through FCT – Foundation for Science and Technology, I.P., in the scope of the framework contract foreseen in the numbers 4, 5 and 6 of the article 23, of the Decree-Law 57/2016, of August 29, changed by Law 57/2017, of July 19. Thanks are also due for the financial support to the PhD grant of S. Coelho (SFRH/BD/137999/2018), and to CESAM (UID/AMB/50017/2019), to FCT/MCTES through national funds, and the co-funding by the FEDER, within the PT2020 Partnership Agreement and Compete 2020.

REFERENCES

[1] EEA (European Environment Agency), Air quality in Europe – 2018 report. EEA Report No 12/2018. Luxembourg: Publications Office of the European Union. ISBN 978-92-9213-989-6, 2018.

[2] EU, Directive (EU) 2016/2284 of the European Parliament and of the Council of 14 December 2016 on the reduction of national emissions of certain atmospheric pollutants, amending Directive 2003/35/EC and repealing Directive 2001/81/EC, 2016.

[3] EC, Communication from the Commission to the European Parliament, the Council, the European Economic and Social Committee and the Committee of the Regions. A Clean Air Programme for Europe. COM(2013) 918 final, 2013.

[4] Borrego, C. et al., Air quality plan for ozone: a case-study in North Portugal. *Air Quality Atmosphere & Health,* **9**(5), pp. 447–460, 2016.

[5] Borrego, C., Monteiro, A., Sá, E., Carvalho, A., Coelho, D., Dias, D. & Miranda, A. I., Reducing NO_2 pollution over urban areas: air quality modelling as a fundamental management tool. *Water, Air, Soil & Pollution,* **223**(8), 5307–5320. 2012.

[6] Duque, L. et al., Evaluating strategies to reduce urban air pollution. *Atmospheric Environment,* **127**, 196–204, 2016.

[7] Monteiro, A. et al., Air quality over Portugal in 2020. *Atmospheric Pollution Research,* **6**(5), pp. 788–796, 2015.

[8] Skamarock, W. C. et al., A Description of the Advanced Research WRF Version 3. NCAR/TN–475+STR Ncar Technical Note. 2008

[9] ENVIRON, CAMx. User's Guide. Comprehensive air quality model with extensions. Vrs. 6.4. 2016.

[10] APA, Portuguese informative inventory report 1990–2015 submitted under the NEC Directive (EU) 2016/2284 and the UNECE convention on long-range transboundary air pollution". Portuguese Environmental Agency. 2017.

[11] DGT, (Direção Geral do Território), 2018. Cartografia de Uso e Ocupação do Solo (COS, CLC e Copernicus). Online. www.dgterritorio.pt. Accessed on: 19 Oct. 2018.

[12] Ferreira, J., Leitão, J., Monteiro, A., Lopes, M. & Miranda, A.I., National Emission Ceilings in Portugal – Trends, compliance and projections. *Air Quality Atmosphere Health,* **10**(9), pp. 1089–1096. 2017.

[13] EEA (European Environment Agency), National Emission Ceilings (NEC) Directive reporting status 2018. Briefing no. 6/2018. www.eea.europa.eu/themes/air/national-emission-ceilings/nec-directive-reporting-status-2018. Accessed on: 23 Apr. 2019

URBAN MOBILITY STRATEGIES TO IMPROVE LOCAL AIR QUALITY: CASE STUDY OF LISBON, PORTUGAL

DIOGO LOPES[1], JOANA FERREIRA[1], SANDRA RAFAEL[1], PATRICIA BAPTISTA[2], MARTA FARIA[3], NUNO CANHA[1,4], SUSAN MARTA ALMEIDA[4] & MARINA ALMEIDA-SILVA[4,5]
[1]Department of Environment and Planning, CESAM, University of Aveiro, Portugal
[2]IN+ Center for Innovation, Technology and Policy Research, Universidade de Lisboa, Portugal
[3]LAETA, IDMEC – Instituto Superior Técnico, Universidade de Lisboa, Portugal
[4]Centre for Nuclear Sciences and Technologies (C2TN), University of Lisbon, Portugal
[5]H&TRC – Health & Technology Research Center, Instituto Politécnico de Lisboa, Portugal

ABSTRACT

Particulate matter concentrations are still exceeding the European air quality limit values, mainly in the urban areas where a large part of the population lives. Despite the implementation of several policies and measures to reduce the atmospheric emissions from road transport, this activity still has high influence in the air quality of the European cities. The main purpose of this study was to assess the potential of urban mobility strategies to reduce the particulate matter concentrations over the main avenue of Moscavide (a parish within the municipality of Loures, very close to the city of Lisbon, Portugal). The proposed strategies were evaluated by applying an air quality modelling system with a high spatial resolution. The results showed a slight air quality improvement (up to 0.9 µg.m^{-3}) for the tested scenarios. However, complementary studies are still necessary to provide a better understanding of the most efficient urban mobility strategies to be applied over the study area.
Keywords: PM_{10}, road transport, urban mobility strategies, air quality modelling.

1 INTRODUCTION

Despite the efforts on reducing particulate matter (PM) atmospheric emissions, the European population lives in urban areas where particles levels with an aerodynamic equivalent diameter less than or equal to 10 µm (PM_{10}) exceed quite often the air quality limit values, [1] with severe implications for human health [2].

Road transport is one of Europe's main sources of PM_{10} [1]. To reduce the atmospheric emissions from this sector, in the last decades, several policies and measures have been implemented such as: (i) definition of European emission standards (e.g. Euro 5 and Euro 6); (ii) modifications in the fuel specifications (e.g. reduction of sulphur content); (iii) development of alternative fuels (e.g. compressed natural gas (CNG)) and devices for treatment of exhaust gases (e.g. diesel particulate filter) [3]; (iv) encouraging a shift among transport modes; (v) land use planning to ensure sustainable transport facilities; and (vi) improving public transport and procurement [1].

The ongoing REMEDIO project (regenerating mixed-use Mediterranean (MED) urban communities congested by traffic through Innovative low carbon mobility solutions, part of Interreg MED Program and co-funded by European Regional Development Fund (ERDF)) aims to strength the capacity of cities to use low carbon transport systems and include them in their mobility plans by testing existing mobility solutions, through an assessment tool and participatory governance schemes that result in an operational path replicable by other MED urban areas with different city sizes. This work is part of this project and aims to evaluate the potential of urban mobility strategies to reduce PM_{10} levels over the main avenue of Moscavide located in Loures municipality (close to Lisbon, Portugal). The baseline case and proposed strategies were evaluated using an air quality modelling system with high temporal and spatial resolution. In addition, data collected in field campaigns conducted for an autumn

WIT Transactions on Ecology and the Environment, Vol 236, © 2019 WIT Press
www.witpress.com, ISSN 1743-3541 (on-line)
doi:10.2495/AIR190251

period, covering traffic counts and air quality monitoring, were used to estimate road transport emissions and to select the simulation period, respectively.

The paper is organised as follow: in Section 2, the air quality modelling setup and configuration are described in detail. Section 3 and Section 4 focus in the simulation episode selection and urban mobility strategies results, respectively. Finally, in Section 5, the main conclusions are presented.

2 AIR QUALITY MODELLING SETUP AND CONFIGURATION

To evaluate the potential of mobility scenarios to improve the air quality in the main avenue of Moscavide, an adequate model able to simulate atmospheric concentrations at street level was selected. The Computational Fluid Dynamic (CFD) model VADIS [4], [5] developed at the University of Aveiro is suitable for the present case study. This model is composed by the FLOW and DISPER modules. The FLOW calculates the wind, pressure, turbulence kinetic energy, turbulent viscosity and temperature three-dimensional (3D) while the DISPER computes the 3D air pollution concentrations using the wind field estimated by the FLOW. The main input data of the VADIS model are the local meteorological conditions (wind speed, wind direction and temperature), road transport emissions and buildings volumetry (buildings coordinates and height).

The VADIS was applied over the main avenue of Moscavide with a domain of 536×536×70 (length×width×height) and a grid resolution of 3×3×3 m^3. The study case consists in one narrow way road of about 400 meters surrounded by buildings with less than 14 meters of height and, consequently, has unfavourable air pollution dispersion conditions. Fig. 1 shows the buildings and road segments (1–5) over the study area.

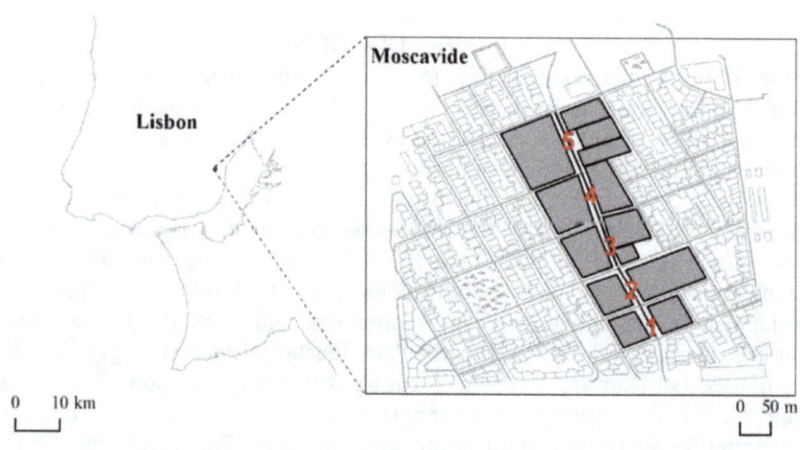

Figure 1: Application area of urban mobility strategies with the buildings and road segments considered.

The meteorological parameters can be obtained either from measurements at a meteorological station or from a meteorological model. For this work the Advanced Research Weather Research and Forecasting model (WRF-ARW) (v.3.7.1) [6] was considered. This model has been extensively used in Portugal and worldwide [7]–[9]. It provides meteorological fields for the region of interest by applying a nesting approach, from global to regional scales, covering the study area. In this work, it was applied with the following nested domains: domain 1 (D1) with a spatial resolution of 125 km, covering the Europe and part of North Africa; D2 with a spatial resolution of 25 km, comprising Iberian peninsula; and D3 with a spatial resolution of 5 km, covering the Portuguese region. The global meteorological fields from the National Center for Environmental Prediction, with 1×1° (≈ 10×10 km^2) of horizontal resolution and 6 hours of temporal resolution, were used to provide initial and boundary conditions for the coarse meteorological domain (D1). The model physical configuration was defined according to previous WRF model simulations over the Portuguese region [10], [11].

The Transport Emission Model for line sources (TREM) [12] was used to estimate road transport emissions for the baseline case and urban mobility strategies. Road traffic counts were used to obtain the traffic flow for the study area. The vehicle fleet composition (vehicle categories and classes) was defined using statistical data from Statistical National Institute (INE) and the Portuguese Automobile Trade Association (ACAP) [13], [14]. For the urban mobility strategies, four scenarios were created and tested over the study area. Scenario 1: vehicles below the Euro 3 emission standard were replaced by newer cars (Euro 4, 5 and 6); Scenario 2: diesel used by the buses was replaced by compressed natural gas (CNG); Scenario 3: maximum vehicle speed limited was increased to 35 km.h^{-1}; and Scenario 4: the road traffic was closed in the road segment 5 (Fig. 1). It should be noted that the road traffic flow was kept the same in all tested scenarios, except for the Scenario 4 where the road segment 5 was closed (without traffic flow). The scenarios of this work have been tested and implemented worldwide, showing for that, their applicability and economic sustainability over the case study [12], [15].

The buildings 3D coordinates input data were provided by the OpenStreetMap dataset [16] and the buildings' shapes were defined according to proximity and geometry criteria (building in 2D are represented in Fig. 1). The building heights were estimated based on number of floors and considering 2.8 meters per floor.

3 SIMULATION EPISODE SELECTION

During November 2016, air quality monitoring and traffic count campaigns over the case study area were conducted. With the main purpose of selecting the hourly episode for the air quality modelling application, the daily profile for the measured PM$_{10}$ levels and traffic flow in the area of interest were analysed (see Fig. 2). The wind speed and wind direction from WRF-ARW cells, located over the modelling domain, for the experimental campaigns were also analysed (Fig. 2).

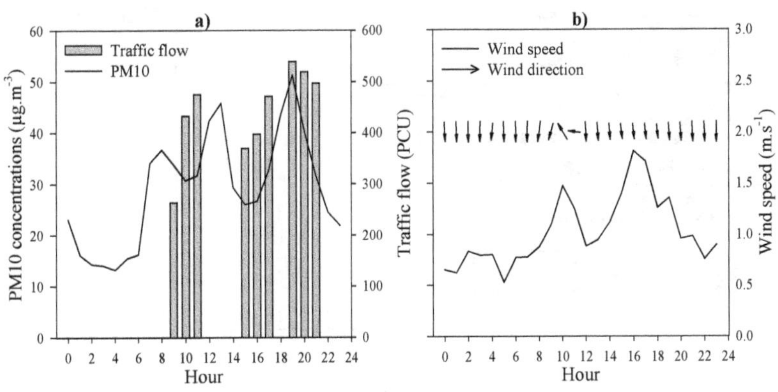

Figure 2: (a) Hourly values of PM$_{10}$ concentrations and traffic flow in the case study; (b) Hourly wind fields at the inlet boundaries of the air quality modelling system.

During the experimental campaign, the PM_{10} levels showed variations consistent with changes in road traffic flow. Three air pollution peaks in the morning (at 9 am), at midday (1 pm) and evening rush hours (at 7 pm) were registered. The highest PM_{10} concentrations (51 $\mu g.m^{-3}$) were recorded at 7 pm when the wind blew from north and wind speed was about 1.4 $m.s^{-1}$. The identification of emission sources of the study area shows that soil, road transport exhaust emissions, non-exhaust emissions, sea salt and fuel-oil contribute to about 36.3, 9.9, 24.8, 17.2 and 11.9% of the measured PM concentrations during the experimental campaign, respectively [17]. Values between 10 and 25 $\mu g.m^{-3}$ were obtained for the off-peak period (0 am–8 am) and the registered PM_{10} levels were lower than the European limit value (daily limit value is 50 $\mu g.m^{-3}$). Regarding the meteorological parameters, the wind speed ranged between 0.5 (at 5 am) and 1.8 $m.s^{-1}$ (at 4 pm). The wind blew from southeast during late morning (10 am and 11 am) and for the remaining hours, the wind direction was mainly from the north.

According to the analysed air pollution levels, traffic flow and meteorological parameters, the hour of 7 pm was selected for the air quality modelling application over the study area. In this sense, the evening rush hour (i.e. 7 pm) was selected for simulation, and thus, in the next sections, the simulated road transport emissions and air pollution levels correspond to the referred period. This analysis is quite important since that is the period when people are most exposed to air pollution, while they are driving, walking on the street or at home.

4 URBAN MOBILITY STRATEGIES

The urban mobility strategies were investigated considering as reference the baseline scenario (current situation). Two different approaches were used to provide this analysis: (i) road transport emissions by road segments to quantify the impact of measures in terms of PM_{10} emissions; (ii) mapping of the hourly PM_{10} levels differences between baseline and mitigation scenarios to understand the spatial variability of the air pollution concentrations.

Fig. 3 shows the road transport emissions by road segment (1–5), for the baseline case (Base case) and urban mobility scenarios (Scenario 1, 2, 3 and 4) for the study episode (7 pm).

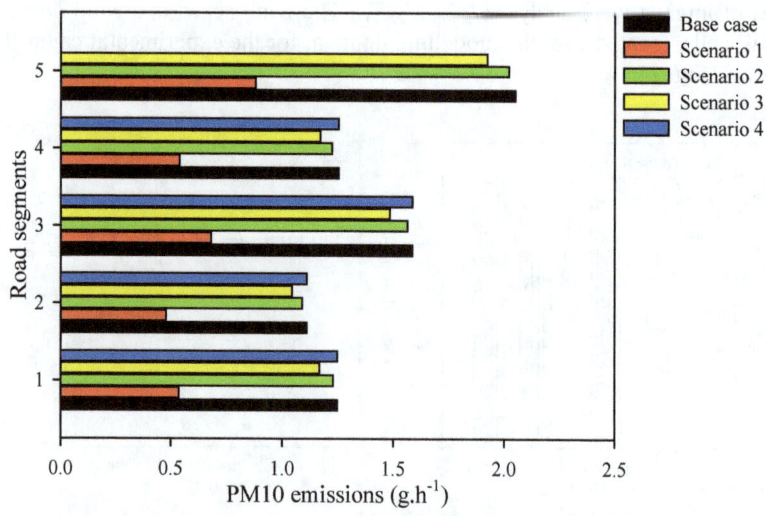

Figure 3: PM_{10} emissions ($g.h^{-1}$), by road segment, for the baseline case and tested scenarios on evening rush hour (7 pm).

For the baseline case, the PM_{10} emissions vary in a range of 1.1 to 2.1 $g.h^{-1}$. Maximum values are obtained for the road segment located further north of the study area (road segment 5). The emissions of the remaining roads represent 50–80% of the total emissions of the main road. Analysing the emissions associated to each urban mobility strategy, the highest impacts are obtained for Scenarios 1 and 4 with a total reduction of about 4.1 (57%) and 2.0 $g.h^{-1}$ (28%), respectively. The replacement of the fuel used by buses (Scenario 2) and the increase of the maximum vehicle speed limits (Scenario 3) allow a total reduction of less than 0.5 $g.h^{-1}$ (6.4%). However, these results only allow to quantify the atmospheric emission impacts of the tested Scenarios. Concerning the air quality impacts, Fig. 4 presents the expected PM_{10} levels reduction (in $\mu g.m^{-3}$) during the study episode (7 pm).

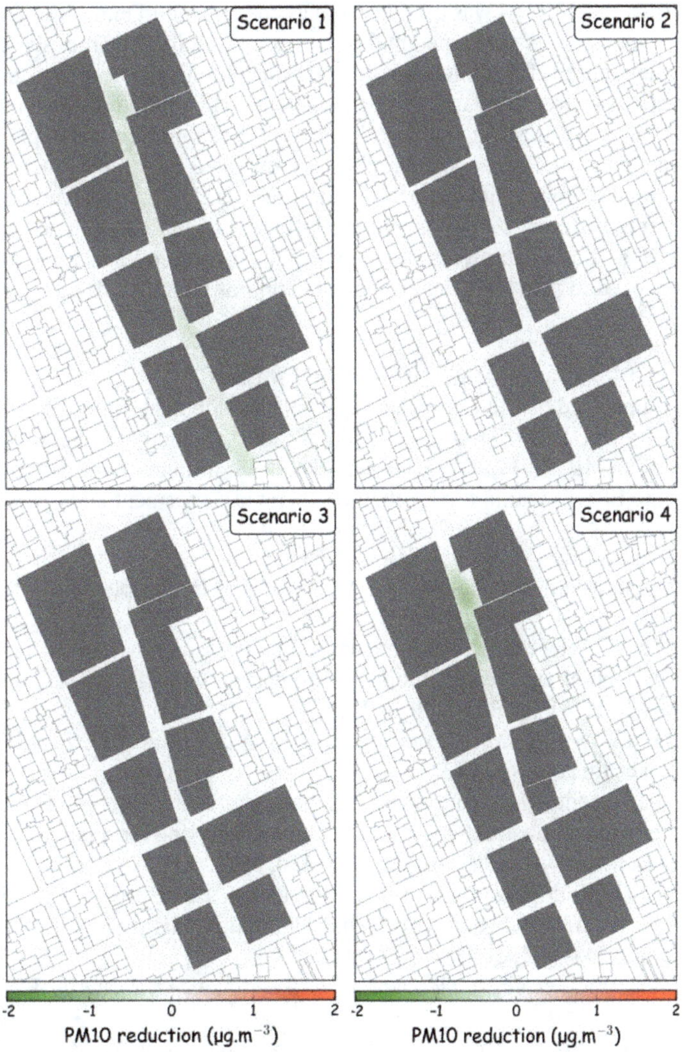

Figure 4: Air pollution reduction of PM_{10} levels based on the difference between each scenario and the baseline case.

It can be observed that the implementation of Scenarios 1 and 2 reduce the PM_{10} concentrations up to 0.9 $\mu g.m^{-3}$ over the study area. The highest air pollution decrease is recorded when the road segment 5 is closed. However, Scenario 1 reveals a larger reduction area with air quality improvements over the entire simulation domain (up to 0.5 $\mu g.m^{-3}$). For the remaining tested Scenarios, the replacement of the buses fuel (Scenario 2) and the increase of the maximum vehicle speed limit (Scenario 3) a slight decrease of PM_{10} concentrations is verified (less than 0.1 $\mu g.m^{-3}$).

5 CONCLUSIONS

The main goal of this work was to evaluate the potential of urban mobility strategies to reduce the PM_{10} levels on a narrow street of an urban area. The proposed strategies were evaluated by the application of the CFD model VADIS, with high spatial and temporal resolution to different scenarios. The TREM was used to estimate road transport emissions for the baseline case, based on traffic counts performed in the study area, and for the proposed scenarios. Air quality monitoring and traffic count campaigns for an autumn period shown that the highest PM_{10} concentrations are recorded on evening rush hours (at 7 pm) when the wind blows from north and the wind speed is about 1.4 $m.s^{-1}$. The tested urban mobility strategies showed that the highest reductions of PM_{10} concentrations are obtained when the Euro 3 vehicles are banned and the road segment with the largest traffic flow is closed. However, complementary studies considering different scenarios and the influence of the non-exhaust emission from the road traffic (this source represents about 25% of the measured PM_{10} concentrations in the study area) is needed to provide a better understanding of the most efficient urban mobility strategies.

ACKNOWLEDGEMENTS

This work was supported by the European Regional Development Fund (ERDF) through the Interreg MED project REMEDIO (Ref. 862). Thanks are due for the financial support to CESAM (UID/AMB/50017/2019), to FCT/MCTES through national funds, and the co-funding by the ERDF, within the PT2020 Partnership Agreement and Compete 2020.

REFERENCES

[1] European Environmental Agency (EEA), Air quality in Europe – 2018. Report, no. 12. 2018.

[2] Costa, S. et al., Integrating health on air quality assessment—Review report on health risks of two major european outdoor air pollutants: PM and NO_2. *J. Toxicol. Environ. Heal. Part B*, **17**(6), pp. 307–340, 2014.

[3] Lopes, D., *Determinação de Fatores de Emissão de Gases de Exaustão de Veículos*, Universidade de Aveiro, 2013.

[4] Rafael, S., Vicente, B., Rodrigues, V., Miranda, A.I., Borrego, C. & Lopes, M., Impacts of green infrastructures on aerodynamic flow and air quality in Porto's urban area. *Atmos. Environ.*, **190**(July), pp. 317–330, 2018.

[5] Amorim, J.H., Rodrigues, V., Tavares, R., Valente, J. & Borrego, C., CFD modelling of the aerodynamic effect of trees on urban air pollution dispersion. *Sci. Total Environ.*, **461–462**, pp. 541–551, 2013.

[6] Skamarock, W.C. et al., *A Description of the Advanced Research WRF Version 3*, 2008.

[7] Lopes, D., Ferreira, J., Hoi, K., Miranda, A.I., Yuen, K.V. & Mok, K.M., Weather research and forecasting model simulations over the Pearl River Delta Region. *Air Qual. Atmos. Heal.*, **12**, pp. 115–125, 2019.

[8] Evans, J.P., Ekstrom, M. & Ji, F., Evaluating the performance of a WRF physics ensemble over South-East Australia. *Clim. Dyn.*, **39**(6), pp. 1241–1258, 2012.
[9] Rama Rao, Y.V., Alves, L., Seulall, B., Mitchell, Z., Samaroo, K. & Cummings, G., Evaluation of the weather research and forecasting (WRF) model over Guyana. *Nat. Hazards*, **61**(3), pp. 1243–1261, 2012.
[10] Carvalho, D., Rocha, A., Gómez-Gesteira, M. & Santos, C., A sensitivity study of the WRF model in wind simulation for an area of high wind energy. *Environ. Model. Softw.*, **33**, pp. 23–34, 2012.
[11] Monteiro, A. et al., Air quality over Portugal in 2020. *Atmos. Pollut. Res.*, **6**(5), pp. 788–796, 2015.
[12] Li, X., Lopes, D., Mok, K.M., Miranda, A.I. & Yuen, K.V., Development of a road traffic emission inventory with high spatial – temporal resolution in the world's most densely populated region – Macau. *Environ. Monit. Assessmen*, pp. 191–239, 2019.
[13] Associação Automóvel de Portugal (ACAP), Portugal Statistics for fleet composition, 2015, http://www.acap.pt/.
[14] Instituto Nacional de Estatística (INE), Statistics Portugal, 2011; CENSUS, 2011-statistical data for Portugal, 2011, Online. http://censos.ine.pt.
[15] Duque, L. et al., Evaluating strategies to reduce urban air pollution. *Atmos. Environ.*, **127**(2), pp. 196–204, 2016.
[16] OpenStreetMap contributors, Planet dump; Data file from $date of database dump$, 2017, Online. https://planet.openstreetmap.org.
[17] Almeida, S.M., Manousakas, M.I., Diapouli, E. & Eleftheriadis, K., Source apportionment in a street canyon : first approach within REMEDIO project. Presented at *5th Iberian Meeting on Aerosol Science and Technology Jointly organised with IMPROVE LIFE13 ENV/ES/263*, Barcelona, Spain, 2017.

SPECIAL SESSION: STRATEGIES FOR HUMAN EXPOSURE REDUCTION

(ORGANISED BY E. C. RADA)

PERSPECTIVES OF STACK AND ENVIRONMENTAL MONITORING IN THE SURROUNDINGS OF A WASTE-TO-ENERGY PLANT

LUCA ADAMI[1], MARCO SCHIAVON[1], MARTINA FERRAI[2], LORIS DALLAGO[2], ELENA CRISTINA RADA[3], MARCO TUBINO[1] & MARCO RAGAZZI[1]
[1]Department of Civil, Environmental and Mechanical Engineering, University of Trento, Italy
[2]ISER Srl, Italy
[3]Department of Theoretical and Applied Sciences, University of Insubria, Italy

ABSTRACT
Waste-to-Energy (WtE) processes respond both to the emerging need for reducing the flow of waste into the environment and, at the same time, to the increasing demand for energy. Despite, these evident advantages, WtE plants still present some critical issues regarding the emissions of heavy metals into the atmosphere, which are regulated by the environmental legislations but, concerning the European Union, are regarded as groups of metals with no consideration of the different carcinogenic potential of each metal. In addition, there are uncertainties on the estimation of the balance of carbon dioxide, which depends on several factors like transportation, type of incoming waste, processes in use and secondary emissions. Despite great improvements in air pollution control from this sector, persistent organic pollutants are still emitted by WtE plants. Therefore, the implementation of adequate environmental monitoring (EM) plans is essential to monitor the impact of WtE plants in their surroundings, especially in the presence of the population, fields and pastures. In view of these considerations, this paper aims to provide guidance on basic and novel approaches that are necessary for a comprehensive monitoring of the impacts of a new WtE plant in terms of air quality and public health. A case study regarding the EM plan proposed for a new WtE plant will also be reported as an example.
Keywords: ambient air, soil, deposition, sediments, dioxin, heavy metals, gasification, waste management.

1 INTRODUCTION

The increasing demand for energy that has taken place since the last century, combined with the increasing need for reducing the flow of waste into the environment, resulted in the great opportunity of producing energy from waste and in the development of the so-called waste-to-energy (WtE) sector [1]. The contribution of this sector towards the achievement of the United Nations' Sustainable Development Goals number seven (Affordable and Clean Energy) and eleven (Sustainable Cities and Community) has already been recognised by the scientific literature in this field [2], [3].

Generally, WtE processes include the direct or indirect combustion of waste or refuse-derived fuel (RDF), which is produced by aerobic treatments of the residual fraction of municipal solid waste (MSW). Being thermal processes, both methods lead to the generation of thermal energy from the combustion of a fuel. The latter can be directly the RDF in the case of direct WtE plants (e.g.: incinerators) or the gas generated by the thermal treatment of waste/RDF, i.e. the syngas [4], [5]. In this last case, WtE plants are commonly regarded as indirect processes and include different gasification treatments, pyrolysis and combinations of these treatments [6]–[8].

MSW incinerators are the first example of WtE technologies [9]. Historically, MSW incinerators were depicted as a threat for human health, especially for their contribution in terms of dioxin in the atmosphere. In the last decades, the WtE sector has undergone rapid changes in the air pollution control lines to limit the emissions of dioxin, heavy metals,

WIT Transactions on Ecology and the Environment, Vol 236, © 2019 WIT Press
www.witpress.com, ISSN 1743-3541 (on-line)
doi:10.2495/AIR190261

particulate matter (PM) in general, nitrogen (NO_x) and sulphur (SO_x) oxides [10]. Besides the use of renewable sources, WtE was recently recognised as one of the cleanest approaches for energy production, thanks to the improvements made in terms of air pollution control [11]. Nowadays, modern WtE plants may include electrostatic precipitators and bag-house filters (for PM removal), multi-stage scrubbing (to remove acids and SO_x, with the optional addition of activated carbon to remove dioxin, hydrocarbons in general and metals), selective or non-selective catalytic reduction (for NO_x removal) and, recently, selective catalytic oxidation (for dioxin removal) [10]. Further improvements are expected by the WtE sector from the general point of view of sustainability, especially concerning the minimisation of the solid residues (e.g., bottom ashes) and the recovery of chemicals [12]–[16]. These expected improvements will allow the WtE sector to comply with a vision of circular economy. Specifically, the literature recently highlighted the positive role of indirect combustion and bottom ash vitrification in a circular economy perspective [17].

However, some potential criticalities still exist, especially concerning the role of heavy metals [18]. In addition, as large combustion installations that partly burn non-biogenic fractions of waste, WtE plants generate net positive emissions of carbon dioxide (CO_2) into the atmosphere, with direct impacts on the climate. Non-biogenic CO_2 contributions also derive from the use of other fossil fuels to ignite the waste and to carry out the vitrification process of the bottom ashes, as an option to landfilling. Methane (CH_4) and coke are usually adopted for those purposes, respectively. Indirect CO_2 net emissions that must be considered to complete the CO_2 balance derive from the transportation (positive contribution to the net CO_2 generation) of waste and from the avoided emissions related to both energy recovery strategies and the choice of performing the co-generation of thermal and electric energy (negative CO_2 contribution). The net CO_2 generation is normally calculated as an estimate based on standard emission factors regarding positive contributions (auxiliary fuel combustion and waste transportation) and the negative contributions (extra thermal energy, obtained by energy recovery strategies, and electric energy production from a reference mix of fuels, which may be compensated by electric energy production by a steam turbine in the plant). The CO_2 generated by waste oxidation can be estimated by the carbon content of the input waste, which requires periodical characterisation. Thus, the amount of CO_2 generated by a WtE plant strongly depends on the input waste, whose composition may not be constant over time. Therefore, the estimation of CO_2 emissions through emission factors is subject to a high degree of approximation. A continuous monitoring system for CO_2 concentration at the stack level would reduce the degree of uncertainty and avoid performing chemical characterisation of the input waste.

In the European Union, the emissions from WtE plants in the atmosphere are regulated by the European Directive 2010/75/EU [19]. The authorisation processes of a WtE plant is regulated by the European Directive 2014/52/EU [20] that reviewed the Directive 2011/92/EU [21] in several aspects: 'purely' procedural requirements, screening, quality and monitoring [22]. Directives (Art. 1 of 16) shall apply to the assessment of the environmental effects of those public and private projects which are likely to have significant effects on the environment. This Directive describes the procedure to evaluate the Environmental Impact Assessment (EIA) of a single project to ensure an integrated pollution prevention and control on the environmental matrices (e.g. soil, water, air, ecosystems). A specific section of the EIA Directive describes the EM plan with the aim to conducting cumulative exposure and risk assessment pre- and post-intervention [23].

An EM plan should monitor the area potentially influenced by the plant and the expected impacts as more completely as possible. All the environmental comparts that could be influenced by the activities of the plant should be correctly identified, as well as the pollutants monitored. To create a reference base to compare the situation when the plant is in operation, it is important to evaluate the pollution level before its construction. If critical impacts cannot be excluded during the plant construction, a monitoring activity should be planned during this phase too.

In the light of the importance of correct EM plans for the WtE sector, the present paper is intended to provide guidance on the basic approaches that are necessary to monitor the impacts of a new WtE plant in terms of air quality. In addition, this work presents other new and unconventional strategies that would allow for a comprehensive monitoring of a WtE plant from the point of view of its impacts on air quality and on the exposure to air pollutants. Such new approach responds to the need for overcoming the uncertainties that characterise this sector in terms of (1) the emissions of heavy metals and persistent organic pollutants (POPs), (2) the role of secondary emissions and (3) its carbon footprint. These aspects are here discussed also by considering the case study of the EM plan proposed for a new WtE plant.

2 BASE APPROACH

Hereby we refer to the case study of a project for a new WtE plant located near the border between the provinces of Trento and Bolzano (North Italy). The project consists in a thermal treatment of industrial waste finalized to produce energy (electricity and heat), through the technology of indirect combustion, with final vitrification of the ashes. Input of the new plant consists in RDF (European Waste Catalogue – EWC – code 19 12 10) and in sorting non-hazardous residues (EWC code 19 12 12) for a total amount of 95,000 t/year. Estimated net production of electric energy is 100 GWh/year, with a maximum thermal power of 69 MW. The main stack of the WtE plant is designed 45 m high, with an internal diameter of 1.8 m. The scheme of the new plant consists in the main following parts: heating furnace, gasifying reactor, temperature monitoring and product analysis. Moreover the thermal line is completed by the presence of a heating recovery system, dry off-gas purification line with catalytic removal of nitrogen oxides and spent reagent storage systems, turbo generator with nominal capacity of 17 MW and connected vapour cycle system. The outgoing airflow rate from the primary stack is 107,000 Nm^3/h.

The EM plan proposed for the case study presented above is partly based on a conventional monitoring scheme that considers classical environmental matrices for air contaminants: exhaust gas, ambient air and soil. Noise pollution and other environmental matrices like superficial water basins and groundwater are considered by the plan but are not reported in this paper. According to the EM plan, ambient air concentrations and soil concentrations should be measured before, during and after the construction of the plant. One critical aspect, which may influence the significance of the monitoring activity, is the location of the sampling points. The most appropriate strategy consists in locating the sampling points where the highest impacts are expected. The choice of the sampling points can then be made on the basis of the results of the dispersion modelling that is part of the environmental impact assessment procedure. In addition to the expected hotspots, especially in the case of soil concentrations, sensitive areas could be selected, such as kindergarten, schools and parks, where children might be exposed to the contaminants by accidental soil ingestion. Regarding the contaminants to monitor, concerning ambient air concentrations, heavy metals and, in general, PM10 are the air pollutants considered by the EM plan of the case study presented here. The choice of these air contaminants to monitor is related to the results of the dispersion modelling and to the health impact assessment, which highlighted that the role of dioxin is negligible with respect to that of heavy metals. The monitoring of soil concentrations concerned a larger variety of pollutants (heavy metals, aromatics, phenols, chlorinated compounds, polycyclic aromatic hydrocarbons, plant protection products, dioxin and other chemicals) according to the legislation on the concentration limit values of pollutants in soils [24].

3 ADVANCED APPROACH

In addition to the classical monitoring approach involving stack emissions, pollutant concentrations in ambient air and soil and noise, other techniques are very useful to characterise the impacts of a new installation and, specifically, a new WtE. Such techniques are not routinely included in EM plans, but their potential in deepening the understanding of how the environment is influenced by the plant is worth consideration for future environmental impact assessment procedures. Some methodologies would provide additional information to measure the footprint of a WtE plant in the surroundings. Other methodologies would overcome issues related to the WtE sector that have been recently pointed out [15].

The contribution of a WtE plant in terms of air pollutants could be evaluated by two unconventional methodologies that have been developed in the last five years. Specifically, two methods are particularly appropriate to monitor the influence of a plant in terms of persistent organic pollutants (POPs). The first method consists in the characterisation of the sediments of a pond, possibly located downwind and within a few kilometres from the plant. By knowing the sedimentation rate of the pond and by collecting a sediment core, it is possible to characterise and quantify the POPs that have deposited as sediments to the bottom of the pond over the years. This method allows reconstructing the history of pollutant contributions in the area and, through the analysis of the fingerprints of chlorinated compounds like dioxin and through the so-called "diagnostic ratios" between different polycyclic aromatic hydrocarbons, the identification of the dominant source(s) of POPs in the area is made possible [25]. Indeed, every source emitting dioxin is characterised by a specific profile of the different congeners in the exhaust off-gas of a combustion plant [26]. In spite of large approximations that must be considered on the fate of POPs in the atmosphere and on their partition between gaseous and particle phases, the detection of a similar profile in other environmental matrices might reveal that the most probable origin of the contamination could be the combustion plant under observation. Another unconventional methodology allows obtaining the same information: the characterisation of the atmospheric depositions to soil. By placing bulk deposition samplers in the surroundings of a combustion plant, the role of the plant can be determined by a comparison between the congener profiles of dioxin and the fingerprint of the specific emitting activity [26]. Deposition samplers should be placed in strategic locations, preferably close to residential areas, schools, hospitals, fields and pastures, in order to cover all the possible routes of exposure to POPs. Indeed, in the case of dioxin, the diet is the dominant exposure route [27], since such group of toxics tend to accumulate in the food chain, specifically in fat tissues. This peculiarity justifies the importance of monitoring pastures and fields [28]. For both methodologies, the quantification of POPs can be made with high-resolution gas chromatography and mass spectrometry, after extraction with hexane and specific purification cycles [25], [26]. Both methodologies are included in the EM plan of the WtE plant presented here as a case study. Figure 1 presents the location of the monitoring points chosen for the measurement of ambient air concentrations, soil concentrations, atmospheric deposition and sediment sampling. The monitoring points were the same for ambient air concentrations and atmospheric depositions (A1 and A2) and are located close to the area of the plant. A1, located in a vineyard, coincides with the cell with the highest concentration of air pollutants that results from the dispersion model used during the environmental assessment study; A2 was chosen to characterise the role of the plant in the industrial area where the plant is located. One monitoring point for soil concentration (SO1) was selected as close to the cell of maximum impact for air pollutants, whereas the second monitoring point (SO2) was chosen to keep a playground under observation. The monitoring point for sediment concentrations (SE1) is located in a nearby pond.

An additional monitoring proposal concerns the speciation of the chromium released by the stack(s) of a WtE plant. At the European level, the environmental legislation established concentration limit values for several heavy metals emitted from incineration plants: cadmium (Cd), thallium (Tl), total chromium (Cr), arsenic (As), nickel (Ni), lead (Pb), vanadium (Va), cobalt (Co), mercury (Hg), manganese (Mn), antimony (Sb) [19]. However, the legislation does not specify limit values for the single heavy metals (apart from Hg), but establishes a limit values for the sum of Cd and Tl, and for the sum of the remaining metals. Some of the aforementioned metals are listed by the International Agency for Research on Cancer (IARC) as carcinogenic (Cd, As, Ni) and possibly carcinogenic (Hg) to humans [29]. In addition, in the emissions from a WtE plant, Cr is composed by Cr III (not carcinogenic) and Cr VI (carcinogen to humans). Among the metals regulated by the legislation, Cr VI and Cd have the highest carcinogenic potentials [30]. This means that, in a worst-case scenario, a WtE plant might emit a large proportion of Cd and Cr VI in relation to the other metals in their respective groups and, in spite of this, would comply with the legislation. In other words, in the absence of specific limits for carcinogenic metals, the environmental authorities of a country/region could authorise a new plant to induce a non-acceptable cancer risk in the population potentially exposed to the emissions.

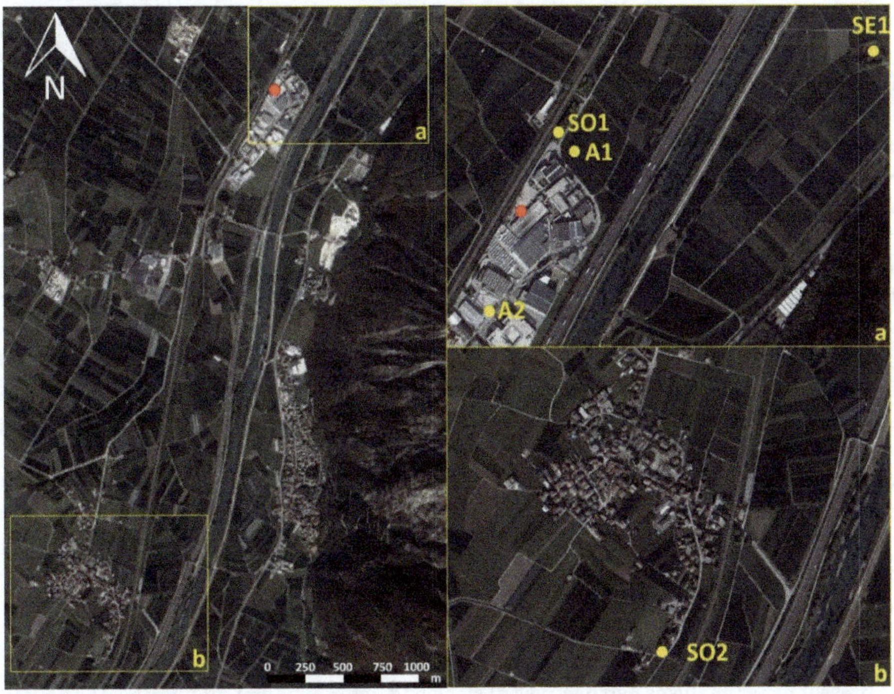

Figure 1: Location of the monitoring points for ambient air concentrations and deposition (A1 and A2), soil concentrations (SO1 and SO2) and sediment sampling (SE1) chosen for the case study; the location of the primary stack is represented by a red dot.

Thus, a monitoring approach consisting in periodical verifications of the concentrations of the specific metals at the stack would help the detection of anomalous contributions of carcinogenic metals in the atmosphere. It is worth reminding that inhalation is not the only route of exposure to metals. Accidental soil ingestion and the diet may give non-negligible contributions to the daily dose too [31]. Therefore, a periodical metal speciation at the stack, which could be carried out by sampling and analysing the particulate matter released by the plant, would allow excluding the occurrence of non-acceptable levels of carcinogenic metals or considering the adoption of improved air pollution control technologies. The quantification of metals could be carried out with inductively coupled plasma optical emission spectrometry on particulate samples. Cr speciation could be determined with ion chromatography-based US EPA method 7199 [32]. This first additional monitoring approach is proposed within the EM plan of the WtE plant presented in Section 2.

The emissions into the atmosphere from a WtE plant usually do not only originate from the main chimney. Secondary emissions may play a significant role in the overall mass flow of the air pollutants released by a WtE plant. In the case study presented here, for instance, the mass flow rate of total suspended particles (TSP) released by secondary emissions are estimated to be almost 28% the TSP mass flow rate from the primary stack. The type of air pollutants released and their mass flow rates vary depending on the type of secondary emission involved. In addition, different secondary emissions may be subject to different limit values, which depend on the nature of the processes and on the fuel used (if any). The characterisation of the emissions from a WtE plant may be more complicated in the presence of diffuse emissions, i.e. the release of air pollutants in the atmosphere without any conveyance systems. However, such conditions are likely to occur in the residual municipal solid waste treatment sector only if the plant includes an aerobic biological stage. In that case, diffuse emissions (represented by an area source) may occur at the outlet of the air treatment line if an open biofilter is used [33]. Given the non-negligible role of secondary emissions, an EM plan should keep the concentrations of air pollutants under control. To this purpose, discontinue samplings at secondary emission points might be sufficient to characterise secondary contributions into the atmosphere.

The conventional monitoring approach usually focuses on local pollutants. However, the carbon footprint of a WtE plant could be also easily assessed by installing a non-dispersive infrared (NDIR) sensor for the monitoring of the concentrations of greenhouse gases at the point of release of the exhaust gas. Given the non-negligible contributions expected from secondary emissions, installing NDIR gas sensors also on secondary stacks would allow for a direct measurement of the carbon footprint of the whole plant. In addition, the direct measurement of greenhouse gases could turn useful to calibrate the average CO_2-equivalent (CO_{2eq}) emission factors that are usually adopted to estimate the carbon footprint from specific processes. The monitoring of CO_{2eq} concentrations (and consequently of the CO_{2eq} mass flow rates, which can be calculated from the airflow rate) at all the emission points of a WtE plant could help detecting priorities in the adoption of strategies to reduce the emissions of greenhouse gases, by focussing, for instance, on the transportation of waste and/or on carbon sequestration processes.

4 CONCLUSIONS

The present paper analyses perspectives of stack and environmental monitoring of a new WtE plant to be built in north Italy, discussing the results of the base approach (due to existing European Directives) and to an advanced approach due to necessary improvements suggested by the specificity of the case studied that can find broader application in the WtE field. The advanced approach proposed includes two unconventional methodologies developed in the last five years to monitor the influence of a WtE plant in terms of persistent organic pollutants (POPs), in particular the characterisation of the atmospheric depositions to soil.

In particular, the first method describes the characterisation of the sediments of a pond, located downwind and in the neighbourhood of the plant. This allow (1) to reconstruct the contribution of pollutants in the surrounding area and (2) to identify the dominant sources of POPs through (1) the analyses of the fingerprints of chlorinated compounds and through (2) the so-called "diagnostic ratios" between different polycyclic aromatic hydrocarbons.

The second methodology proposed consists in placing bulk deposition samplers in strategic positions (e.g. public areas and buildings like schools, hospitals and parks, residential areas, fields and pastures) in the surroundings of the plant, comparing the congener profiles of dioxin and the fingerprint of the specific emitting activity from the stacks.

In addition to the two unconventional methodologies we propose (1) the speciation of the chromium released by the stacks of the WtE plant with the aim to quantify the cancerogenic effect due to the presence of Cr VI in the total Cr emitted and (2) the specific analyses of the emissions from secondary chimneys.

Although the application of first unconventional method is limited by the presence of a pond in the neighbourhood of the plant, the second methodology as well as the speciation of the chromium and the analyses of the secondary points of pollution emission can be considered innovative approaches for the characterization of a WtE plant that aim to minimize its environmental footprint.

ACKNOWLEDGEMENT

The authors wish to thank PA Holding for the financial support to the research that was the starting point of the present article.

REFERENCES

[1] Brebbia, C.A., Telles, J.C.F. & Wrobel, L.C., (eds), *Boundary Element Techniques*, Springer-Verlag: Berlin and New York, pp. 11–13, 1984.

[2] Osifchin, N. & Vau, G., Power considerations for the modernization of telecommunications. *Proceedings of the Fourth Annual Portable Design Conference*, pp. 137–142, 1997.

[3] Test Methods for Evaluating Solid Wastes, Physical/Chemical Methods; U.S. Environmental Protection Agency, Office of Solid Wastes, SW-846. www.epa.gov/epaoswer/hazwastes/test/main.htm. Accessed on: 23 Jun. 2015.

[4] Bratanow, T. & De Grande, G., Numerical analysis of normal stress in non-Newtonian boundary layer flow. *Engineering Analysis*, **6**(2), pp. 20–25, 1985.

[5] Klinghoffer, N.B., Themelis, N.J. & Castaldi, M.J., Waste to energy (WTE): an introduction. *Waste to Energy Conversion Technology*, eds N.B. Klinghoffer & M.J. Castaldi, Woodhead Publishing: Cambridge, pp. 3–14, 2013.

[6] AlQattan, N., Acheampong, M., Jaward, F.M., Ertem, F.C., Vijayakumar, N. & Bello, T., Reviewing the potential of Waste-to-Energy (WTE) technologies for Sustainable Development Goal (SDG) numbers seven and eleven. *Renewable Energy Focus*, **27**, pp. 97–110, 2018.

[7] United Nations, About the Sustainable Development Goals. www.un.org/sustainabledevelopment/sustainable-development-goals/. Accessed on: 21 Dec. 2018.

[8] Costa, M., Curcio, C., Piazzullo, D., Rocco, V. & Tuccillo, R., RDF incineration modelling trough thermo-chemical conversion and gaseous combustion coupling. *Energy*, **161**, pp. 974–987, 2018.
[9] Bosmans, A., Vanderreydt, I., Geysen, D. & Helsen, L., The crucial role of Waste-to-Energy technologies in enhanced landfill mining: A technology review. *Journal of Cleaner Production*, **55**, pp. 10–23, 2013.
[10] Ribeiro, A., Soares, M., Castro, C., Mota, A., Araujo, J., Vilarinho, C. & Carvalho, J., Waste-to-energy technologies applied for refuse derived fuel (RDF) valorization. *Lecture Notes in Electrical Engineering*, **505**, pp. 641–647, 2018.
[11] Lombardi, L., Carnevale, E. & Corti, A., A review of technologies and performances of thermal treatment systems for energy recovery from waste. *Waste Management*, **37**, pp. 26–44, 2015.
[12] Ragazzi, M. & Rada, E.C., Multi-step approach for comparing the local air pollution contributions of conventional and innovative MSW thermo-chemical treatments. *Chemosphere*, **89**, pp. 694–701, 2012.
[13] Xu, H., Lin, W.Y. & Romagnoli, A., Technological review on enhancing the energy efficiency of MSW incineration plant. *Proceedings of the 2018 Asian Conference on Energy, Power and Transportation Electrification*, 8610756, 2019.
[14] Makarichi, L., Jutidamrongphan, W. & Techato, K., The evolution of waste-to-energy incineration: A review. *Renewable and Sustainable Energy Reviews*, **91**, pp. 812–821, 2018.
[15] Carneiro, M.L.N.M. & Gomes, M.S.P., Energy, exergy, environmental and economic analysis of hybrid waste-to-energy plants. *Energy Conversion and Management*, **179**, pp. 397–417, 2019.
[16] Ardolino, F., Berto, C. & Arena, U., Environmental performances of different configurations of a material recovery facility in a life cycle perspective. *Waste Management*, **68**, pp. 662–676, 2017.
[17] Rada, E.C., Ragazzi, M., Torretta, V., Castagna, G., Adami, L. & Cioca, L.I., Circular economy and waste to energy. *AIP Conference Proceedings*, **1968**, 030050, 2018.
[18] Schiavon, M., Ragazzi, M., Rada, E.C., Magaril, E. & Torretta, V., Towards the sustainable management of air quality and human exposure: Exemplary case studies. *WIT Transactions on Ecology and the Environment*, vol. 230, WIT Press: Southampton and Boston, pp. 489–500, 2018.
[19] European Union, Directive 2010/75/EU of the European Parliament and of the Council of 24 November 2010 on industrial emissions (integrated pollution prevention and control). *Official Journal of the European Union*, Document 32010L0075.
[20] European Union, Directive 2014/52/EU of the European Parliament and of the Council of 16 April 2014 amending Directive 2011/92/EU on the assessment of the effects of certain public and private projects on the environment Text with EEA relevance. *Official Journal of the European Union*, Document 32014L0052.
[21] European Union, Directive 2011/92/EU of the European Parliament and of the Council of 13 December 2011 on the assessment of the effects of certain public and private projects on the environment Text with EEA relevance. *Official Journal of the European Union*, Document 32011L0092.
[22] Arabadjieva, K., 'Better Regulation' in environmental impact assessment: The amended EIA directive. *Journal of Environmental Law*, **28**(1), pp. 159–168, 2016.
[23] Barzyk T.M., Conlon K.C., Chahine T., Hammond D.M., Zartarian V.G. & Schultz B.D., Tools available to communities for conducting cumulative exposure and risk

assessments. *Journal of Exposure Science and Environmental Epidemiology*, **20**, pp. 371–384, 2010.

[24] European Union, Directive 2004/35/CE of the European Parliament and of the Council of 21 April 2004 on environmental liability with regard to the prevention and remedying of environmental damage. *Official Journal of the European Union*, Document 32004L0035.

[25] Argiriadis, E., Rada, E.C., Vecchiato, M., Zambon, S., Ionescu, G., Schiavon, M., Ragazzi, M. & Gambaro, A., Assessing the influence of local sources on POPs in atmospheric depositions and sediments near Trento (Italy). *Atmospheric Environment*, **98**, pp. 32–40, 2014.

[26] Rada, E.C., Ragazzi, M. & Schiavon, M., Assessment of the local role of a steel making plant by POPs deposition measurements. *Chemosphere*, **110**, pp. 53–61, 2014.

[27] Sasamoto, T., Ushio, F., Kikutani, N., Saitoh, Y., Yamaki, Y., Hashimoto, T., Horii, S., Nakagawa, J. & Ibe, A., Estimation of 1999–2004 dietary daily intake of PCDDs, PCDFs and dioxin-like PCBs by a total diet study in metropolitan Tokyo, Japan. *Chemosphere*, **64**(4), pp. 634–641, 2006.

[28] Schiavon, M., Ragazzi, M. & Rada, E.C., A proposal for a diet-based local PCDD/F deposition limit. *Chemosphere*, **93**(8), pp. 1639–1645, 2013.

[29] International Agency for Research on Cancer, IARC Monographs, Classifications. http://monographs.iarc.fr/ENG/Classification/latest_classif.php. Accessed on: 20 Dec. 2018.

[30] U.S. Environmental Protection Agency, Integrated Risk Information System (IRIS). www.epa.gov/IRIS/. Accessed on: 20 Dec. 2018.

[31] Yousaf, B., Amina, Liu, G., Wang, R., Imtiaz, M., Rizwan, M.S., Zia-ur-Rehman, M., Qadir, A. & Si, Y., The importance of evaluating metal exposure and predicting human health risks in urban–periurban environments influenced by emerging industry. *Chemosphere*, **150**, pp. 79–89, 2016.

[32] United States Environmental Protection Agency, Method 7199 – Determination of hexavalent chromium in drinking water, groundwater and industrial wastewater effluents by ion chromatography. www.epa.gov/sites/production/files/2015-12/documents/7199.pdf. Accessed on: 20 Dec. 2018.

[33] Schiavon, M., Ragazzi, M., Torretta, V. & Rada, E.C., Comparison between conventional biofilters and biotrickling filters applied to waste bio-drying in terms of atmospheric dispersion and air quality. *Environmental Technology*, **37**(8), pp. 975–982, 2016.

CO_2 MEASUREMENTS FOR UNCONVENTIONAL MANAGEMENT OF INDOOR AIR QUALITY

MARCO RAGAZZI[1], ROSSANO ALBATICI[1], MARCO SCHIAVON[1], NAVARRO FERRONATO[2] & VINCENZO TORRETTA[2]
[1]Department of Civil, Environmental and Mechanical Engineering, University of Trento, Italy
[2]Department of Theoretical and Applied Sciences, University of Insubria, Italy

ABSTRACT

Carbon dioxide (CO_2) is a known global pollutant and is responsible for the global warming that the planet has been experiencing for the last few decades. On a local scale, outdoor CO_2 does not pose any risk for the environment and humans. Risks are usually far from involving human beings in the majority of indoor environments, although high CO_2 concentrations may entail temporary adverse effects that are similar to the typical symptoms of the so-called "sick-building syndrome". Such effects become even more important on workplaces or at school/university, since high CO_2 levels may negatively influence the productivity and the learning capability of individuals. To understand the magnitude of the problem, a monitoring campaign was carried out in four classrooms and in a library of a university in northern Italy. Three of the classrooms under observation were not equipped with an air extraction system. The CO_2 concentration was monitored with low-cost non-dispersive infrared digital sensors in two periods of the year: between February and March and at the end of May. During those periods, the number of occupants, temperature and relative humidity were also monitored, as well as any opening of the windows (where available). The results showed that, where automatic air extraction is not available, CO_2 concentrations can exceed 5,000 ppm. In general, the lower the ratio between the room volume and the number of occupants, the higher the concentrations achieved. The installation of low-cost sensors might prove useful to prevent the negative effects from the exposure to high CO_2 levels and help achieve more sustainable conditions in indoor spaces, since the sensors could inform lecturers and students on the need for opening doors and/or windows when air extraction is not provided.
Keywords: carbon dioxide, indoor air quality, comfort, exposure, sustainability, ventilation, air quality monitoring.

1 INTRODUCTION

According to the World Health Organization (WHO), on average, people spend about 90% of their daily time in indoor environments [1]. In the light of such outcome, the air quality in schools, workplaces, houses and other indoor spaces is then crucial to limit the personal exposure to air contaminants [2]–[5]. In addition, cohort studies revealed that indoor air pollution is correlated with an increase in mortality, due to respiratory morbidity [6]. Poor indoor air quality is at the basis of the so-called sick-building syndrome (SBS), i.e. an ensemble of symptoms (e.g. loss of attention, fatigue, pains and allergic symptoms) that are associated with the staying of an individual in his/her workplace or house [7], [8]. Contrarily to the outdoor space, indoor environments, if not adequately built, may contribute to increase the concentrations of air contaminants that comes from outside [9] or that are generated indoors, especially if the ventilation and the rate of exchange of the indoor air are weak. Indeed, similarly to outdoor environments, ventilation acts as the main dispersion mechanism in indoor spaces [10].

The indoor air can enrich with several air pollutants: particulate matter [6], volatile organic compounds [11], nitrogen dioxide [12] and ozone [13] are some of the most common air contaminants that have been studied in indoor environments so far. Such pollutants can enter an indoor space from outside by infiltration through doors and windows or simply when doors and windows are opened. If the outdoor concentration of contaminants is higher than

WIT Transactions on Ecology and the Environment, Vol 236, © 2019 WIT Press
www.witpress.com, ISSN 1743-3541 (on-line)
doi:10.2495/AIR190271

the indoor concentration, the latter increase. If no forced aeration system is present, the contaminants may stagnate inside the indoor environment. Consequently, the human exposure to air pollutants may increase with respect to outdoors [14]. As an example, Gonzalez-Flesca et al. [15] found out that the personal exposure to benzene measured in four French urban areas was 3.5 times higher than the mean outdoor concentration.

In addition to external contributions, the indoor air is strongly influenced by indoor sources. Cooking, biomass burning for heating and cooking purposes, natural-gas burners, cigarette smoke and new furniture are known sources of particulate matter, carbon monoxide, nitrogen dioxide, polycyclic aromatic hydrocarbons, dioxin and volatile organic compounds [16]. Besides those air contaminants, in the recent years the scientific community has paid growing attention to a substance that has not been traditionally accounted for when investigating the air quality of confined environments: carbon dioxide (CO_2). The global effects of such substance on the environment are well known. On a local scale, with the exception of specific workplaces (e.g. fermentation tanks), CO_2 does not pose direct risks for human health. However, in indoor environments, CO_2 may assume the features of an indoor pollutant, since it promotes the occurrence of some of the symptoms of the SBS. Indeed, high concentrations of CO_2, within certain limits, induce negative (though reversible) effects on humans, such as decrease of attention, reduced productivity and physical discomfort [17].

The concentration of CO_2 in the outdoor environment is normally in the range 300–400 ppm. The generally weak ventilation that characterises indoor environments with occupants leads to higher values. According to the German Indoor Air Hygiene Commission and the Working Group of the Supreme Health Authorities of the Federal States, CO_2 concentrations may be regarded as "hygienically insignificant" (<1,000 ppm), "hygienically evident" (1,000–2,000 ppm) and "hygienically unacceptable" (>2,000 ppm) [18]. The German Federal Environment Agency's Indoor Air Hygiene Commission declares that rooms exceeding a CO_2 concentration of 1,000 ppm should require an exchange of air. Such an advice becomes an obligation if the concentration exceeds the value of 2,000 ppm [19]. Further works concluded that concentrations above the "hygienically insignificant" range might negatively influence the learning ability [20]. Headache, loss of attention and sleepiness are symptoms related to higher concentrations, up to 5,000 ppm [21]. As previously mentioned, mortality is associated with extreme conditions, i.e. where CO_2 is so abundant that oxygen becomes limited [22]. Although such extreme levels are not proper of houses, offices or schools, other symptoms like headache, sleepiness and decreased attention may often occur in crowded and non-ventilated public places.

The dependence of CO_2 concentrations on the air exchange rate of a room made some authors conclude that CO_2 can also be a good indicator of the presence of other substances [17] and, according to Fanger [23], "may in many cases also provide a first indication of a possible health risk" from toxic air contaminants. Monitoring CO_2 concentrations in indoor spaces can help highlighting critical situations requiring an exchange of air (where automatic ventilation systems are absent) or an increase in the air exchange rate (where automatic ventilation systems are present), in order to reduce the exposure of the occupants to potentially toxic substances and decrease the level of discomfort directly induced by CO_2 inhalation.

Schools and universities, especially, have the role of educating the future ruling class. High CO_2 concentrations in crowded rooms may influence the learning quality and may pose a risk to the achievement of this goal. Recent researches carried out in primary and secondary schools showed that CO_2 could achieve mean concentrations higher than 1,000 ppm, which thus highlight a "hygienically evident" problem [24], [25]. Other studies showed that peak concentrations close to [26] or higher than 4,000 ppm [27] may be achieved in schools, and highlighted the positive role of natural ventilation in taking the concentrations back to acceptable levels. In spite of the importance of keeping the learning capability at high standards, the number of studies on the CO_2 concentrations in schools and, especially, in universities is still low.

In the light of the previous considerations, this paper wants to share the results of a monitoring campaign of indoor CO_2 carried out in an Italian university and highlight the higher levels of concentrations that may be achieved in classrooms when no automatic ventilation systems are installed. This work is intended to shed a light on an underestimated factor that negatively affects the sustainability in working and teaching environments. Continuous CO_2 monitoring can turn useful to plan renovations in school and university environments or to inform teachers/professors on the need for exchanging the air (e.g. by opening doors and/or windows) when air extraction is not provided.

2 MATERIALS AND METHODS

2.1 Air quality monitoring campaign

The air quality monitoring campaign was carried out in a building of the University of Trento. Trento is a town with about 117,000 inhabitants located in an Alpine valley in northern Italy. The building considered in this study is located on a hill, about 80 m above the town, in an area that could be regarded as an urban background zone from the point of view of air quality, the main emission source being a secondary street (lowest distance: 91 m) and the university parking (Fig. 1). The building hosts a total of 24 classrooms with different size. Four classrooms and the library were selected to be representative of all the typologies of rooms present in the building and frequented by students. The layouts of the selected rooms are presented in Fig. 2. All the rooms, with the exception of room T3, lack of an automatic air extraction system, but are equipped with air conditioning. Only the doors of room M2 have a small grid in the lower part that connect the rooms with their respective corridor. Only rooms M2, D1 and D2 have windows that can be opened. The features of the rooms monitored are reported in Table 1. The monitoring sessions lasted between 8 and 11 days and took place in two periods: the classrooms were monitored between February and March 2017, whereas the measurements in the library were carried out at the end of May 2017. During each session, the number of occupants, as well as the opening of windows (in rooms M2, D1 and D2), were also monitored. The number of occupants was evaluated on a 1-hour time resolution, by counting the people inside the rooms. For each classroom, the occupants were the professor and the students attending their classes. In the case of the library, the occupants were mainly students, with the occasional presence of a few researchers and professors.

2.2 Instrumentation

The CO_2 measurements were carried out by using two identical portable real-time monitoring devices (ENERair v7.0, Enerconsult srl, Italy) equipped with a non-dispersive infrared digital sensor with self-calibration. The sensor is able to measure CO_2 concentrations in the range 0–5,000 ppm, with an accuracy of ± 30 ppm (± 3%) and a resolution of 1 ppm. The device contains an in-built datalogger, which locally stores the data on a micro-SD card and transmits the measurements to a central server via Ethernet connection or WiFi. The device is also equipped with additional sensors that turn useful to monitor other parameters for indoor air quality, such as temperature and relative humidity, which were measured by a

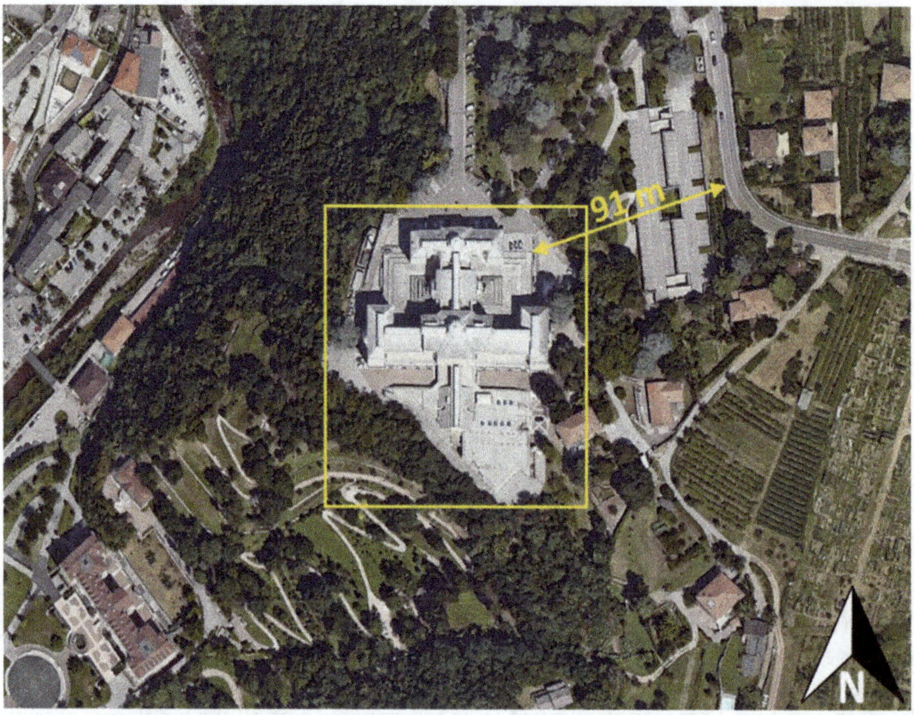

Figure 1: Map of the building with detail of the distance from the closest emission source (a secondary street).

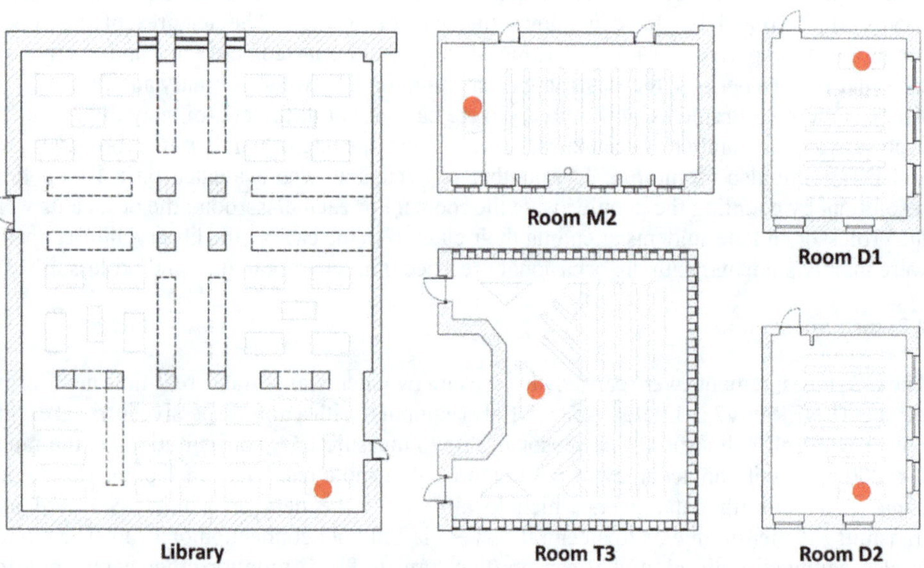

Figure 2: Layouts of the rooms monitored. The red dots indicate the location of the monitoring device.

Table 1: Features of the rooms monitored.

	Library	Room T3	Room M2	Room D2	Room D1
Net surface (m^2)	520	145	85	62	62
Net volume (m^3)	1,460	760	280	205	205
Floor	0	0	2	2	1
Orientation	SE	E	S	SE	SE
Maximum number of seats	162	98	88	47	47
Windows	Cannot be opened	Cannot be opened	Can be opened	Can be opened	Can be opened
Sampling period	02.08–02.17	02.17–02.26	02.17–02.26	02.27–03.08	02.27–03.08

semiconductor I2C-bus sensor. The sampling time can be personalised and was set as equal to 15 s for the whole duration of the tests. In all the classrooms, the device was placed on the desk, with exception of the library, where the device was located on a table of the room.

3 RESULTS AND DISCUSSION

The trends of CO_2 concentration, temperature (T), relative humidity (RH) and the number of occupants during the monitoring campaign that took place in the library are presented in Fig. 3. The trends of T and CO_2 concentration follow that of the number of occupants. As expected, the trend of RH is antithetical with respect to T. During the whole monitoring period, the CO_2 concentration remained in the acceptable range (<1,000 ppm), although no air extraction is present and in spite of the relatively low value of the minimum ratio between volume and number of occupants (13.6 m^3/person). This can be explained by the fact that the library is widely used during the whole day by students that often open the main door to enter/exit the room, and by the fact that the airtightness of the building envelope is quite low.

The situation gets slightly worse when considering the largest classroom (T3). Here, on the eighth day from the beginning of the monitoring campaign, the CO_2 concentration reached a peak of 1,507 ppm after four hours of lessons (Fig. 4). That day, the minimum ratio between volume and number of occupants was 16.5 m^3/person, i.e. similar to the value measured in the library. This difference in the CO_2 peak concentration is probably due to the fact that this room has been built after the main building, in place of an original inner courtyard, so with higher attention to constructive quality (better airtightness) but with supply and return air conditioning elements mainly concentrated in the upper part of the space (the roof is not horizontal but presents a slight sloping and a maximum inner high of about 6.50 m) and non at the users level, so with non-optimal air fluxes.

Higher concentration values were measured in classroom M2 (Fig. 5). Here the CO_2 peak concentration (1,791 ppm) was achieved on the fifth day of measurements, in the correspondence of the highest number of occupants (30 people). In this situation, the ratio between volume and number of occupants was 9.3 m^3/person. Room M2, like rooms D1 and D2, has windows that can be opened by the occupants. During this peak episode, the windows were opened and the CO_2 concentration soon decreased to <700 ppm.

The situation was even worse when considering the smallest classrooms monitored. Room D2 showed a maximum concentration of 3,855 ppm on the third day of measurements (Fig. 6). In that occasion, the number of people was 44, corresponding to a ratio between volume and number of occupants of 4.7 m^3/person. After this peak episode, the windows

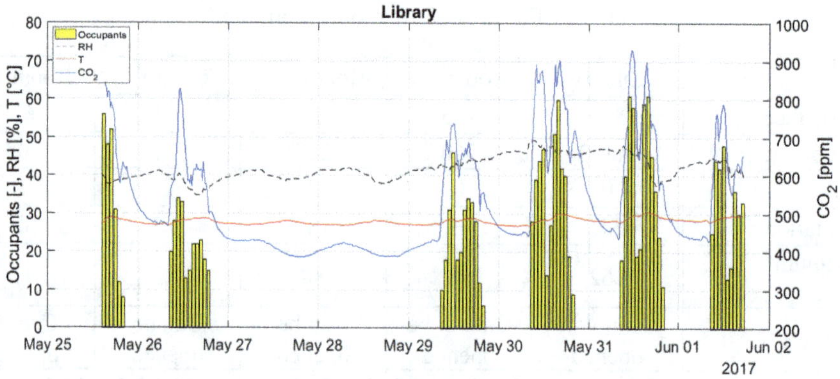

Figure 3: Trends of CO_2 concentrations, T, RH and number of occupants in the library during the monitoring campaign.

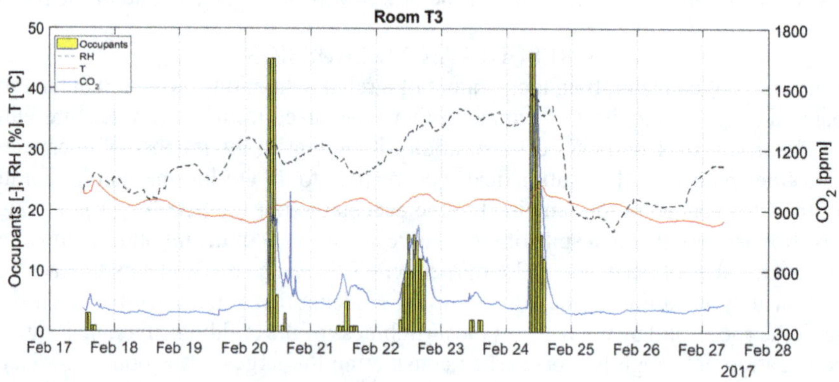

Figure 4: Trends of CO_2 concentrations, T, RH and number of occupants in room T3 during the monitoring campaign.

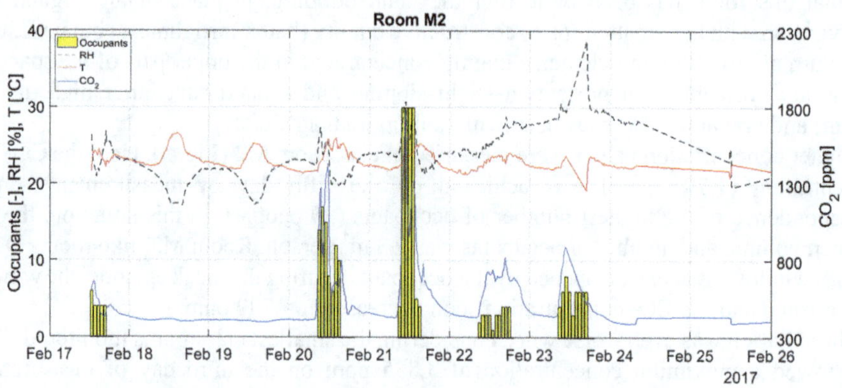

Figure 5: Trends of CO_2 concentrations, T, RH and number of occupants in room M2 during the monitoring campaign.

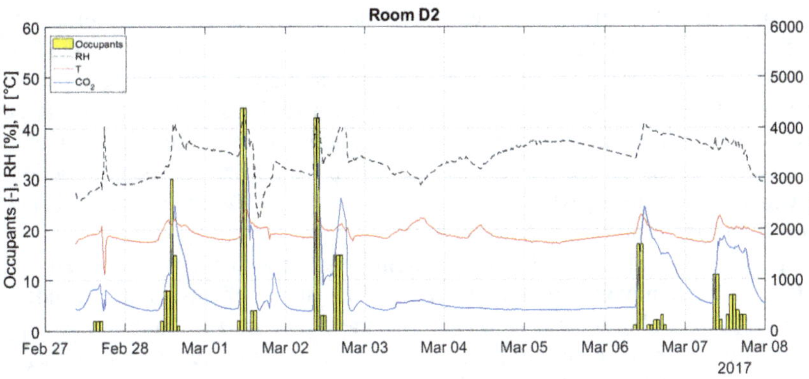

Figure 6: Trends of CO_2 concentrations, T, RH and number of occupants in room D2 during the monitoring campaign.

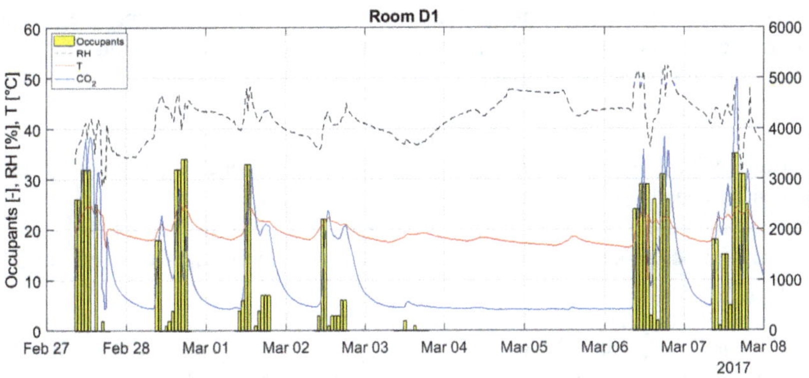

Figure 7: Trends of CO_2 concentrations, T, RH and number of occupants in room D1 during the monitoring campaign.

were opened by the occupants. The highest concentration value was recorded in room D1 on the last day of measurements (Fig. 7), when the value reached the instrumental end of scale (5,000 ppm). The number of occupants in that situation was 30. The corresponding ratio between volume and number of occupants was 6.8 m^3/person. This peak episode was recorded after five hours of lessons without opening the windows. The high concentration values made the occupants open the windows to refresh the air. Following windows opening, the concentration dropped to about 1,000 ppm. Contrarily to room M2, rooms D1 and D2 are not equipped with any grid on the entrance door and this may partly explain the higher concentrations measured.

4 CONCLUSIONS

The monitoring campaign carried out on university rooms highlighted that high concentrations of CO_2, generated by the human metabolism, might be easily achieved during classes in the absence of air extraction systems. The critical levels of CO_2 concentrations (>1,000 ppm) that can be achieved in such situations entail adverse (though reversible) effects on the occupants, who might experience decreased their learning capability, fatigue, sleepiness, loss of attention, headache and other symptoms that recall those of the SBS. The use of low-cost equipment, such as non-dispersive infrared digital sensors, could prevent the occurrence of CO_2 levels that might become critical to students, lecturers and, in general, employees working in relatively crowded rooms and/or in the absence of air extraction systems. Low-cost sensors could make people aware of the need for simple actions, like opening windows and doors, which could rapidly take the CO_2 concentration back to the "hygienically insignificant range" (<1,000 ppm).

In this paper, CO_2 is thus seen from a different perspective, i.e. not as a global pollutant, but as a local contaminant that humper the achievement of sustainable conditions in workplaces and school buildings. Conversely, the low-cost CO_2 monitoring approach presented in this paper can help the identification of critical environments that may require renovations (e.g. installation of automatic ventilation systems) or can alert the occupants when it is opportune to favour the natural ventilation of indoor spaces (e.g. by opening windows and/or doors).

ACKNOWLEDGEMENTS

The authors wish to thank Mr Andrea Marchiori, for his support during the monitoring campaign, Enerconsult srl and, especially, arch. Basilio Guerra, for the monitoring apparatus and the support to the data management.

REFERENCES

[1] World Health Organization, Regional Office for Europe, Combined or multiple exposure to health stressors in indoor built environments. www.euro.who.int/__data/assets/pdf_file/0020/248600/Combined-or-multiple-exposure-to-health-stressors-in-indoor-built-environments.pdf. Accessed on: 20 Nov. 2018.

[2] Camacho-Montano, S.C., Wagner, A., Erhorn-Kluttig, H., Mumovic, D. & Summerfield, A., Clearing the air on EU guidance projects for school buildings. *Building Research and Information*, **47**(5), pp. 624–634, 2019.

[3] Kelly, F.J. & Fussell, J.C., Improving indoor air quality, health and performance within environments where people live, travel, learn and work. *Atmospheric Environment*, **200**, pp. 90–109, 2019.

[4] Almeida-Silva, M., Wolterbeek, H.T. & Almeida, S.M., Elderly exposure to indoor air pollutants. *Atmospheric Environment*, **85**, pp. 54–63, 2014.

[5] Schiavon, M., Rada, E.C., Ragazzi, M. & Antolini, D., Indoor measurements of particulate matter during steak cooking under different conditions. *WIT Transactions on Ecology and the Environment*, vol. 176, WIT Press: Southampton and Boston, pp. 255–264, 2013.

[6] Kumar, R., Nagar, J.K. & Gaur, S.N., Indoor air pollutants and respiratory morbidity: A review. *Indian Journal of Allergy, Asthma and Immunology*, **19**, pp. 1–9, 2005.

[7] Norhidayah, A., Chia-Kuang, L., Azhar, M.K. & Nurulwahid, S., Indoor air quality and sick building syndrome in three selected buildings. *Procedia Engineering*, **53**, pp. 93–98, 2013.

[8] Hansel, N.N., McCormack, M.C. & Kim, V., The effects of air pollution and temperature on COPD. *COPD: Journal of Chronic Obstructive Pulmonary Disease*, **13**(3), pp. 372–379, 2016.

[9] Cioca, L.-I., Ivascu, L., Rada, E.C., Torretta, V. & Ionescu, G., Sustainable development and technological impact on CO2 reducing conditions in Romania. *Sustainability*, 7(2), pp. 1637–1650, 2015.

[10] Schiavon, M., Rada, E.C., Ragazzi, M., Antognoni, S. & Zanoni, S., Domestic activities and pm generation: A contribution to the understanding of indoor sources of air pollution. *International Journal of Sustainable Development and Planning*, **10**(3), pp. 347–360, 2014.

[11] Goodman, N.B., Wheeler, A.J., Paevere, P.J., Selleck, P.W., Cheng, M. & Steinemann, A., Indoor volatile organic compounds at an Australian university. *Building and Environment*, **135**, pp. 344–351, 2018.

[12] Andrade, A. & Dominski, F.H., Indoor air quality of environments used for physical exercise and sports practice: Systematic review. *Journal of Environmental Management*, **206**, pp. 577–586, 2017.

[13] Salonen, H., Salthammer, T. & Morawska, L., Human exposure to ozone in school and office indoor environments. *Environment International*, **119**, pp. 503–514, 2018.

[14] Villanueva, F., Tapia, A., Amo-Salas, M., Notario, A., Cabañasa, B. & Martínez, E., Levels and sources of volatile organic compounds including carbonyls in indoor air of homes of Puertollano, the most industrialized city in central Iberian Peninsula: Estimation of health risk. *International Journal of Hygiene and Environmental Health*, **218**, pp. 522–534, 2015.

[15] Gonzalez-Flesca, N. et al., Personal exposure of children and adults to airborne benzene in four French cities. *Atmospheric Environment*, **41**, pp. 2549–2558, 2007.

[16] Schiavon, M., Ragazzi, M., Rada, E.C., Magaril, E. & Torretta, V., Towards a sustainable management of air quality and human exposure: exemplary case studies. *WIT Transactions on Ecology and the Environment*, vol. 230, WIT Press: Southampton and Boston, pp. 489–500, 2018.

[17] Ragazzi, M., Rada, E.C., Zanoni, S., Passamani, G. & Dalla Valle, L., Particulate matter and carbon dioxide monitoring in indoor places. *International Journal of Sustainable Development and Planning*, **12**(6), pp. 1032–1042, 2017.

[18] Bekanntmachung des Umweltbundesamtes, Gesundheitliche Bewertung von Kohlendioxid in der Innenraumluft (Health evaluation of carbon dioxide in indoor air). *Bundesgesundheitsblatt – Gesundheitsforschung – Gesundheitsschutz, 51*, pp. 1358–1369, 2008.

[19] German Federal Environment Agency's Indoor Air Hygiene Commission, *Guidelines for Indoor Air Hygiene in School Buildings*, KOMAG mbH: Berlin, 2008.

[20] Satish, U. et al., Is CO_2 an indoor pollutant? Direct effects of low-to-moderate CO_2 Concentrations on human decision-making performance. *Environmental Health Perspectives*, **120**(12), pp. 1671–1677, 2012.

[21] NASA, In-flight carbon dioxide exposures and related symptoms: associations, susceptibility and operational implications, TP–2010–216126. https://ston.jsc.nasa.gov/collections/trs/_techrep/TP-2010-216126.pdf. Accessed on: 21 Nov. 2017.

[22] Langford, N.J., Carbon dioxide poisoning. *Toxicological Reviews*, **24**(4), pp. 229–235, 2005.

[23] Fanger, P.O., Sensory characterization of air quality and pollution sources. *Chemical, Microbiological, Health and Comfort Aspects of Indoor Air Quality – State of the Art in SBS*, H. Knöppel & P. Wolkoff, eds, Kluwer Academic Publishers: Dordrecht, pp. 59–71, 1992.

[24] Wang, Y. et al., Effects of solar collectors on indoor air quality in junior classrooms in winter. Presented at *Building a Better New Zealand Conference*, Auckland, New Zealand, 2013.

[25] Allen, J.G., MacNaughton, P., Satish, U., Santanam, S., Vallarino, J. & Spengler, J.D., Associations of cognitive function scores with carbon dioxide, ventilation, and volatile organic compound exposures in office workers: A controlled exposure study of green and conventional office environments. *Environmental Health Perspectives*, **124**, pp. 805–512, 2016.

[26] Schibuola, L., Scarpa, M. & Tambani, C., Natural ventilation level assessment in a school building by CO_2 concentration measures. *Energy Procedia*, **101**, pp. 257–264, 2016.

[27] Stabile, L., Dell'Isola, M., Russi, A., Massimo, A. & Buonanno, G., The effect of natural ventilation strategy on indoor air quality in schools. *Science of The Total Environment*, **595**, pp. 894–902, 2017.

INTEGRATED METHODOLOGY FOR THE MANAGEMENT OF HUMAN EXPOSURE TO AIR POLLUTANTS

MARCO SCHIAVON[1], ELENA CRISTINA RADA[2], LUCA ADAMI[1], FEDERICA FOX[1] & MARCO RAGAZZI[1]
[1]Department of Civil, Environmental and Mechanical Engineering, University of Trento, Italy
[2]Department of Theoretical and Applied Sciences, University of Insubria, Italy

ABSTRACT
The assessment of the impacts of future civil/industrial plants on air quality is traditionally based on the environmental assessment procedure, which, however, may lead to different conclusions depending on the choice of the target area and the emission sources already present, and may underestimate the global impacts on health, society and economy. The paper proposes an organic multi-step methodology to carry out health risk assessments on a target area, to assess the incidence of emissive activities under evaluation for authorisation, and, as an additional matter of novelty, to study mitigation/compensation measures to take their future impacts back to acceptable levels, if their contributions are expected to exceed acceptable risk parameters. Criteria for the definition of target areas, emission calculation, dispersion modelling, the health risk assessment procedure, the verification of the risk acceptability and the compensation approach are described in detail. The current potential limitations of the approach are discussed and solutions to overcome such weak points are proposed. In view of a sustainable integration of new activities within local contexts, this methodology aims at guiding decision makers to carry out the screening of projects in terms of health impacts, according to a standardised approach.
Keywords: dispersion modelling, health risk, emissions, air quality, cancer risk, hazard index, environmental impact, external costs, environmental sustainability.

1 INTRODUCTION

Air pollution is the first environmental cause of worsening of human health and/or mortality on Earth, causing about 7 million deaths per year [1]–[4]. In the 28 countries of the European Union, following the exposure to particulate matter (PM), ozone (O_3) and nitrogen dioxide (NO_2), air pollution is believed to be responsible for the death of 483,400 people per year [5]. Among other environmental issues, the traditional environmental impact assessment procedure evaluates the compatibility of future activities with the air quality of the area where those would be located and the compliance with health risk parameters of acceptability. Indeed, health protection should be the ultimate purpose of the authorisation procedure for environmental impact assessment. However, the conventional approach presents some limitations [4], [6]–[10]:

1. the health risk may significantly vary within the area of study, depending on the local prevailing mechanisms of dispersion of the air pollutants, on the spatial distribution of the residents and on the other emission sources that may be present in the area;
2. the definition of the study area itself requires a standardised approach, since the impacts of one or more emission sources quantitatively depend on the extension and location of the target area;
3. although a new activity may comply with health risk acceptability parameters, it may also entail increased social costs, since health impacts, though considered as acceptable, might increase;
4. the conventional approach is based on the use of different methodologies, which requires different input data and formats, and does not follow a standard procedure.

WIT Transactions on Ecology and the Environment, Vol 236,
www.witpress.com, ISSN 1743-3541 (on-line)
doi:10.2495/AIR190281

The novelty of this paper consists in the proposal for an organised and organic multi-step methodology that allows local administrators and decision makers to:

1. carry out the assessment of the risk for health related to air pollution in a target area through an integrated approach;
2. evaluate scenarios that consider the presence of additional emissive activities and the acceptability of their supplemental impacts; and
3. consider possible mitigation measures to compensate the authorisation of a new activity, in order to improve (or, at least, to keep as constant) the status of health in the area.

A scheme of the multi-step methodology that will be discussed in the paper is presented in Fig. 1. The health risk assessment is carried out through the traditional methods of air dispersion modelling and the estimation of the intake of contaminants through the relevant routes of exposure, but the procedure is presented in a unified way and with a logical structure, to facilitate the work of local authorities. The evaluation of scenarios including the activity of additional emission sources allows assessing the local incidence of a new planned industrial/civil plant with respect to the current situation. Finally, the compensation approach should be used when the new activity, in spite of adopting the best available techniques, is expected to induce a health risk that exceeds the criteria of acceptability in the target area. The compensation procedure may translate, for instance, into obliging the owners of the new activity to invest in enhanced air pollution control systems for plants that had been already authorised in the area and that have room for improvement in air pollution control or energy saving policies. As another example, in the case of new waste-to-energy plants, the compensation approach may consist in using the electric energy produced by the plant and ask the owners of the plant to install electrical heating systems in dwellings, as integration (partial substitution) of domestic boilers fed by fossil fuels or wood stoves, the latter being known sources of PM [11], [12].

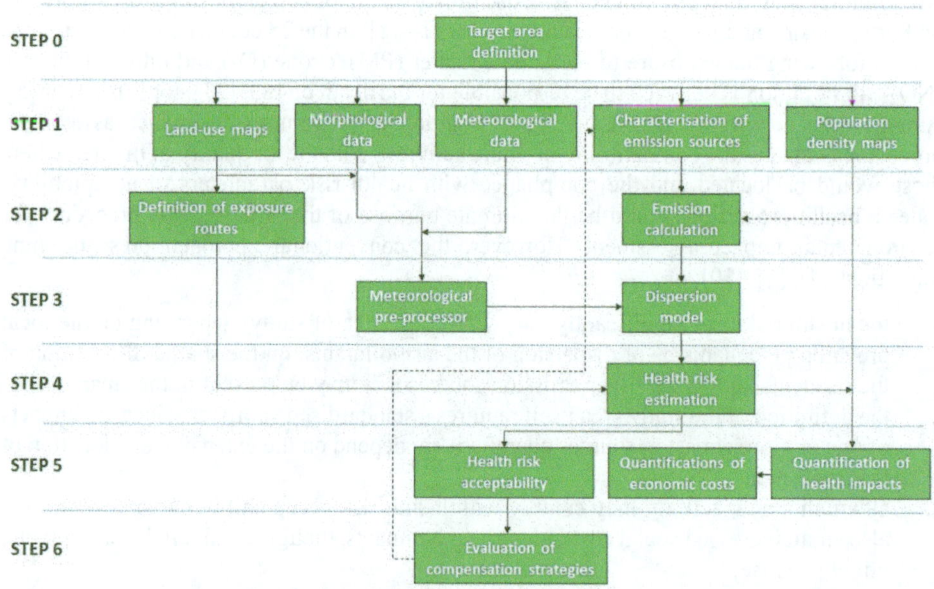

Figure 1: Conceptual scheme of the integrated methodology for the management of human exposure to air pollutants.

2 METHODOLOGY

2.1 Selection of the target area

The definition of the target area is a necessary step to identify the dominant routes of exposure, to define the limits of the computation domain of the dispersion model, to estimate the impacts on health, to evaluate the influence of additional emission sources and of possible mitigation strategies. To quantify the impacts, the target area cannot be limited to the area including the industrial sources to investigate, but should be extended to the adjacent municipalities or residential areas, and up to a variable distance from the point of interest that depends on the possible presence of orography, which may impede the dispersion of pollutants. As a reference criterion, the target area should include all the municipalities/ districts contained in a radius of 5 km from the industrial source (or, if more than one, from the barycentre of the industrial sources). In addition, if important external emission sources are present (e.g., an industrial hub), the target area should include a buffer containing the municipalities/districts adjacent to the area defined by the previous criterion, in order to account for the possible influence of external sources on the air quality of the inner target area. Concerning the horizontal grid resolution of the dispersion model, the higher the resolution, the higher the accuracy of the heath impact estimation. Cells with sides of 100 m could be a reasonable compromise between accuracy of the results and computational efforts. If the target area includes the buffer, the latter can have a lower horizontal resolution (i.e., higher cell dimensions), which can be imposed with a nesting procedure in the dispersion model.

2.2 Data collection and emission calculation

The collection of data is functional to the simulations of the dispersion of air pollutants and to the health risk quantification. The management of data must be carried out on a georeferenced base, i.e., in a geographical information system (GIS) environment.

In addition to direct intake of air pollutants via inhalation, other routes of exposure may be significant in determining the daily dose of air contaminants: the diet, the dermal contact with contaminated food, the dermal contact with soil and the accidental soil ingestion. If the target area includes cultivated lands, pastures and/or places where kids can get into contact with soil (e.g., schools and playgrounds), the air pollutants that deposit to soil may activate those additional exposure routes. Accidental soil ingestion may be a significant route of exposure for children, especially if schools or parks are present in the area [13]. A basic knowledge of the area and the consultation of land-use maps, such those created by the CORINE Land Cover project [14], help identifying the relevant exposure routes to consider. Every land-use class of the CORINE maps also reports a value of the surface roughness, which is an essential parameter to the meteorological pre-processor of the selected dispersion model for the calculation of the wind field in the domain. Depending on the size of the computation domain of the dispersion model, the urban fabric might be explicitly defined rather than parametrising it by defining surface roughness values compatible with the land use [10]. In this case, an elevation map of the buildings must be retrieved or specifically developed.

If orography is present in the target area, a digital elevation model (DEM) must be included in the dispersion model. DEMs are available as georeferenced raster maps on various online databases. The DEM should have a horizontal resolution lower or equal to that of the computation domain of the dispersion model.

The meteorological pre-processors of dispersion models require hourly meteorological data relative to one or more weather stations in the area or to a regional/global scale model (e.g., WRF). Depending on the dispersion model chosen, the essential data may vary. When the purpose is the assessment of the impacts on health, as in the present case, meteorological data should cover a whole solar year of observations.

It is then essential to retrieve data on the emissions of the largest possible number of pollutants from the largest possible number of sources. In the case of point sources (e.g., chimneys of mills or centralised district heating systems), the data consists in: the location and elevation of the source, the temperature of the gas emitted, the airflow rates, the concentration of pollutants at the stack level, the possible presence of buildings near the source. Those data can be retrieved through the environmental assessment reports. If accurate emission values can be estimated for activities subject to environmental impact assessment, the estimation of the emissions from other sources has a much higher degree of uncertainty. This is the case of road traffic, whose emissions must be estimated based on: a characterisation of the vehicle fleet (usually available from national statistics databases), on the vehicle fluxes along the road network (usually available from local authorities) and on emission factors available in the literature. At a European level, emissions factors for road traffic are provided by the European Environment Agency (EEA) [15] and can be calculated by applying the COPERT 5 algorithm [16]. Roads are usually treated as linear sources in dispersion modelling, thus by approximation to polylines in GIS environments. If aggregate emission factors are available, some activities are conveniently modelled as area sources, due to their ubiquity [17]. This is the case, for instance, of small industries and diffuse sources (e.g., de-centralised domestic heating), which could be approximated to polygons. The emissions from domestic heating could be weighed on the resident population, by retrieving population density maps from regional/national statistics institutions. Regional air pollutant inventory emissions, like the case of INEMAR in Italy [18], turn particularly useful to this purpose. The emissions from other specific areal sources (waste landfills or wastewater treatment plants) may be estimated with the use of national or international databases, like EEA emission inventory [15]. To reduce the uncertainties related to this approximated approach, daily, weekly and yearly patterns of emissions should be assumed. All the mass flow rates of pollutants should cover the same period of meteorological data. Consistently with meteorological data, dispersion models require hourly values of the mass flow rates of the pollutants considered. Given a specific dispersion model, emissions and meteorological data play the most important role in determining the accuracy of the simulation results. Thus, the more emission values are accurate, the higher is the accuracy of the output.

Besides the emission sources located in the target area (with the possible inclusion of the buffer area), it is important to consider the background annual mean concentrations of as more pollutants as possible, by retrieving the data measured by rural background air quality stations within or close to the computation domain. Interpolation methods, like the ordinary Kriging method [19], could be used to obtain more realistic background concentration maps if more than one rural background stations are available.

In addition to weighing the emissions from diffuse sources, population density maps are essential to quantify the impacts in terms of the number of cases of cancer, hospitalisation and mortality expected on an annual basis, and to estimate the economic costs for the society.

2.3 Dispersion modelling

The meteorological data, recorded by one or more meteorological stations within the target area or obtained by regional-scale models, are the input data to the meteorological pre-processor of the selected dispersion model. Some manipulations are necessary to convert the data into the format required by the pre-processor. Depending on the dispersion model, vertical profiles of wind, temperature and pressure may be requested as input data. The pre-processor computes the horizontal and vertical wind fields that are necessary to calculate the dispersion of air pollutants within the atmosphere. Generally, the output from the dispersion model consists in maps of annual mean and maximum concentrations and, in the case of macropollutants, concentrations compatible with the limit values defined by the environmental legislation. The concentrations simulated for each pollutant must be added to the background concentrations of the same pollutants (if available) measured at rural background sites. In the case that other exposure routes are relevant for the target area, in addition to inhalation, atmospheric depositions should be selected besides air concentrations as the second desired output of the dispersion model.

2.4 Health risk estimation

Diet, dermal contact and accidental soil ingestion are relevant pathways for inorganic (heavy metals) and organic (dioxin and polycyclic aromatic hydrocarbons) micropollutants, which tend to accumulate in soil and in vegetable and animal tissues, due to their physical–chemical properties. Such pollutants may of course enter the human body through inhalation too. The estimation of the health risk by exposure to micropollutants is carried on the basis of the guidelines developed by the United States Environmental Protection Agency (US-EPA) for the calculation of the cancer and non-cancer risks [20]. The US-EPA guidelines focus on the long-term effects of micropollutants. All micropollutants may induce toxic non-carcinogenic effects, whereas only some of them are carcinogens: dioxin, polycyclic aromatic hydrocarbons, some volatile organic compounds like benzene and formaldehyde, and some metals including cadmium, arsenic, mercury, nickel and chromium VI [21]. Through specific relations that depend on the exposure route, the guidelines allow estimating the daily intake of each contaminant and the associated risk, by starting from annual mean concentrations and deposition values. The calculation of the exposure through the diet requires considering the statistics on food consumption for the target population, which are usually elaborated by national statistics institutes, like the Italian National Institute for the Research on Food and Nutrition (INRAN) [22]. The estimation of the overall risk is carried out through the superposition principle, thus by assuming that the risk related to a single contaminant can be added to the risks related to the others. By applying the US-EPA guidelines to each cell of the computation domain, it becomes possible to obtain maps of the individual risk (related to the single pollutant) for cancer or non-cancer effects, or a map of the overall (by summing all contributions) cancer risk. As a precautionary measure, the present methodology presented here suggests assuming that all the population living in the target consume only locally-grown food, thus subject to the potential contamination that is the object of the study.

Although the US-EPA guidelines do not consider macropollutants (NO_2, PM, O_3 and carbon monoxide), the risk related to the exposure of the individuals to macropollutants can be anyway estimated through the several cohort studies available in the literature, which allow obtaining the so-called relative risk (*RR*), i.e., the incremental probability that an effect occurs in an exposed population with respect to the probability that the same effect occurs in an unexposed one [23]. The literature provides *RR* values concerning the incidence of specific non-cancer pathologies that require hospitalisation (e.g., respiratory and cardiovascular diseases) and the incidence of mortality. *RR* values are inserted in parametrised dose-response functions of the following type:

$$\Delta y = -P \cdot y_0 \cdot (e^{-RR \cdot \Delta C} - 1), \tag{1}$$

where Δy is the excess annual incidence of a specific effect, P is population exposed to the increase of the annual mean concentration of a specific air pollutant (ΔC) with respect to the unexposed population and y_0 is the baseline prevalence of the effect per year [24]. ΔC and RR are normally expressed as $\mu g/m^3$ and its reciprocal, respectively. Values of y_0 may be obtained either from the same cohort studies reporting the RR values chosen or, if available, from national or international statistics database. The more specific the statistics are for the local population, the higher is the expected accuracy of the results.

2.5 Verification of the exposure acceptability

In the case of cancer risk, no threshold values are considered below which the impact can be considered as null, since a single damaged cell might be sufficient to initiate the development of cancer. However, the cancer risk may be considered as acceptable if comprised between 10^{-6} and 10^{-4} [25]. In the Italian legislation, 10^{-6} is the acceptable cancer risk related to the emission of a single pollutant from a single source, while this value is increased to 10^{-5} when considering more pollutants from a single source [26]. Considered the purpose of the present methodology, 10^{-5} is selected as the reference value to assess the acceptability of the overall cancer risk in the area, and the consequent possibility to start new industrial activities. This value is precautionary, since the methodology considers the as wider as possible number of emission sources in the area. Concerning the non-carcinogenic risk, the doses of each contaminant are divided by a contaminant-specific reference dose established by the US-EPA [21], below which no effects occur. This ratio is the so-called hazard index (HI). HI < 1 indicates the absence of toxic non-cancer impacts for health.

To quantify the impacts related to the inhalation exposure to macropollutants, the following procedure can be applied:

- convert the population density vector map into a raster map reporting the number of inhabitants in each cell (the cell size should equal that used in the dispersion model);
- for each pollutant and for the corresponding effects of interest, apply the dose-response function to each cell, now characterised by the pollutant concentration and the number of inhabitants; depending on the effects, it could be necessary to use the maximum concentrations instead of the annul mean ones;
- cell by cell and for each pollutant, quantify the effects in terms of hospitalisation for different health outcomes and of mortality;
- calculate the total number of cases within the computational domain;
- estimate the economic costs for the society, for instance through the cost functions defined by the ExternE methodology [27].

By applying the population raster map to the cancer risk map, it is possible to estimate the potential annual cancer cases related to air pollution within the target area. In this case, the economic quantification of the impacts is not feasible, due to the variability of the course of the disease among individuals.

To evaluate the incidence of one or more additional emission sources in the target area, the same methodology must be reapplied by replacing the current emission sources with the additional ones. By adding up the risk maps related to the current situations and the corresponding risk maps related to the additional sources, the new risk maps will take into account the contribution of the additional emission sources. If the new scenario did not comply with one of the acceptability criteria, or if the additional sources implied unacceptable sanitary impacts and social costs (according to criteria to be defined with the local authorities), it would be necessary to consider strategies to compensate the impacts or, eventually, deny the authorisation to the new activities.

2.6 Evaluation of compensation measures

To evaluate compensation strategies, it is necessary to define alternative emissive scenarios and apply again the methodology on the dominant emission sources of the area. Once the health risk will have been estimated depending on the air pollutants (macropollutants, carcinogenic micropollutants or toxic non-carcinogenic micropollutants), the acceptability of the risk must be verified.

3 CRITICAL ASPECTS

The methodology presented in Section 2 follows a multi-step approach that makes use of several procedures. Generally, at present, the latter are poorly harmonised and require a huge amount of input data, both in terms of quantity and in terms of variety. Therefore, efforts are necessary to facilitate the retrieving of the data of interest and to make them compatible with the procedures adopted. A set of tools could be developed by the scientific community to facilitate the fast conversion of data into the format required by who applies the methodology. However, the public sector (e.g., environmental agencies) should ease the operation of data collection by making all the data of interest available to the public and reachable on a dedicated web portal. In addition, the following critical aspects are worth consideration:

1. the background contributions of heavy metals, dioxin, benzene, polycyclic aromatic hydrocarbons and carbon monoxide would be neglected in the assessment of the health impacts if those pollutants were not measured by the air quality monitoring stations located in the target area;
2. at present, an institutional methodology to estimate the transfer of toxic and persistent pollutants from the environment to animals and from animals to meat, dairy products and eggs is still absent;
3. the different databases required to extract relevant emission factors might consider different pollutants and, in some cases, pollutants that are aggregated in different manners.

The first point would require specific monitoring campaigns covering the unmonitored pollutants. However, two assumptions could be formulated on this aspect:

- the background concentrations of benzene and carbon monoxide, whose presence is generally limited to the proximity with emission sources [28], [29], might be neglected;
- the possible underestimation of persistent pollutants might be compensated by the precautionary strategy of considering that the diet is 100% based on locally-grown food.

The second point implies that, in case farm animals present in the target area, the estimation of the health impacts should consider animal- and pollutant-specific food-chain models available in the literature. As an example, McLachlan [30] developed a model to assess the accumulation of dioxins and furans through the feed-cattle-beef/milk chain.

The third point entails the need for harmonising different databases, introducing hypotheses or neglecting some pollutants.

In the light of the previous considerations, the role of public institutions is crucial in view of a homogenisation of data and procedures that would avoid wrong conclusions in a health impact assessment or approximations that might underestimate the real impacts.

4 CONCLUSIONS

The present paper proposes a methodology aimed at assessing the health risk related to the air quality in a target area and a criterion to evaluate the acceptability of the additional incidence of a new emissive activity. The criterion is based on the concepts of acceptable cancer risk, unitary HI and the compliance with local sanitary and economic targets, respectively considering the effects of carcinogenic micropollutants, the toxic non-carcinogenic effects of micropollutants and macropollutants. The methodology actually consists in a sequence of consolidated procedures (emission calculation, dispersion modelling, health risk estimation, external cost estimation) and novel steps (compensation criterion), which were here presented in a unified manner, in order to serve as guidelines to local administrations. The implementation of this methodology could facilitate the decision-making process concerning the authorisation procedure of new industrial/civil activities, but requires efforts by the public sector at local and national/international levels. Specifically, additional monitoring campaigns in the areas under investigation might be necessary to retrieve background contributions of air pollutants that are not monitored by the air quality stations in the target areas. In addition, at a national/international level, the public sector should support the harmonisation of different sources of data to facilitate the application of the complete methodology presented here.

ACKNOWLEDGEMENT

The authors wish to thank the ATS Consortium for the financial support to the research that was the starting point of this article.

REFERENCES

[1] UN Environmental Assembly, Air. http://web.unep.org/environmentassembly/air. Accessed on: 30 Jan. 2019.

[2] De Marco, A. et al., Impacts of air pollution on human and ecosystem health, and implications for the National Emission Ceilings Directive: Insights from Italy. *Environment International*, **125**, pp. 320–333, 2019.

[3] Schiavon, M., Ragazzi, M., Rada, E.C., Magaril, E. & Torretta, V., Towards the sustainable management of air quality and human exposure: Exemplary case studies. *WIT Transactions on Ecology and the Environment*, vol. 230, WIT Press: Southampton and Boston, pp. 489–500, 2018.

[4] Rada, E.C., The sustainable city and air pollution. *WIT Transactions on Ecology and the Environment*, vol. 191, WIT Press: Southampton and Boston, pp. 1369–1380, 2014.

[5] European Environment Agency, Air Quality in Europe – 2018 Report. www.eea.europa.eu//publications/air-quality-in-europe-2018. Accessed on: 30 Jan. 2019.

[6] Tian, Y., Yao, X. & Chen, L., Analysis of spatial and seasonal distributions of air pollutants by incorporating urban morphological characteristics. *Computers, Environment and Urban Systems,* **75**, pp. 35–48, 2019.

[7] Cai, J. et al., Association between airborne fine particulate matter and residents' cardiovascular diseases, ischemic heart disease and cerebral vascular disease mortality in areas with lighter air pollution in China. *International Journal of Environmental Research and Public Health*, **15**(9), p. 1918, 2018.

[8] Murakami, K., Itsubo, N., Kuriyama, K., Yoshida, K. & Tokimatsu, K., Development of weighting factors for G20 countries. Part 2: Estimation of willingness to pay and

annual global damage cost. *International Journal of Life Cycle Assessment*, **23**(12), pp. 2349–2364, 2018.
[9] Pimpin, L. et al., Estimating the costs of air pollution to the National Health Service and social care: An assessment and forecast up to 2035. *PLoS Medicine,* **15**(7), e1002602, 2018.
[10] Schiavon, M. et al., Assessing the air quality impact of nitrogen oxides and benzene from road traffic and domestic heating and the associated cancer risk in an urban area of Verona (Italy). *Atmospheric Environment*, **120**, pp. 234–243, 2015.
[11] Swab, C., Allen, P., Armitage, S. & Biberic, A., 2014 residential wood combustion survey: Results overview and spatial allocation of emissions estimates. *Atmospheric Environment,* **198**, pp. 12–22, 2019.
[12] Ionescu, G., Apostol, T., Rada, E.C., Ragazzi, M. & Torretta, V., Critical analysis of strategies for PM reduction in urban areas. *UPB Scientific Bulletin, Series D: Mechanical Engineering*, **75**, pp. 175–186, 2013.
[13] Rodrigues, S.M. & Römkens, P.F.A.M., Human health risks and soil pollution. *Soil Pollution*, eds A.C. Duarte, A. Cachada & T. Rocha-Santos, Elsevier Academic Press: London, pp. 217–250, 2018.
[14] Copernicus Land Monitoring Service, CLC 2012. http://land.copernicus.eu/pan-european/corine-land-cover/clc-2012/view. Accessed on: 28 Jan. 2019.
[15] European Environment Agency, EMEP/EEA air pollutant emission inventory guidebook 2016. www.eea.europa.eu/publications/emep-eea-guidebook-2016. Accessed on: 23 Jan. 2019.
[16] EMISIA, COPERT 5 – Introduction. https://copert.emisia.com/. Accessed on: 23 Jan. 2019.
[17] Tomasi, E., Antonacci, G., Giovannini, L., Zardi, D. & Ragazzi, M., Atmospheric dispersion modelling with AERMOD for comparative impact assessment of different pollutant emission sources in an Alpine valley. *WIT Transactions on Ecology and the Environment*, vol. 198, WIT Press: Southampton and Boston, pp. 431–442, 2015.
[18] INEMAR, XWiki. www.inemar.eu/xwiki/bin/view/Inemar/WebHome. Accessed on: 24 Jan. 2019.
[19] Wackernagel, H., *Multivariate Geostatistics*, Springer: Berlin, Heidelberg, pp. 74–81, 1995.
[20] U.S. Environmental Protection Agency, Office of Solid Waste and Emergency Response, Human health risk assessment protocol for hazardous waste combustion facilities (final), Report no. EPA530-R-05-006.
[21] U.S. Environmental Protection Agency, Integrated risk information system (IRIS). www.epa.gov/IRIS/. Accessed on: 20 Dec. 2018.
[22] INRAN, National investigation on food consumption in Italy. http://nut.entecra.it/710/I_consumi_alimentari__INRAN-SCAI_2005-06.html. Accessed on: 17 Jan. 2019. (In Italian.)
[23] National Cancer Institute, Definition of relative risk. www.cancer.gov/publications/dictionaries/cancer-terms/def/relative-risk. Accessed on: 30 Jan. 2019.
[24] Logue, J.M., Price, P.N., Sherman, M.H. & Singer, B.C., A method to estimate the chronic health impact of air pollutants in U.S. residences. *Environmental Health Perspectives*, **120**(2), pp. 216–222, 2012.
[25] U.S. Environmental Protection Agency, Residual risk report to congress. www.epa.gov/sites/production/files/2013-08/documents/risk_rep.pdf. Accessed on: 23 Jan. 2019.

[26] Italian Republic, Decree 4/2008 – Additional corrective and integrative measures to the decree 152/2006 on environmental legislation. *Official Journal of the Italian Republic*, **24**, 2008.
[27] EMRC, Cost-benefit analysis of final policy scenarios for the EU clean air package. http://ec.europa.eu/environment/air/pdf/TSAP%20CBA.pdf. Accessed on: 24 Jan. 2019.
[28] Skorokhod, A.I., Berezina, E.V., Moiseenko, K.B., Elansky, N.F. & Belikov, I.B., Benzene and toluene in the surface air of northern Eurasia from TROICA-12 campaign along the Trans-Siberian Railway. *Atmospheric Chemistry and Physics*, **17**, pp. 5501–5514, 2017.
[29] Zhang, K. & Batterman, S., Near-road air pollutant concentrations of CO and $PM_{2.5}$: A comparison of MOBILE6.2/CALINE4 and generalized additive models. *Atmospheric Environment*, **44**, pp. 1740–1748, 2010.
[30] McLachlan, M.S., A simple model to predict accumulation of PCDD/Fs in an agricultural food chain. *Chemosphere*, **34**, pp. 1263–1276, 1997.

INFLUENCE OF KEY PARAMETERS ON THE REMOVAL EFFICIENCY OF AIR POLLUTANTS BY A BIOTRICKLING FILTER

VINCENZO TORRETTA[1], MARCO SCHIAVON[2], PAOLO CARUSON[3] & MARCO RAGAZZI[2]
[1]Department of Theoretical and Applied Sciences, University of Insubria, Italy
[2]Department of Civil, Environmental and Mechanical Engineering, University of Trento, Italy
[3]Air Clean Srl, Italy

ABSTRACT

A chemical scrubber followed by a biotrickling filter (BTF) was integrated in the air treatment line of a composting plant, originally equipped with an open-bed biofilter. Such integration is motivated by the poor performance of the biofilter in the removal of odorants like ammonia (NH_3), sulphides and volatile organic compounds (VOCs). The scrubber and BTF units were monitored during two summer periods in two consecutive years (08.10.2016–09.13.2016 and 07.26.2017–09.20.2017) by analysing key parameters such as airflow rate, metabolite concentrations and, especially, the temperature of the circulation water of the BTF. During the second period, a 17% decrease in the waste-gas flow rate and a consequent increase in the waste-gas temperature were measured. The control of temperature proved to be crucial for the removal of the three groups of pollutants investigated in this study. The occurrence of thermophilic conditions in the BTF inhibited nitrification and generally lowered VOC biodegradation. The increased NH_3 removal rate that was observed during the second period can only be explained with the increase in the residence time in the scrubber, as an analysis of the nitrate and ammonium concentrations in the circulation water of the BTF revealed. While adverse temperature conditions may be compensated by increased residence time for NH_3, temperature is crucial for VOCs, which were only slightly affected by the increased residence time. On the other hand, a relative increasing trend of VOC removal rate in the BTF was observed in the thermophilic range. In addition, the removal efficiency of dimethyl sulphide clearly increased during the second period. More research is needed to understand if such a positive effect can be related to the increased residence time in the pre-treatment line or if thermophilic conditions can help the biodegradation of VOCs and sulphides.
Keywords: biofiltration, air pollution control, absorption, volatile organic compounds, dimethyl sulphide, ammonia.

1 INTRODUCTION

In spite of the evident advances in the emission control from industrial and civil activities, air pollution is still an alarming issue related to the continuous economic growth of developed and developing countries around the world, which determines an increasing number of activities that release air pollutants into the atmosphere [1]. Ozone, particulate matter (PM) and ammonia (NH_3) act on a regional scale, the first two being transported over hundreds of kilometres and the last one promoting secondary PM formation [2]–[4]. Ozone also induce global warming on a global scale [5] PM also acts on a local scale, inducing respiratory and cardiovascular diseases [6], as well as nitrogen dioxide, carbon monoxide, volatile organic compounds (VOCs) and micropollutants. VOCs may be directly harmful to humans through inhalation and may promote ozone formation [7]. In addition, NH_3 and other nitrogen-based compounds, several VOCs and sulphur-based compounds may be perceived by the sense of smell at low concentrations, thus inducing odour nuisance to people living near activities that emit such substances [8].

Physical and chemical methods (e.g., thermal or catalytic oxidation, absorption, catalytic reduction and adsorption) have been largely and conveniently used to remove gaseous air pollutants when the polluted airflow is characterised by high pollutant concentrations

WIT Transactions on Ecology and the Environment, Vol 236, © 2019 WIT Press
www.witpress.com, ISSN 1743-3541 (on-line)
doi:10.2495/AIR190291

(> 100 ppm) and low-to-medium flow rates (< 10^4 Nm^3/h) [9], [10]. The treatment of large airflow rates characterised by lower pollutant concentrations have proved to be more conveniently achieved by methods that limit the use of fuels, electric energy and chemicals [11]. This is the case, for instance, of biofiltration, which is a widespread, simple and cost-effective technology that is especially convenient to remove a large variety of air pollutants at low concentrations (generally < 100 ppm) but with medium-high airflow rates (> 10^4 Nm^3/h) [12]. Such operating conditions are typical of mechanical-biological treatments (MBTs) of municipal solid waste like composting. In fact, biofiltration has become a fundamental component of the air pollution control lines of MBTs.

Composting and other aerobic MBTs (e.g., biostabilisation and biodrying) emit a large variety of compounds compared to specific industrial activities. The major constituents of the process air of aerobic MBTs include volatile organic compounds (VOCs), nitrogen- and sulphur-based compounds. Among VOCs, the main groups of compounds in the process air are aromatics, ketones, aldehydes, terpenes, volatile fatty acids and alcohols [13]. Nitrogen- and sulphur-based compounds are mainly ammonia (NH_3), ammines, hydrogen sulphide, dimethyl sulphide (DMS) and dimethyl disulphide [14], [15]. All VOCs except aromatics are generated by incomplete aerobic biodegradation, while sulphides are formed by local conditions of anaerobic biodegradation that can develop in the waste piles the porosity of waste is low [13], [16]. Ammonia and ammines are formed as degradation products of proteins [16]. Aromatic VOCs and other micropollutants (e.g., dioxin) may be released by MBT facilities too. Contrarily to the previous compounds, such substances are not formed during the biodegradation process but are already present in the waste. The presence of aromatics may be due to the disposal of solvents in the waste, while the presence of dioxin may be related to the presence of contaminated food waste [17]. Indeed, it is known that about 90% of the human daily exposure to dioxin occurs through the diet [18].

Traditionally, biofiltration is carried out by open-bed biofilters, which requires large areas to ensure satisfying residence time and homogeneous conditions along the vertical direction. More recently, as an alternative to traditional biofilters, biotrickling filters (BTFs) have gained consideration because of their more compact size, lower pressure drop, effective pH control and due to the possibility of removing different pollutants that require different pH conditions and bacterial strains [19]. Indeed, contrarily to biofilters, BTFs can be easily divided into two or more stages operating under different pH conditions and inoculated with the most suitable microorganisms for the removal of the pollutants.

The control of temperature in a BTF is particularly important because temperature influences the reproduction of the microorganisms that carry out the biodegradation process. Mesophilic conditions, with temperatures in the range 10–40°C, are the most suitable for most of the bacterial strains and, consequently, for most of the pollutants that are removed by biofiltration [20]. Generally, temperatures in the thermophilic range (> 45°C) may cause the death of most of the microogranisms. This should be taken into account even when operating a BTF under mesophilic conditions, since biodegradation is an exothermic process [19]. Nevertheless, thermophilic conditions are favourable in specific situations, e.g. when the target is to remove nitrogen oxides from an airflow [21]–[23]. On the other hand, temperatures in the cryophilic range (< 20°C) may inhibit the microbial activity [24].

The present paper describes the operation of an air treatment system used to pre-treat the air extracted from the biocells of a composting plant treating the organic fraction of municipal solid waste (OFMSW). The latter is located in northern Italy and was object of numerous complaints by the local population, due to the odour nuisance perceived in the area, which was related to a malfunctioning of the existing open-bed biofilter, caused by the high concentrations of NH_3 in the waste-gas flow and by the high waste-gas temperature. In addition, high concentrations of VOCs and sulphides were detected in the waste gas, which were generated by anaerobic zones in the biocells. Sulphides like DMS are characterised by lower odour threshold levels compared to other compounds that are generally formed during composting [7]. To reduce the concentrations of odorants in the waste gas upstream of the existing biofilter, a pre-treatment line was installed. The latter is composed of a chemical scrubber and a BTF. In the attempt to decrease the waste-gas temperature, the gas flow rate to the pre-treatment line was increased by drawing the air of the compartment assigned to receive the incoming waste.

In view of the history of the composting plant, the aim of this paper is two-fold:

- to evaluate the technical convenience of pre-treating the heavily polluted air coming from the biocells of a composting plant through advanced biological systems, alone or in combinations with chemical scrubbers;
- to investigate the impact of different process parameters (including the waste-gas temperature) on the performance of the pre-treatment stage, especially on the capability of removing NH_3, VOCs and DMS.

2 MATERIALS AND METHODS

2.1 Description of the air pre-treatment line

The waste gas coming from the biocells for the accelerated oxidation of OFMSW (about 8,550 m^3/h at about 65°C) is mixed with the waste gas coming from the reception compartment (28,575 m^3/h at about 30°C) in the attempt to reduce the gas temperature. The waste gas is then sent to the chemical scrubber operated with a washing solution (1,400 L/min) based on sulphuric acid (H_2SO_4) to abate NH_3. The scrubber consists in a column composed of two chambers filled with spheres of polyethylene.

The waste gas is then sent to a BTF, designed to operate at a temperature of 40°C, for the removal of peaks of NH_3, VOCs and sulphides. The filling material consists in shells, which ensures pH stability in case of production of acidic by-products during biodegradation. In addition, the BTF, in an appropriate range of temperature between 25 and 35°C [25], is capable of operating the nitrification of the residual NH_3 coming from the chemical scrubber. The BTF ensures a residence time of 16 s at the total airflow rate (37,125 m^3/h at 38°C). The washing solution is recirculated over the filling material with a flow rate of 160 L/min.

The biofilter, originally treating a waste-gas flow rate of 84,000 m^3/h coming from the biocells and the maturation stage, is now requested to treat the additional 28,575 m^3/h from the reception compartment. Consequently, the specific volumetric load increased from 44 to 60 m^3/h per m^3 of filling material.

2.2 Air sampling

The efficiency of the air treatment stage of the composting facility was monitored through the analysis of air samples collected by seven sampling points along the air treatment line. Specifically, three air sampling ports are located at the inlet of the chemical scrubber (A), at the inlet of the BTF (B), at the outlet of the BTF and downstream of the blower (C). Fig. 1 presents a schematic view of the layout of the air treatment stage of the composting facility and of the locations of the three sampling points.

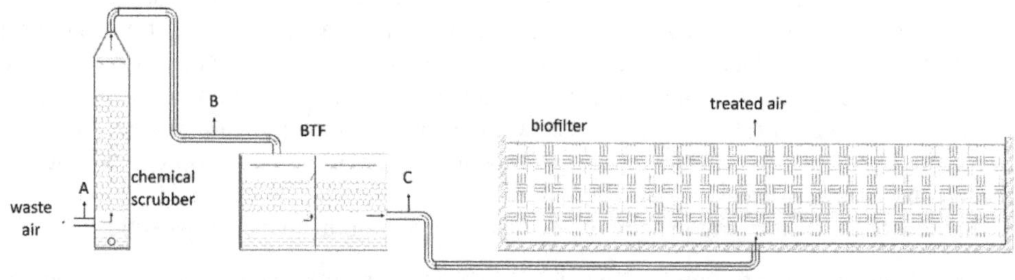

Figure 1: Layout of the air treatment line of the composting facility.

2.3 Analytical methods

The concentrations of NH_3 and DMS in the process air were measured by means of colorimetric vials (Drägerwerk AG & Co. KGaA, Germany), inserted into the sampling ports A, B and C, allowing for measuring NH_3 in the ranges 2–30 ppm and 5–100 ppm, and DMS in the range 0–15 ppm. The total VOC concentration in the process air was monitored by a photoionisation detector (MiniRAE Lite, RAE Systems, USA), allowing for measuring VOC concentrations in the range 0–5,000 ppm in the sampling points A, B and C. A hot-wire anemometer (HD 2303.0, Delta OHM Srl, Italy) was used to measure the temperature and the flow rate of the process air in the sampling point A. In addition, the temperature and the concentration of ammonium (NH_4^+) and nitrates (NO_3^{2-}) ions were monitored in the circulation water of the BTF. Finally, the pH of the washing solution of the scrubber and the pressure drop between the BTF were respectively monitored by a pH-meter and digital differential manometers. Two measurement campaigns were carried out in two consecutive years: the first one took place in the period 08.10.2016–09.13.2016 and the second one was performed in the period 07.26.2017–09.20.2017.

3 RESULTS AND DISCUSSION

3.1 NH_3 removal

The removal of NH_3 from the process air of the composting facility is preliminarily carried out by the scrubber through chemical absorption. Generally, the absorption of NH_3 increases both with the contact time of the pollutant with the washing solution and with decreasing the pH of the washing solution.

The effect of the contact time is visible by comparing the removal efficiency of NH_3 between the two study periods, since the flow rate of the process air was decreased by 17% between 2016 and 2017 by the company that manages the composting plant (Fig. 2). During the whole observation period, the pH remained almost constant and comprised between 4.0 and 4.7. The NH_3 concentration at the inlet of the scrubber varied between 6 and 60 ppm during the two monitoring periods. Overall, the chemical scrubber allowed obtaining NH_3 removal efficiencies > 85%, determining an NH_3 concentration at the inlet of the BTF < 2.5 ppm (mean value: 0.6 ppm).

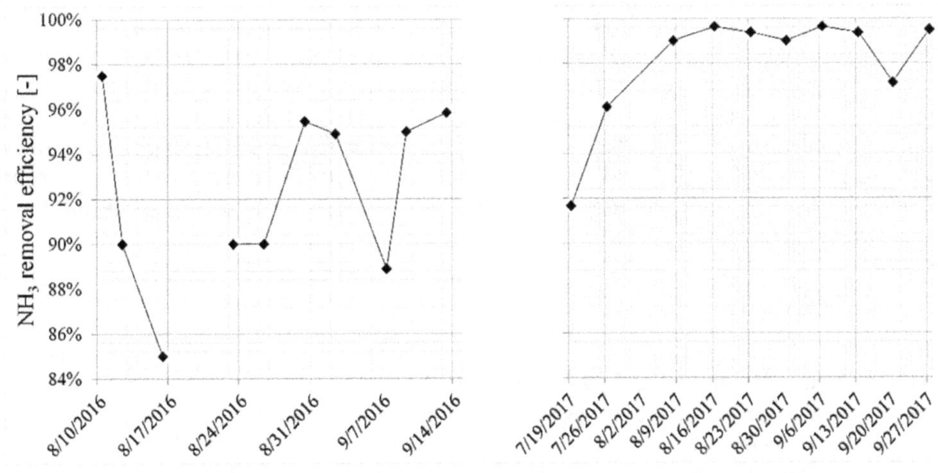

Figure 2: Trends of the removal efficiencies of NH_3 by the chemical scrubber during the first and the second monitoring period.

The removal performance of NH_3 by the BTF was not as constant as that of the chemical scrubber for the same observation period. These results are confirmed by the trend of the concentrations of NH_4^+ and NO_3^{2-} in the circulation water of the BTF. Specifically, NO_3^{2-} concentration in the circulation water is a good indicator of the biological activity of the microorganisms present in the BTF, since a well-performing biological oxidation process transforms NH_4^+ into NO_3^{2-} through nitrification. Therefore, by assuming the airflow rate, the circulation water and nutrient flow rates and the inlet NH_3 concentration as constant, a decrease in the NO_3^{2-} concentration in the circulation water over time would indicate that the biological activity of the microorganisms is weakening. This would be accompanied by an increase in the NH_4^+ concentration in the circulation water. The concentration trends of NH_4^+ and NO_3^{2-} in the circulation water are reported in Fig. 3.

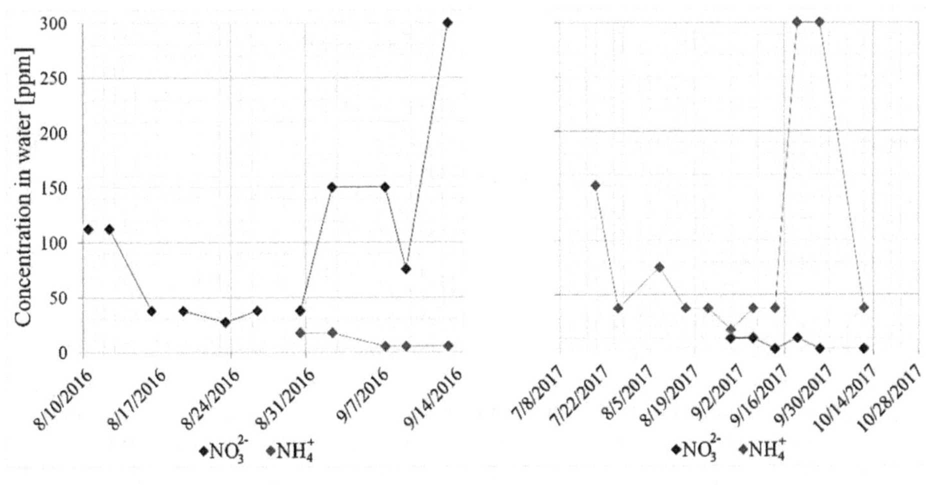

Figure 3: Trends of the concentrations of NO_3^{2-} and NH_4^+ in the circulation water of the BTF during the first and the second monitoring periods.

The concentration trends of NH_4^+ and NO_3^{2-} are not constant over time. This behaviour could be partly related to the high temperature of the circulation water, which reaches peaks > 45°C, especially during summer 2017. Nitrification is known to take place under mesophilic conditions [25]. Under thermophilic conditions, nitrification is inhibited. This may explain the low concentrations of NO_3^{2-} in the circulation water. The high temperature of the washing solution of the BTF experienced in September 2017 explains the significant increase in the NH_4^+ concentration. This event is associated with an increase in the NH_3 concentration at the outlet of the BTF.

3.2 DMS removal

An increase in the overall removal efficiency of DMS was observed during the second monitoring period (summer 2017) with respect to the first one (Fig. 4). This increase may be due to the increase in the residence time caused by the decrease of the airflow rate that occurred in the second year. An additional explanation could be related to the increase in the temperature of the circulation water of the BTF that might have created favourable conditions in the BTF. However, this hypothesis is not supported by the scarce number of scientific papers in the literature [27]. To prove this hypothesis, the concentration of DMS should be measured in B. However, such measurements are not available at present.

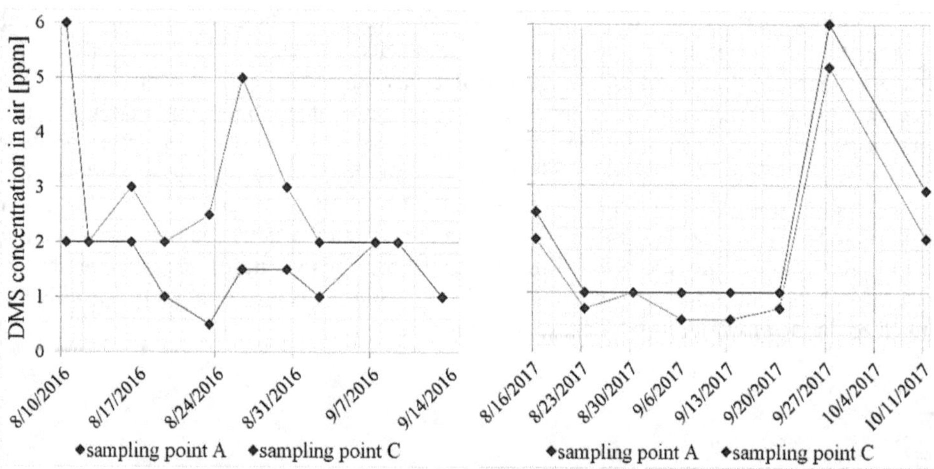

Figure 4: Trends of the DMS concentrations measured in the sampling points A and C during the first and the second monitoring periods.

3.3 Total VOC removal

During the two monitoring periods, the VOC concentrations at the inlet of the scrubber varied between 5.9 and 51.2 ppm (Fig. 5). A decrease in the VOC removal efficiency of the scrubber-BTF combined system can be observed in 2017 compared to the previous year. Such lower performance may be due to the transition to thermophilic conditions in the BTF, which may have slowed down the activity of the microorganisms performing VOC biodegradation. The very low removal efficiencies in the scrubber indicate that most of the removal process occurs in the BTF. In fact, absorption is known not to represent an efficient option for VOC removal, due to the generally low solubility of VOCs in water [26].

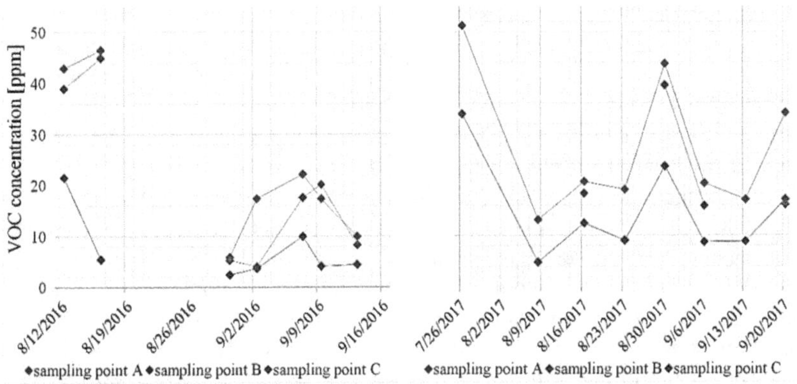

Figure 5: Trends of VOC concentrations during the first and the second monitoring periods.

3.4 Effect of temperature

In the attempt of understanding how the temperature can influence the removal of the three investigated categories of pollutants, Fig. 6 reports the relationships between the removal efficiencies of NH_3, DMS and VOCs as a function of the temperature of the circulation water in the BTF, by assuming a temperature of 40°C as a threshold value for the transition between mesophilic and thermophilic conditions.

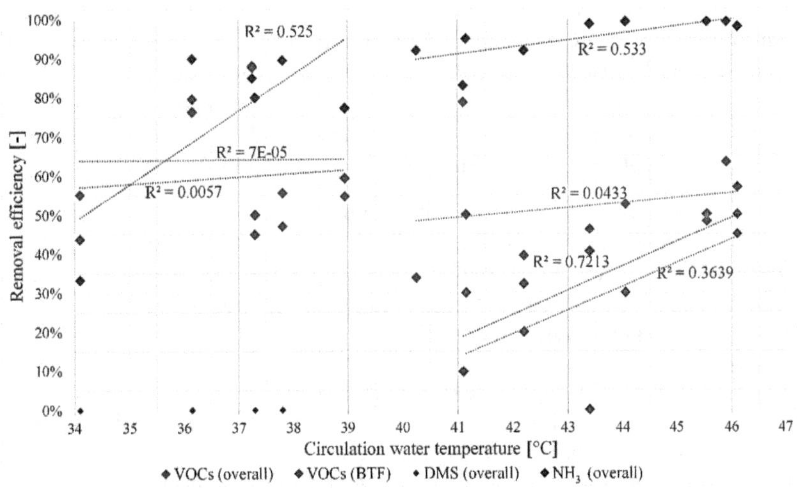

Figure 6: Relationships between the circulation water temperature in the BTF and the removal efficiencies of NH_3, DMS and VOCs; overall removal efficiencies refer to the abatement by the scrubber-BTF system.

Concerning VOCs, the available data allow calculating both the removal efficiencies of the scrubber-BTF system and the removal efficiencies of the BTF alone. Regarding DMS, due to the lack of data in B, only the removal efficiencies of DMS by the scrubber-BTF system were included in Fig. 4. Similarly, the relationship between temperature and NH_3 removal efficiency is presented only with respect to the scrubber-BTF system. A decreasing trend in the removal efficiencies of total VOCs is visible at increasing temperature in the circulation water of the BTF. The highest temperature generally occurs during 2017, when the waste-gas flow rate was 17% lower than 2016 on average. This confirms that, in spite of an increase in the residence time, an increase in temperature negatively affects total VOC removal. Concerning the only BTF, the VOC removal efficiency seems to increase after transition to thermophilic conditions. However, the removal efficiencies obtained under such circumstances are lower than those achieved in the mesophilic range of temperature. Additional tests should be carried out at higher temperatures to confirm the positive effect of temperature. Regarding NH_3, the increase in the overall removal efficiency with temperature is only apparent, since nitrification is inhibited under thermophilic conditions. Therefore, the overall improvements in the NH_3 removal efficiency may be related to the increase in the contact time in the chemical scrubber, due to the decrease in the airflow rate that occurred in 2017, the same period during which the temperature of the circulation water increased. The abatement performance of DMS seems to increase at increasing the temperature of the washing solution of the BTF.

This behaviour can be explained both by the increased residence time in 2017 (when higher temperatures were measured) and by a possible favourable situation to sulphide-degrading microorganisms determined by thermophilic conditions. To confirm this effect, measurements of the DMS concentration in B are needed. If this effect were confirmed, thermophilic conditions could facilitate the removal of sulphides.

If the main target is to reduce NH_3 and VOC concentrations upstream of a biofilter, two alternative options are available:

- the company that manages the composting plant could insert a heat exchanger upstream of the pre-treatment line to cool down the waste-gas flow rate, if its temperature is incompatible with the optimal conditions for nitrification in a BTF;
- the waste-gas flow rate could be further increased by an amount of colder air coming from outside or from internal compartments with low pollutant concentrations (e.g., the reception compartment or the last maturation stages).

Both the options must be evaluated by the company managing the plant, since additional costs would be involved: the first option would imply the condensation of part of the waste gas, which would require additional operational costs to treat the generated wastewater; the second option would entail the enlargement of the treatment line (e.g., an additional stage in the BTF), with additional investment costs.

4 CONCLUSIONS

The air pollution treatment line of a composting facility was monitored during two observation periods. The line comprises a chemical scrubber and a BTF upstream of a biofilter. Special attention was given to assess the performance of the first two elements of the treatment line, which, if properly operated, would ensure lower pollutant loads to the biofilter and, consequently, higher overall removal rates. In view of improved performance, the control of the temperature of the system proved to be crucial. Specifically, the establishment of thermophilic conditions showed adverse effects on the biodegradation of NH_3 in the BTF and a reduced performance in the overall VOC abatement. A decrease in the airflow rate occurred during the second period of observation and this may explain the increase in the removal rate of NH_3, thanks to a consequent increase in the residence time. However, the temperature of the circulation water increased at the same time and this could be considered as a positive factor towards the removal of DMS that was observed in the thermophilic range. Additional confirmations of the effects of high temperature on DMS removal would be available if the concentrations of DMS were measured upstream of the BTF (sampling point B). If the positive effect of temperature were confirmed, the increase in the DMS removal during the second monitoring period could be related to the occurrence of thermophilic conditions and this could open to additional options in the field of air pollution control. The increase in the residence time did not have positive effects on VOC removal and this behaviour may be explained with the low solubility of VOCs in water. While adverse temperature conditions may be compensated by increased residence time for NH_3, temperature is a crucial factor for VOC removal. The complete monitoring of key parameters (pollutant concentrations in air and in water, metabolite concentrations, temperature, pH and airflow rate) in all the sampling points located upstream and downstream of each component would allow for the correct management of a multi-step air pollution treatment line and would give definitive indications on their role.

REFERENCES

[1] Wu, Y., Zhu, Q. & Zhu, B., Comparisons of decoupling trends of global economic growth and energy consumption between developed and developing countries. *Energy Policy*, **116**, pp. 30–38, 2018.

[2] Querol, X. et al., On the origin of the highest ozone episodes in Spain. *Science of the Total Environment*, **571**, pp. 379–389, 2016.

[3] Garg, S. & Sinha, B., Determining the contribution of long-range transport, regional and local source areas, to PM10 mass loading in Hessen, Germany using a novel multi-receptor based statistical approach. *Atmospheric Environment*, **167**, pp. 566–575, 2017.

[4] Erisman, J.W. & Schaap, M., The need for ammonia abatement with respect to secondary PM reductions in Europe. *Environmental Pollution*, **129**, pp. 159–163, 2004.

[5] Mickley, L.J., Murti, P.P., Jacob, D.J. & Logan, J.A., Radiative forcing from tropospheric ozone calculated with a unified chemistry-climate model. *Journal of Geophysical Research*, **104**, pp. 30153–30172, 1999.

[6] Rodopoulou, S., Samoli, E., ChalbotbIlias, M.-C.G. & Kavouras, G., Air pollution and cardiovascular and respiratory emergency visits in Central Arkansas: A time-series analysis. *Science of The Total Environment*, **536**, pp. 872–879, 2015.

[7] Schiavon, M. et al., Characterisation of volatile organic compounds (VOCs) released by the composting of different waste matrices. *Environmental Pollution*, **231**, pp. 845–853, 2017.

[8] Aatamila, M. et al., Odour annoyance and physical symptoms among residents living near waste treatment centres. *Environmental Research*, **111**, pp. 164–170, 2011.

[9] Pourmohammadbagher, A., Jamshidi, E., Ale-Ebrahim, H., Dabir, B. & Mehrabani-Zeinabad, M., Simultaneous removal of gaseous pollutants with a novel swirl wet scrubber. *Chemical Engineering and Processing: Process Intensification*, **50**, pp. 773–779, 2011.

[10] Ragazzi, M., Tosi, P., Rada, E.C., Torretta, V. & Schiavon, M., Effluents from MBT plants: Plasma techniques for the treatment of VOCs. *Waste Management*, **34**, pp. 2400–2406, 2014.

[11] Fridman, A., *Plasma Chemistry*, Cambridge University Press: Cambridge, 2008.

[12] Schiavon, M., Ragazzi, M., Torretta, V. & Rada, E.C., Comparison between conventional biofilters and biotrickling filters applied to waste bio-drying in terms of

atmospheric dispersion and air quality. *Environmental Technology*, **37**, pp. 975–982, 2016.

[13] Pagans, E., Font, X. & Sánchez, A., Emission of volatile organic compounds from composting of different solid wastes: abatement by biofiltration. *Journal of Hazardous Materials*, **131**, pp. 179–186, 2006.

[14] Dorado, A.D., Husni, S., Pascual, G., Puigdellivol, C. & Gabriel, D., Inventory and treatment of compost maturation emissions in a municipal solid waste treatment facility. *Waste Management*, **34**, pp. 344–351, 2014.

[15] Federal Environment Agency – Austria, *Exhaust Emissions from Mechanical-Biological Waste Treatments in Austria*, Umweltbundesamt: Vienna, 1998. (in German.)

[16] Shao, L.-M., Zhang, C.-Y., Wu, D., Lü, F., Li, T.-S. & He, P.-J., Effects of bulking agent addition on odorous compounds emissions during composting of OFMSW. *Waste Management*, **34**, pp. 1381–1390, 2004.

[17] Rada, E.C., Ragazzi, M., Panaitescu, V. & Apostol, T., The role of bio-mechanical treatments of waste in the dioxin emission inventories. *Chemosphere*, **62**, pp. 404–410, 2006.

[18] Eduljee, G.H. & Gair, A.J., Validation of a methodology for modelling PCDD and PCDF intake via the food-chain. *Science of The Total Environment*, **187**, pp. 211–229, 1996.

[19] Schiavon, M., Ragazzi, M., Rada, E.C. & Torretta, V., Air pollution control through biotrickling filters: a review considering operational aspects and expected performance. *Critical Reviews in Biotechnology*, **36**, pp. 1143–1155, 2016.

[20] Schnelle, K.B.J & Brown, C.A., *Air Pollution Control Technology Handbook*, CRC Press: Boca Raton, 2001.

[21] Liang, W., Huang, S., Yang, Y. & Jiang, R., Experimental and modeling study on nitric oxide removal in a biotrickling filter using Chelatococcus daeguensis under thermophilic condition. *Bioresource Technology*, **125**, pp. 82–87, 2012.

[22] Yang, Y., Huang, S., Liang, W., Zhang, Y., Huang, H. & Xu, F., Microbial removal of NO_x at high temperature by a novel aerobic strain Chelatococcus daeguensis TAD1 in a biotrickling filter. *Journal of Hazardous Materials*, **203–204**, pp. 326–332.

[23] Xu, F., Huang, S., Liu, Y. & Chen, S., Comparative study on the production of poly(3-hydroxybutyrate) by thermophilic Chelatococcus daeguensis TAD1: A good candidate for large-scale production. *Applied Microbiology and Biotechnology*, **98**, pp. 3965–3974, 2014.

[24] Hui, Y.H. & Sherkat, F., *Handbook of Food Science, Technology, and Engineering – 4 Volume Set*, CRC Press: Boca Raton, 2005.

[25] Cho, K.H., Kim, J.-O., Kang, S., Park, H., Kim, S. & Kim, J.M., Achieving enhanced nitrification in communities of nitrifying bacteria in full-scale wastewater treatment plants via optimal temperature and pH. *Separation and Purification Technology*, **132**, pp. 697–703, 2014.

[26] Dumont, E., Couvert, A., Amrane, A., Couriol, C., Darracq, G. & Le Cloirec, P., Equivalent Absorption Capacity (EAC) concept applied to the absorption of hydrophobic VOCs in a water/PDMS mixture. *Chemical Engineering Journal*, **287**, pp. 205–216, 2016.

[27] Luvsanjamba, M., Sercu, B., Van Peteghem, J. and Van Langenhove, H., Long-term operation of a thermophilic biotrickling filter for removal of dimethyl sulfide. *Chemical Engineering Journal*, **142**, pp. 248–255, 2008.

SPECIAL SESSION: NATURALLY OCCURRING IONIZING POLLUTANTS

(ORGANISED BY R. IPPOLITO)

RADON ENTRY MODELS INTO BUILDINGS VERSUS ENVIRONMENTAL PARAMETERS, BUILDING SHAPE AND TYPES OF FOUNDATION

ROSARIA IPPOLITO & ROMOLO REMETTI
Department of Basic and Applied Sciences for Engineering, Sapienza – University of Rome, Italy

ABSTRACT

Many studies have been carried out around the world to identify the carcinogenic risk associated with human exposure to air pollution and, nowadays, epidemiological evidences are the way to characterize the risk. Human exposure to naturally occurring ionizing sources is one of the main risks highlighted by the WHO. There is still a great need to make the population aware of this risk, to avoid exposure. For decades, the radon entry has been studied through diagnostic measurement techniques, designing efficient mitigation systems. The rate of radon entry into indoor air also varies with climatic conditions, such as rain, which alters the soil conditions and thus the flow of gas through the soil to the building shell. Environmental parameters and building specific shapes need to be examined to quantify their influence for radon entry. Several mechanisms are responsible: the dominant ones are the "stack effect" driven by temperature differences between the indoors and the outdoors, the effect of wind on the building shell, the operation of mechanical ventilation systems which distribute heated or cooled air throughout the house and each types of foundation which connects the building to the ground. Italy is a geologically fragile country, constantly hit by earthquakes; this allowed to develop over the centuries, safer building strategies. Most of the Italian building heritage, in over 7,900 municipalities, consists of masonry buildings, often made up of local materials with a high radium content. The purpose of this article is to evaluate and analyse how environmental, anthropic and constructive factors can influence radon entry models into buildings. Understandings of the various mechanisms that drive radon into buildings permit the development of specific technologies aimed to limit the radon entry rate and satisfy the Council Directive 2013/59/EURATOM requirements.

Keywords: radon, indoor air pollution, IAQ, case studies, aerosols and particles, environment, air pollution modelling, ventilation, ionizing radiations.

1 INTRODUCTION

The Italian building heritage is mainly made up of load-bearing masonry constructions. This is due to the history of the Country, and above all to the culture of conservation handed down over the centuries, which advises to restore and maintain buildings, not only historical ones, rather than demolish and rebuild with innovative technologies, materials and systems. Many earthquakes, over the centuries and in recent years have struck Italy.

The building stock present in the regions classified as seismic risk consists of 11.1 million buildings, of which 60% are residential buildings made of load-bearing masonry, about 5.2 million buildings built before 1981. In this study, we are dealing with masonry buildings built before 1946, which currently represent 26.4% of the Italian building heritage in seismic risk areas [2]. Most of these buildings were built with local or volcanic materials, with a high content of radium-226, such as those studied, for our case, in the Lazio region, and in particular in the area affected by the last eruption of the Latium Volcano, now Albano Lake, that occurred more than 5,000 years ago. By adding contributions from the soil, building materials and other factors, the population is unknowingly exposed to radioactive doses that far exceed the reference values imposed by 2013/59 / EURATOM [1].

WIT Transactions on Ecology and the Environment, Vol 236, © 2019 WIT Press
www.witpress.com, ISSN 1743-3541 (on-line)
doi:10.2495/AIR190301

2 BACKGROUND

Radon, its harmful and oncological effects due to the gas itself and its decay products are often less considered than other indoor pollutants, particularly in Italy, due to regulatory gaps.

^{238}U, progenitor of radon, constitutes 99.3% of the mixture of natural uranium and, with a half-life of about 4.5 billion years, it is widespread from the origins of the earth; you can find it in all soils and building materials produced directly or indirectly from soils, rocks, in particular from granite and volcanic ones.

Radon, a noble and chemically stable gas, has a half-life of 3.82 days; therefore, the most inhaled fraction of this gas is expelled without contributing significantly to the expected respiratory damage. Its decay products, ^{218}Po, ^{214}Pb, ^{214}Bi, and ^{214}Po cause the major health problems. These radionuclides supply the dose of alpha radiation to the bronchial tissue and increase the risk of developing lung cancer.

The environmental aerosol, both for transport and for respiratory deposition, influences the behavior of the attached fraction of radon decay products. These particles include natural materials such as pollens, road dust as well as anthropogenic emissions. Many factors can modify the internal aerosol such as cooking, smoking and other occupants' lifestyles.

A new WHO *Handbook on Indoor Radon* [3] recommends that Countries adopt reference levels of the gas of 100 Bq/m^{-3}, lowering the values recommended by 2013/59 / EURATOM set at 300 Bq/m^{-3} for annual average indoor radon concentrations.

3 INFLUENCING FACTORS ON INDOOR RADON CONCENTRATION

Many and concurrent factors influence the entrance of radon into buildings: soil physical characteristics, building materials, shape and relationships among the height H_B, length L_B and width W_B of the building, foundation type, environmental parameters, human behavior and lifestyle (Fig. 1).

The main principles of radon transport in materials are due to diffusion and convection.

Diffusion, due to the difference between the concentration of radon in the soil or in building materials, is a process generated by concentration gradients. The gas extends, according to Fick's law, which connects flux density and concentration gradient, to distribute homogeneously between the layers (eqn (1)):

$$J_d(Rn) = -D_e(Rn)grad\, C(Rn), \quad (1)$$

where:

- *Jd (Rn)* is the diffusion flux density of the radon activity (Bq/m^2s);
- *D_e (Rn)* is the actual diffusion coefficient obtained from the ratio between the diffusion coefficient *D* and the porosity of the layer e which influences the gas diffusion rate;
- *C (Rn)* is the concentration in terms of radon activity (Bq/m^3).

Convection, a phenomenon regulated by Darcy's law, is due to differences of the pressure gradient (eqn (2)):

$$\boldsymbol{v(Rn)} = \frac{-\boldsymbol{K}}{\boldsymbol{\mu}}\Delta \boldsymbol{P}, \quad (2)$$

where:

- ***v(Rn)*** is the surface velocity vector;
- ***K*** is the permeability coefficient (m^2);
- m is the viscosity of the fluid;
- **P** is the atmospheric pressure in Pa.

Figure 1: Summary of the interaction among the numerous parameters that affect indoor radon concentration levels.

The transport mechanism is caused by the pressure difference between external and internal air, normally lower, and finds privileged flow channels in the joints of the foundations, of the water and natural gas pipes and in the cracks of the basements.

The *chimney effect* is due to the difference in temperature between the building's interior and exterior: the lighter warm air tends to rise and this flow involves the depressurization inside the building and, consequently, an air recall from the outside and from the ground.

Convection is generated by wind, houses heating, atmospheric pressure or mechanical ventilation, that cause pressure differences between the outdoor and indoor of the dwellings. The amount of the pressure drop caused by wind depends on the shape of building and on the wind speed. The shape of the building affects the concentration of the internal radon. Fig. 1 summarizes the geometry of the building and relates the dimensions, height H_B, width W_B and length LB. Experimental studies have shown that tower buildings ($H_B > L_B$, $H_B > W_B$) have high concentrations of radon even on floors higher than fifth, thanks to the chimney effect. Low buildings with a large base on the ground floor have high concentrations due to the great contribution made by the contact with the ground and the internal depression which increases the entry of radon from the subsoil.

The pressure drop is transmitted with a certain speed and this influences the rate of radon entry into the building. The parameter that increases or decreases this velocity is the permeability of the ground. Advection is even more effective in case of cracks in the soil.

The "*wind effect*" due to the difference in air speed between the internal and external building exerts a force on the walls and on the ground in the direction of the wind, which pushes the radon into the building.

Infiltration indicates the exchange of air between the indoor and the outdoor of buildings, as doors and windows, ventilation and gaps and openings through the shell of the buildings. Other factors facilitate the migration of radon from the ground to the building, such as rain or ice layers: in fact, the rainwater that saturates the soil prevents radon from being released into the atmosphere and conveys it to the foundations area.

It is very important to know the building foundation system (Fig. 2). In masonry constructions, the continuous foundation often does not allow a ventilated crawl space and the lower floor of the building rests directly on the ground. Because of these construction characteristics, radon between the floor and the crawl space enters the building through all the joints and cracks of the floor and piping systems, due to the difference in atmospheric pressure between inside and outside.

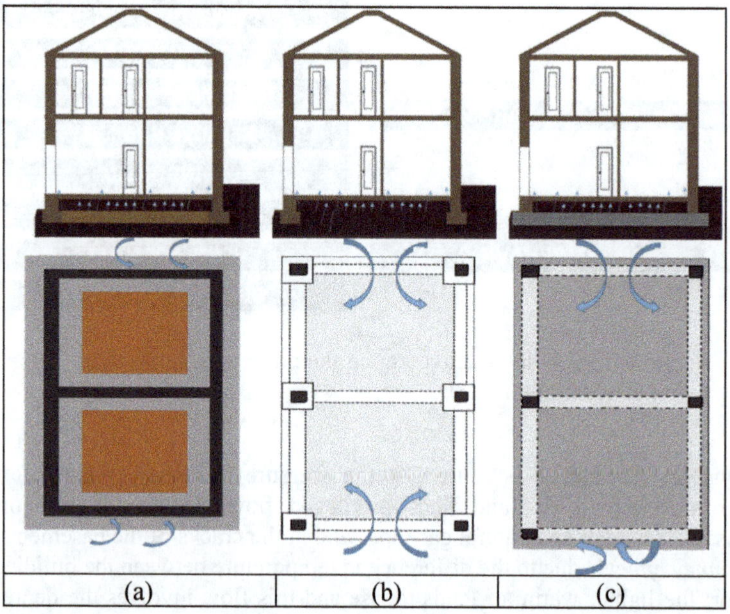

Figure 2: Plant and elevation of shallow foundation types. (a) Strip footings consist of load-bearing walls. The ventilation of the crawl space is difficult; (b) Spread footings are the most common and consist of single or combined columns. Sometimes the floor rests directly on the ground; and (c) Raft footings are used to spread the load from a structure over the entire area of the building, because of soft or loose soils. Drawings © [4].

4 MATERIALS AND METHODOLOGY

4.1 The area

Santa Maria delle Mole, Municipality of Marino, is located south of Rome, between the Appia Antica Park and the base of the ancient Latium Volcano, now Lake of Albano. There are many underground water springs and high gaseous emissions of carbon dioxide CO_2 and sulfur dioxide SO_2, a toxic gas with a nauseating smell. This area is not completely urbanized and will be the subject of new constructions; because of its geophysical characteristics, it is very important to check the concentration levels of radon gas and other pollutants.

4.2 Building characteristics

The building studied has some interesting characteristics that make it right as an experimentation model. The shape of the building follows the land morphology, with considerable differences in height. In particular, the building under test is located on a deep lava flow of black leucitite [5]. This construction, dating back to the Second World War, 1939, has a supporting structure in volcanic tuff masonry. The building has mixed foundations due to an old structural failure on shallow foundations, strip footing type on the side of bedroom and spread type on the living room side (Fig. 3). The most significant rooms, objects of different measurement campaigns in recent years, are the living room and the bedroom. The two rooms have different characteristics: the living room rests on an almost underground room; the bedroom instead rests directly on the rock and is slightly lower than the external one. Both rooms have an inter-floor height of 4 m, the living room is 50 m^2 and the bedroom are about 25 m^2.

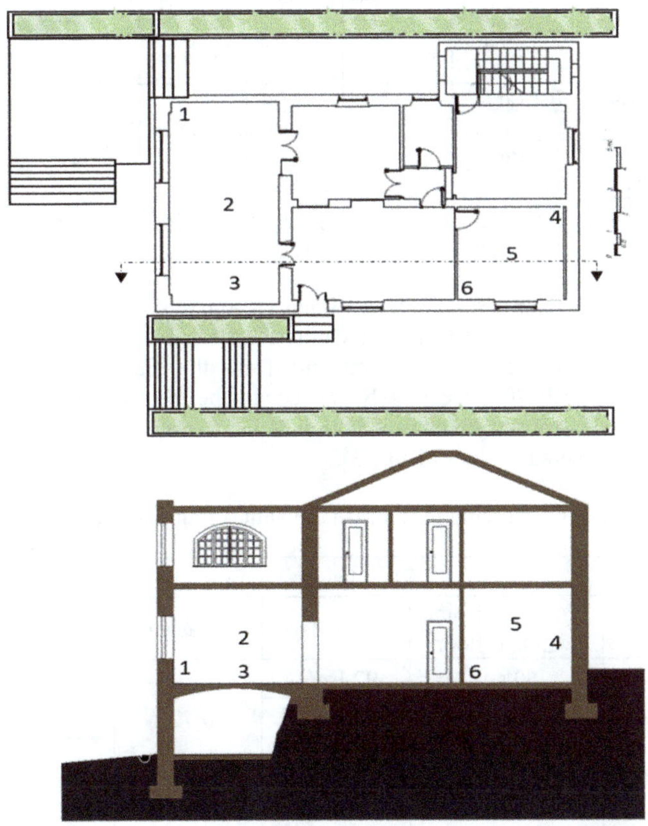

Figure 3: Plant and elevation. Numbers represent the measurement points. Drawings © [4].

4.3 Measurement methodology

Test 1: January–March 2018; July–September 2018
Detector: Solid-state nuclear tracks detector (SSTD CR39)

Test 2: January–March 2019
Detectors: Solid-state nuclear tracks detector (SSTD CR39)
MR1 ZnS(Ag) scintillation cell with sensor of humidity, temperature, atmospheric pressure
TESTO anemometer

Some passive dosimeters were placed in different positions in each room. Measurements were made in the coldest winter months and in the hottest summer ones. Three dosimeters were positioned at different heights, compatible with the lifestyle of the occupants. The measurements inside the same room differ between the min and the max: in the living room even more than 35%, in the bed area around 28% on radon concentration levels (Table 1).

Table 1: 2018 Measurements.

SSNTD CR 39	2018 Measurements						
	Distance from wall and floor	Jan.–Mar.	Δ measure	Jul.–Sep.	Δ measure	Average measure per point	Average measure per room
1 L	60 cm	761 Bq/m^3	35.98%	293 Bq/m^3	29.69%	527 Bq/m^3	472 Bq/m^3
2 L	160 cm	523 Bq/m^3		206 Bq/m^3		364 Bq/m^3	
3 L	120 cm	817 Bq/m^3		231 Bq/m^3		524 Bq/m^3	
4 B	120 cm	834 Bq/m^3	28.12%	220 Bq/m^3	25.75%	527 Bq/m^3	509 Bq/m^3
5 B	180 cm	644 Bq/m^3		196 Bq/m^3		420 Bq/m^3	
6 B	60 cm	896 Bq/m^3		264 Bq/m^3		580 Bq/m^3	

The average of the measured values is around those fixed by Italian Legislative Decree 230/95, 500 Bq/m^3, which recommends repeating the measurements in case of values close to the set limits, but the levels recommended by the European Council Directive 2013/59/Euratom, 300 Bq/m^3, are largely exceeded. The measurements were repeated in 2019. Due to different causes affecting variation of radon concentration such a time span is necessary to obtain average values (Table 2).

Table 2: The measurements confirm the values found in the previous year.

2019 Measurements				
SSNTD CR 39	Distance from wall and floor	Jan.–Mar.	Δ measure	Average measure per room
1 L	60 cm	692 Bq/m^3	30.72%	
2 L	160 cm	548 Bq/m^3		677 Bq/m^3
3 L	120 cm	791 Bq/m^3		
4 B	120 cm	690 Bq/m^3	17.79%	
5 B	180 cm	596 Bq/m^3		670 Bq/m^3
6 B	60 cm	725 Bq/m^3		

5 MATERIALS AND METHODOLOGY

Radon measurements were performed with a scintillation cell detector. The difference in concentration, thanks to the measurement of the airflow speed, shows that the dosimeters must be positioned compatibly with the lifestyle of the occupants (Table 3).

Table 3: 2019 Measurements by MR1 and anemometer.

2019 Measurements					
MR1	Distance from wall and floor	2–3 Feb.	Δ measure	Airflow speed by infiltration	Average measure per room
1 L	60 cm	702 Bq/m^3		0.00 m/s	
2 L	160 cm	612 Bq/m^3	12.82%	0.12 m/s	664 Bq/m^3
3 L	120 cm	680 Bq/m^3		0.07 m/s	
4 B	120 cm	713 Bq/m^3		0.00 m/s	
5 B	180 cm	694 Bq/m^3	14.84%	0.22 m/s	740 Bq/m^3
6 B	60 cm	815 Bq/m^3		0.00 m/s	

6 CONCLUSION

After all these analyses we can conclude that to make a good evaluation, we need to keep in mind many factors and that it is difficult to generalize because too many parameters are at stake. New elements are suggested for a good diagnosis of the building. Occupant lifestyles play a very important role. The objective measurement does not guarantee the occupant from unconscious exposure. As seen that from experimental data the radon concentration differences can be even higher than 35%, that in terms of radiation protection means increasing the cancer risk. Measurement differences within the same room, suggests the operator to take into account also the furnishing of the house. First of all the bedroom design, where every human being spends at least 8 hours a night. Portions of interior rooms may remain completely unventilated and sometimes right on the bedside.

This study shows how many topics are to be examined for a correct analysis for radon entry model into buildings:

- area on which the building stands;
- geological analysis and anthropic transformations;
- building materials and outdoor flooring;
- construction technology, foundation type;
- plant systems, cooling and heating;
- building design, shape and openings that regulate the correct dilution of indoor air;
- careful study of environmental, meteorological and microclimatic parameters;
- interior design that can influence the lifestyle of the occupants (Fig. 4).

All these issues should be treated by various specialists, each in his own field and work as a team for a better result.

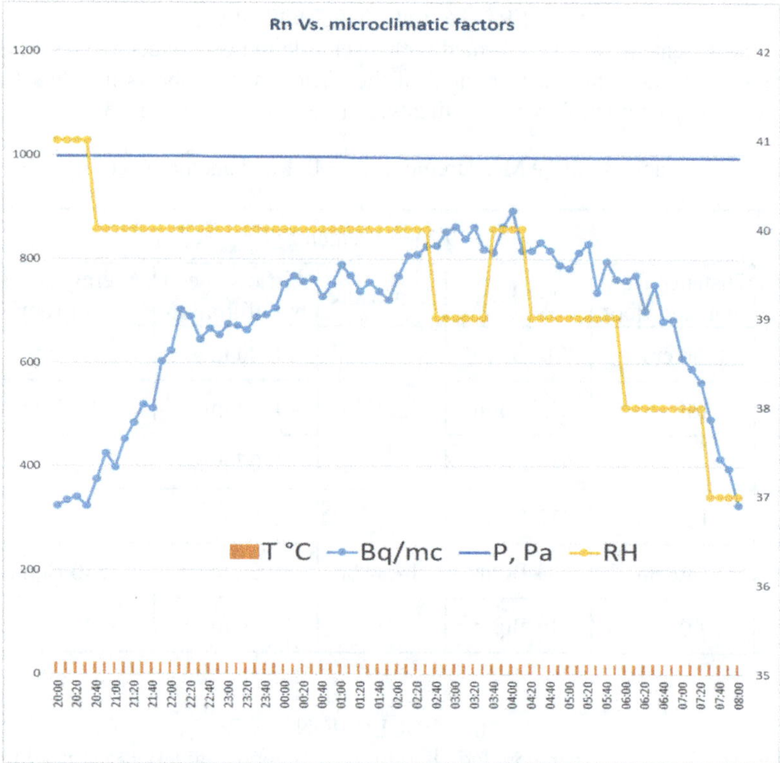

Figure 4: The average value of indoor radon concentration is quite high (Table 3). The graph shows how, as the outdoor temperature decreases at night, the concentration rises.

REFERENCES

[1] European Commission, Council Directive 2013/59/Euratom of 5 December 2013 laying down basic safety standards for protection against the dangers arising from exposure to ionising radiation, and repealing Directives 89/618/Euratom, 90/641/Euratom, 96/29/Euratom, 97/43/Euratom a, *Off. J. Eur. Union*, no. December 2003, pp. 1–73, 2014

[2] ANCE (Associazione Nazionale Costruttori Edili), Il mercato della sicurezza e dell'efficienza energetica della casa, Direzione Affari Economici, Centro Studi e Area Tecnologie, Qualità e Sicurezza delle Costruzioni Roma 26 settembre 2017. www.ance.it/docs/docDownload.aspx?id=40567.

[3] WHO, *Handbook on Indoor Radon: A Public Health Perspective*, World Health Organization, Geneva, pp. x–xi, 2009.

[4] Studio Ippolito and Muceli, Original CAD drawings ©. www.studioippolitomuceli.com.

[5] Carta Geologica d'Italia 1:50.000 Foglio Geologico 387, Albano Laziale, Servizio Geologico D'Italia, Ispra (The Italian Institute for Environmental Protection and Research).

INDOOR RADON SURVEY IN UNIVERSITY BUILDINGS: A CASE STUDY OF SAPIENZA – UNIVERSITY OF ROME

CHRISTIAN DI CARLO[1,2], ROMOLO REMETTI[2], FEDERICA LEONARDI[3], ROSABIANCA TREVISI[3], LUIGI LEPORE[2] & ROSARIA IPPOLITO[4]
[1]National Center for Radiation Protection and Computational Physics, Italian National Institute of Health, Italy
[2]Department of Basic and Applied Sciences for Engineering, Sapienza – University of Rome, Italy
[3]Department of Occupational Hygiene, Italian National Workers Compensation Authority (INAIL), Italy
[4]Faculty of Architecture, "Valle Giulia", Sapienza – University of Rome, Italy

ABSTRACT

The indoor radon concentration in underground workplaces pertaining to Sapienza – University of Rome have been monitored since the 1990s according to the prescription of the Italian Legislative Decree 230/95. In the last few years, the recommendations contained in the Council Directive 2013/59/Euratom have shifted the focus to all indoor exposure situations by promoting actions to identify workplaces and dwellings with radon concentrations exceeding the reference level of 300 Bq/m^3. In response to the upcoming transposition into national legislation, Sapienza has promoted the first Italian survey addressing workplaces in university buildings, regardless of the position with respect to the ground floor. The survey has interested more than 300 workplaces, i.e. administration and professors' offices, research and educational laboratories, conference rooms and classrooms, distributed in 15 different buildings. Places monitored are strongly heterogeneous in terms of users' habit, occupancy pattern and building characteristics. The influence of these parameters into seasonal variation have been addressed by organizing the survey into four quarters. The indoor radon concentration is measured by solid state nuclear track detectors, CR39. The aim of the paper is to present features, methods and intermediate results of the survey. The work, relying on the analysis of previous measurements interesting underground workplaces, focuses on methodology followed during all the preliminary and preparatory phases: active measurements by ionization chamber radon continuous monitor, radon progeny equilibrium factor estimations by radon daughters monitor, strategies for occupants' awareness, positioning protocol and provisions to maximize representativity of results.

Keywords: indoor radon, survey, university, radon in school, radon in workplace.

1 INTRODUCTION

Inhalation is the primary patter for public exposure to natural radiation [1]. The main contribution is given by isotope 222 of radon (in the following referred to simply as "radon" or "Rn") and its progeny [2]. This nuclide is showed to be the primary cause of lung cancer among people who have never smoked (i.e., the second cause of lung cancer after cigarette smoke). The percentage of lung cancer attributable to radon exposure is estimated to range between 3% and 14% [3]: in Italy, the annual average of lung cancer deaths attributable to radon is 3,326 (1,118–5,882), ranging between 3% and 18% of total lung cancer deaths [4].

Radon concentration in Italy has been addressed, from 1989 to 1998, by a representative National Survey aiming to evaluate the exposure in dwellings of all the 21 Italian regions. According to the results of such survey, Lazio was found to be one of the region with the highest mean radon concentration (>100 Bq/m^{-3}) [5]. Since then, several studies have been carried out in Italian workplaces and dwellings (e.g., [6]).

The current Italian Legislative Decree 230/95, transposing recommendation of Directive 96/29 Euratom, demands radon measurements in underground workplaces only. Nevertheless, exposure in schools (i.e., kindergarten, primary, and secondary school) has been studied due to the peculiar building characteristics (schools are generally at the ground

WIT Transactions on Ecology and the Environment, Vol 236, © 2019 WIT Press
www.witpress.com, ISSN 1743-3541 (on-line)
doi:10.2495/AIR190311

and first floor) and due to the age of population generally frequenting such buildings (e.g., [7]–[10]).

The need of performing and documenting design and methods of a survey addressing university buildings arises from: (i) no case study interesting such buildings in Italy, and very few in other countries (e.g., [11]), are available; (ii) due to the dimension and the spatial distribution of such buildings, specific protocols should be considered; and (iii) Council Directive 2013/59/Euratom (transposition into national legislation is in progress) will require measurements in workplaces and buildings with public access without restriction on floors [12].

In this paper, the survey on indoor radon concentration in Sapienza – University of Rome will be described in detail. The materials and methods will be discussed in depth with respect to radon detectors used, number of buildings (and rooms) involved, and deployment–removal protocols. The inner representativity, in terms of intended use and occupancy factor, of such survey will be analyzed. Moreover, the results of measurements addressing radon equilibrium factor F and unattached fraction f_p will be reported.

2 MATERIALS AND METHODS

The survey has been completely designed and managed by Sapienza – University of Rome and Italian National Institute for Insurance against Accidents at Work (INAIL). The indoor radon measurements interested 18 buildings, all of pertinence of the Department of Basic and Applied Sciences for Engineering (SBAI).

2.1 Rooms census

A comprehensive and systematic census of rooms to be considered within the survey has been carried out. Such analysis is mandatory to allow a proper estimation of dose for students and workers. Ten intended uses have been considered, each to be associated to a correspondent occupation factor, T (expressed in terms of fraction of weekly working hours, i.e., 40 h, for the occupant maximally exposed to indoor radon). For offices, workshops, porter's lodges, and research laboratories, T=1 is considered. Reading rooms, libraries, classrooms, and teaching laboratories are assumed to have T=0.5. Storage rooms, kitchens, archives, and copy rooms are all considered unattended rooms (T=1/20). Public toilets, stairways, corridors and elevators have been excluded from the survey due to the low occupancy factor (generally lower than 1/40) and the high air exchange rate.

The results of the census are reported in Table 1.

A total of 335 rooms have been censed: as expected, offices (156) and research laboratories (111) cover the large majority of rooms. As a consequence of the census, three buildings (12, 13 and 16) have been excluded from the survey due to the lack of rooms satisfying requirements about minimum occupancy factors.

2.2 Radon detectors

CR-39 solid state nuclear track detectors (SSNTD) have been used. The sizes of the plastic polymer chosen, manufactured by TASL Ltd, are 37x13 mm. The CR-39 has been coupled with NRPB/SSI holder made of conductive plastic.

Results of studies by Tokonami et al. [13], [14] show a thoron sensitivity for such holders of 0.1 cm^{-2} kBq^{-1} m^3 h^{-1} (0.05 in relative terms). Thus, the Rn-220 penetration into NRPB/SSI holders should not be neglected when deploying the passive dosimeter in rooms. If no

Table 1: Results of intended use census for building interested by the survey. Columns two and three report the distribution of rooms according to the intended use. Columns four and five report the maximum value for occurrences (absolute and percentage respectively) of the intended use in row in the 18 buildings. Columns six and seven report the mean value for the same occurrences. The minimum values are not listed because equal to zero for each row.

	No.		Max.		Mean	
	(#)	(%)	(#)	(%)	(#)	(%)
Office	156	47	33	96	9	38
Workshop	1	0	1	3	0	0
Reading room	8	2	2	13	0	2
Lodge room	5	1	1	13	0	1
Kitchen	1	0	1	3	0	0
Copy room	1	0	1	14	0	1
Archive	3	1	1	3	0	0
Classroom	25	7	11	100	1	28
Teaching lab	24	7	6	100	1	21
Research lab	111	33	93	84	6	8

attention is paid to the position of detectors, this may result in incorrect estimates. From several scientific findings, in fact, it was clarified that radon readings by radon detectors were overestimated by the presence of thoron [15]. If the indoor radon activity concentration is roughly uniform in rooms, the spatial distribution of thoron decreases exponentially from the wall surface [16]–[18]: according to this, a minimum distance of 30 cm between CR-39 and the closer wall has been adopted during the current survey.

According to scientific findings concerning CR-39 detectors background [19], a chemical etching of the detectors has been performed prior to their exposure to radon in order to make easier and more accurate the discrimination of true tracks (i.e., due to interaction of radon and its progeny) from those of the background (i.e., due to surface defects, micro-voids and alpha interactions before exposure).

2.3 Design of the survey

A long-term periodicity in indoor radon concentration is acknowledged [20]: radon follows usually a seasonal cycle with higher levels in cooler months and lower levels in warmer months. The magnitude of the resulting cycle depends on temperature and weather condition of the area. Such periodicity may be modified by: living habits (i.e., changes in the house's ventilation rate), occupancy pattern and building characteristics. Thus, the correlation between Rn concentration and meteorological parameters is far from simple. Results from the Italian National Survey [5] showed several evidences of winter to summer Rn concentration ratio lesser than one. In order to verify if such *inverse* seasonal variation could affect the buildings addressed, it has been decided to divide the survey into four 3 month periods, nearly coincident with the four seasons. The measurements started at the beginning of spring 2019.

Such approach has also the aim of collecting data to estimate the average (with respect to the site) seasonal correction factors. These coefficients allow to perform measurements shorter than one year (e.g., three months [21]), thus estimating the annual mean on the basis of previous observation of seasonal variations in the same area. This approach, followed by several Countries, provides strong reduction of time and cost efforts to manage regulatory requirements. The same approach could be applied to Italy: the current Italian Legislation expresses, in fact, the radon limit in term of annual mean concentration (i.e., 500 Bq/m^{-3}).

2.4 Dosimeters deployment

Dosimeters have been deployed by a trained team according to the following main criteria: (i) the instrument should be placed in a safe position to reduce the risk of manumissions, accidental falls or hits; ii) the instrument should be placed far from heat sources, including surfaces with direct solar radiation; (iii) the instrument should be placed far from air conditioner, windows or doors; and (iv) the instrument should be placed at least 30 cm far from the bearing walls and 1 m from floors and ceilings [22].

In each room one CR-39 detector has been placed, except for rooms larger than 100 m^2 whose radon concentration is being monitored by two dosimeters.

During the deployment procedures, the content of the holder has been showed to the rooms occupants. Such an expedient is aimed to avoid manumissions of the instruments during the measurements period, whose main cause is represented by the desire to know what is inside the holder.

The dosimeters have been coupled with a flyer containing concise information about the nature of measurements taking place and the contacts of survey directors always available for answering questions. On or two reference persons for each building (according to the number of rooms and occupants) have been informed and trained about radon risk, regulatory legislation, and kind of measurements being performed. Then, the information has been transmitted to all the workers by the reference persons themselves. All these precautions are aimed to make the workers feel confident about the measurements taking place in their rooms. In fact, occupant's cooperation is mandatory in checking and saving the dosimeter during the measuring period, so reducing the number of missing results. The cleaning company has been informed about the rooms interested by such measurements together with the positions of each dosimeter deployed.

2.5 Equilibrium factor and unattached fraction

The evaluation of dose arising from radon progeny inhalation is often related to the gas exposure only due to the supposed compensation between doses from attached and unattached fraction. [23] shows that such a self-compensation is not fully realized, and for the same radon gas concentration, the effective dose can vary by a factor up to 2: thus, equilibrium factor and unattached fraction should be evaluated for proper doses evaluation.

According to the ICRP recommendations, an equilibrium factor of 0.4 can be assumed. However, since this factor depends largely on human-related and environmental conditions (e.g., living habits, occupancy pattern, humidity), 7 rooms, contemporary monitored by CR-39 detectors, have been also sampled to host measurements of equilibrium factor and unattached fraction. Rooms choice aimed to make the samples representative with respect to intended use.

Two instruments have been used to evaluate such parameters: an active ionization chamber radon detector (AlphaGUARD DF2000), and a double channel radon progeny monitor (Tracerlab BWLM-2S PLUS). The potential alpha energy concentration (PAEC) of each radon daughter has been determined by alpha spectrometry considering attached and unattached fraction individually.

Each measurement lasted more than 24 hours in order to superimpose short term aperiodic fluctuations and periodic diurnal cycle of radon-related parameters.

3 RESULTS AND DISCUSSION

Dosimeters have been deployed in a total of 178 rooms, with a prevalence of offices (66%), research and teaching laboratories (8% and 16%, respectively). The distribution of CR-39 detectors, according to intended uses of monitored rooms, is reported in Table 2.

Table 2: Summary of rooms whose indoor radon concentration is being measured, i.e., CR-39 detectors have been placed inside actually. Columns two and three report the distribution of rooms according to the intended use. Columns four and five report the maximum value for occurrences (absolute and percentage respectively) of the intended use in row in the 18 buildings. Columns six and seven report the mean value for the same occurrences. The minimum values are not listed because equal to zero for each row.

	No.		Max.		Mean	
	(#)	(%)	(#)	(%)	(#)	(%)
Office	118	66	33	100	7	47
Workshop	1	1	1	3	0	0
Reading room	5	3	2	13	0	2
Lodge room	2	1	1	13	0	1
Kitchen	1	1	1	3	0	0
Copy room	3	2	1	14	0	2
Archive	1	1	1	3	0	0
Classroom	5	3	4	27	0	3
Teaching lab	14	8	6	100	1	31
Research lab	28	16	14	100	2	12

Four other buildings have been excluded from the survey during this phase: they are all mainly hosting classrooms characterized by large areas (more than 150 Bq/m^{-3}) and high air exchange rates.

Inner representativity of rooms sampling during the procedure of dosimeter deployment is discussed with respect to two parameters: intended use (Table 3) and floor level (Table 4).

Rooms actually being measured are well representative of the whole “population” of rooms censed, considering both the occupancy factor and the floor level. Tables 3 and 4 show maximum differences of 4% and 6%, respectively.

3.1 Equilibrium factor and unattached fraction

As shown in Table 5, the average equilibrium factor for offices is 0.52 which is nearly equal to the suggested value, 0.5, of [24], but 30% higher than the traditionally assumed value, 0.4, of [25] (still considered appropriate by [2], showing a range from 0.1 to 0.9). The F value increases for libraries, usually characterized by small surface to volume ratios, and decreases for archives and storage rooms due to their very small free inner volume.

Table 3: Analysis of inner representativeness with regards to rooms occupancy factor expressed in terms of hours spent there in one year: (i) offices, workshops, porter's lodges, and research labs, 2000 h year^{-1}; (ii) reading rooms, libraries, classrooms, and teaching labs, 1000 h year^{-1}; (iii) storage rooms, kitchens, archives, and copy rooms, 100 h year^{-1}.

	Rooms censured		Rooms measured	
	(#)	(%)	(#)	(%)
2000 h/year	273	81%	149	84%
1000 h/year	57	17%	24	13%
100 y/year	5	1%	5	1%

Table 4: Analysis of inner representativeness with regards to floor levels of rooms hosting indoor radon concentration measurements.

	Rooms censured		Rooms measured	
	(#)	(%)	(#)	(%)
PS1	68	10	40	11
PTE	55	8	48	13
PR1	115	17	39	11
P01	77	11	51	14
P02	4	1	0	0
P03	5	1	0	0
P04	11	2	0	0

Table 5: Summary of measurements results about equilibrium factor F and unattached fraction f_p during the first three month period (corresponding to spring).

	Equilibrium factor		Unattached fraction	
	Mean	σ (k=1)	Mean	σ (k=1)
Office 1	0.49	30%	0.09	68%
Office 2	0.39	32%	0.14	49%
Office 3	0.51	33%	0.14	51%
Office 4	0.48	26%	0.13	28%
Porter's lodges	0.52	29%	0.12	76%
2000 h year^{-1}	0.48	10%	0.12	16%
Library	0.66	34%	0.08	28%
1000 h year^{-1}	0.66	34%	0.08	28%
Archive	0.33	15%	0.20	21%
100 h years^{-1}	0.33	15%	0.20	21%

The average unattached fraction for the offices considered is 0.12, greater than the central value of the unattached fraction in houses, 0.05, reported by [2]. Despite what expected, measurements in archives and storage rooms, usually characterized by higher aerosol concentrations (so attachment rates), return higher values of unattached fraction (0.2). Similar considerations should be extended to the reduction (with respect to 0.4) of the equilibrium factor in such places: indeed, a value of 0.33 is generally associated to lower aerosol concentration than the average house.

4 CONCLUSION

A survey addressing measurements of radon concentration in university buildings (at Sapienza – University of Rome) has been described and discussed in depth. The survey aims to provide information about radon exposure in workplaces (e.g., offices, research laboratories) and places attended by students (e.g., classrooms, teaching laboratories and libraries).

The sample of rooms being measured has been proved to be representative of all the rooms censed at the beginning of the survey. Two main variables have been considered in representativity analysis as parameters potentially influencing indoor radon concentration: occupancy factor coupled with the intended use for rooms (i), and the floor level (ii).

Measurements are well under way since April 2019 being designed to be divided according to seasons in order to appreciate the periodicity, direct or inverse, in radon seasonality.

The paper also shows the results of a preliminary study addressing the evaluation of equilibrium factor F and unattached fraction f_p in 7 rooms sampled within the places interested by radon measurements. Such measurements will be repeated in the following three month monitoring periods (i.e., summer, autumn, and winter) extending the number of rooms considered.

REFERENCES

[1] United Nations Scientific Committee on the Effects of Atomic Radiation, *Sources and Effects of Ionizing Radiation*, 2008.

[2] United Nations Scientific Committee on the Effects of Atomic Radiation, *Sources and Effects of Ionizing Radiation*, 2000.

[3] World Health Organization, Indoor radon: A public health perspective. *Int. J. Environ. Stud.*, **67**(1), p. 108, 2009.

[4] Bochicchio, F., Antignani, S., Venoso, G. & Forastiere, F., Quantitative evaluation of the lung cancer deaths attributable to residential radon: A simple method and results for all the 21 Italian regions. *Radiat. Meas.*, **50**, pp. 121–126, 2013.

[5] Bochicchio, F. et al., Annual average and seasonal variations of residential radon concentration for all the Italian regions. *Radiat. Meas.*, **40**(2–6), pp. 686–694, 2005.

[6] Bochicchio, F., Ampollini, M., Antignani, S., Bruni, B., Quarto, M. & Venoso, G., Results of the first 5 years of a study on year-to-year variations of radon concentration in Italian dwellings. *Radiat. Meas.*, **44**(9–10), pp. 1064–1068, 2009.

[7] Gaidolfi, L. et al., Radon measurements in kindergartens and schools of six italian regions. *Radiat. Prot. Dosimetry*, **78**(1), pp. 73–76, 1998.

[8] Tollefsen, T. et al., Radon in indoor air of primary schools: a systematic survey to evaluate factors affecting radon concentration levels and their variability. *Indoor Air*, **24**(3), pp. 315–326, 2013.

[9] Trevisi, R., Leonardi, F., Simeoni, C., Tonnarini, S. & Veschetti, M., Indoor radon levels in schools of South-East Italy. *J. Environ. Radioact.*, **112**, pp. 160–164, 2012.

[10] Trevisi, R. et al., A pilot study on natural radioactivity in schools of south-east Italy. *Environ. Int.*, **36**(3), pp. 276–280, 2010.

[11] Kobayashi T. & Takaku, Y., Intermittent measurements of 222Rn and 220Rn progeny in air for four years. *Radioisotopes*, **46**(9), pp. 603–614, 1997.

[12] European Commission, Council Directive 2013/59/Euratom of 5 December 2013 laying down basic safety standards for protection against the dangers arising from exposure to ionising radiation, and repealing Directives 89/618/Euratom, 90/641/Euratom, 96/29/Euratom, 97/43/Euratom. *Off. J. Eur. Union*, no. December 2003, pp. 1–73, 2014.

[13] Tokonami, S., Takahashi, H., Kobayashi, Y., Zhuo, W. & Hulber, E., Up-to-date radon-thoron discriminative detector for a large scale survey. *Rev. Sci. Instrum.*, **76**(11), pp. 1–5, 2005.

[14] Tokonami, S. et al., Influence of environmental thoron on radon measurements and related issues. *Proceedings of the Natural Radiation Environment – S International Symposium*, pp. 145–149, 2008.

[15] Tokonami, S., Why is thoron measurements important? *Raduation Protection Dosimetry*, **141**(4), pp. 335–339, 2017.

[16] Hosoda, M. et al., Characteristic of thoron (220Rn) in environment. *Appl. Radiat. Isot.*, **120**(October 2016), pp. 7–10, 2017.

[17] Doi, M., Fujimoto, K., Kobayashi, S. & Yonehara, H., Spatial distribution of thoron and radon concentrations in the indoor air of a traditional Japanese wooden house. *Health Phys.*, **66**(1), pp. 43–49, 1994.

[18] Tschiersch, J. & Meisenberg, O., The HMGU thoron experimental house: A new tool for exposure assessment. *Radiat. Prot. Dosimetry*, **141**(4), pp. 395–399, 2010.

[19] Mishra, R., Orlando, C., Tommasino, L., Tonnarini, S. & Trevisi, R., A better understanding of the background of CR-39 detectors. *Radiat. Meas.*, **40**(2–6), pp. 325–328, 2005.

[20] Bossew, P. & Lettner, H., Investigations on indoor radon in Austria, Part 1: Seasonality of indoor radon concentration. *J. Environ. Radioact.*, **98**(3), pp. 329–345, 2007.

[21] Miles, J.C.H., Howarth, C.B. & Hunter, N., Seasonal variation of radon concentrations in UK homes. *J. Radiol. Prot.*, **32**(3), pp. 275–287, 2012.

[22] Agenzia per la protezione dell'ambiente e per i servizi, *Linee guida per le misure di radon in ambienti residenziali*, 2004.

[23] Nikezic, D. & Yu, K.N., Are radon gas measurements adequate for epidemiological studies and case control studies of radon-induced lung cancer. *Radiat. Prot. Dosimetry*, **113**(2), pp. 233–235, 2005.

[24] United Nations Scientific Committee on the Effects of Atomic Radiation, *Sources, Effects and Risks of Ionizing Radiation*, 1988.

[25] United Nations Scientific Committee on the Effects of Atomic Radiation, *Sources and Effects of Ionizing Radiation*, 1993.

AN INEXPENSIVE AND CONTINUOUS RADON PROGENY DETECTOR FOR INDOOR AIR-QUALITY MONITORING

CHRISTIAN DI CARLO[1,2], LUIGI LEPORE[2], LUCA GUGLIERMETTI[2] & ROMOLO REMETTI[2]
[1]Italian National Institute of Health, National Center for Radiation Protection and Computational Physics, Italy
[2]Department of Basic and Applied Sciences for Engineering, Sapienza University of Rome, Italy

ABSTRACT

A silicon photodiode-based inexpensive detector working as a counter and spectrometer for alpha particles has been conceived, designed, constructed and analyzed in depth. Monte Carlo simulations by means of MCNPX ver. 2.7.0 code have been carried out to select the most suitable sensitive element for the intended applications. The detecting unit has been coupled to an Arduino board and tested for low-rate alpha-particle counting and spectroscopy. Results demonstrate a maximum count rate of $4 \cdot 10^3$ s^{-1}, an energy resolution corresponding to a full width at half maximum of 160 keV over the entire energy range of measured alpha (namely 4 ÷ 6.5 MeV), and the sensitive element's intrinsic efficiency of about 100%. Being the detector capable of distinguishing alpha energy associated to decays of radon daughters, its applications include ^{222}Rn progeny monitoring. The air sampling system has been realized by a volumetric micro-pump forcing the air-flow through a millipore filter. By knowing the air-flow rate processed and the corresponding alpha energy spectrum measured, the concentrations of ^{218}Po, ^{214}Po and ^{210}Po are determined. The potential alpha energy concentration-in-air is inferred, and effective dose evaluated. Calibration and testing measurements have been carried out by comparing the obtained results to the outputs of professional and expensive radon progeny monitor. The detector capability of "following" radon progeny concentration-in-air vs. time has been demonstrated. The device studied here can be configured as a prototype for an inexpensive radon progeny sensor to be potentially suitable for indoor air-monitoring in residential buildings, evaluating people's exposures to radon and initiating corrective actions (e.g., mechanical ventilation) if necessary.

Keywords: radon exposure, alpha spectrometry, silicon photodiode, Arduino, PAEC.

1 INTRODUCTION

From 2013 on, the EU Member States have been busy in the transposition of Council Directive 2013/59/Euratom [1] into national legislation. Among novelties introduced, increased attention is paid to exposures from natural sources of radiation, and radon-related aspects particularly.

Updated primary risk model (BEIR VI [2] and Czech–French models [3]) studies regarding epidemiological approach to radon effects on health encouraged the review of the probability coefficient for radon- and radon-progeny-induced lung cancer from 2.8 E-4 to 5 E-4 per WLM [4]. The application of general biokinetic models (mainly the Human Respiratory Tract Model, HRTM, from ICRP 66 [5]) to radon exposures (including the progeny intakes in lungs and whole body) results in a values of effective dose per unit exposure ranging from 10 to 20 mSv per WLM. Because of that, with respect to the preceding legislation, the 2013/59/EURATOM established a "Reference Level" of 300 Bq m^{-3} for annual average indoor radon concentrations (including both occupational and public exposures), overcoming the 500 Bq m^{-3} "Action Level" of the 1996/29/EURATOM (regarding occupational exposures only).

It should be considered that the 2013/59/EURATOM Directive advises Member States that National Reference Level for radon "*shall not be set higher than 300 Bq m^{-3}*", encouraging to choose lower values in situations involving long-term exposures to radon, i.e., workplaces and dwellings. Moreover, being the 300 Bq m^{-3} value conceived as a "reference level" instead of an "action level", radon exposures are subjected to the ALARA

WIT Transactions on Ecology and the Environment, Vol 236, © 2019 WIT Press
www.witpress.com, ISSN 1743-3541 (on-line)
doi:10.2495/AIR190321

principle (As Low As Reasonably Achievable) and the annual average radon indoor concentration should be kept below 300 Bq m^{-3} as more as reasonably achievable.

With all these premises and the potential market related to radon issues and radiation protection aspects, the increasing number of new radon detectors developed in last years and service companies available for radon measurements can be understood. However, regardless any future legal implication, manufacturers or measurements providers can rely on the fact that radon is acknowledged as the second cause of lung cancer after cigarette smoking [6], [7]: in many cases it is a sufficient expedient to promote and diffuse their products to the public without any legal obligation.

The current and common tendency for manufacturers is to conceive "smart" devices even for radon measurements. e.g., Wi-Fi sensors with associated database and detectors which can be geo-localized allowing measurements linked to the specific features of the territory. Further, any member of the public could be able to measure indoor radon concentration by itself and receive notifications about high radon concentration with a message on his smart phone, etc.

A "smart" radon detector could be useful to people truly, as an alternative solution to traditional remedial action applied to "radon-rich" locations. Currently, in most Member States, when at a specific location the annual radon indoor concentration overcomes the "action level", remedial actions need to be applied to reduce such concentration. Typical actions are ventilation with external air, isolation of the building from foundations, "pressure traps" to intercept radon paths to the interior of the building, etc. Actions like these are usually applied continuously, regardless the effective indoor radon concentration. Here instead, a "smart" radon detector could be suitable to control, suggest, and optimize the remedial action operation, e.g., powering fans and active parts only when needed (only when people are there and concentrations are exceeded), or warning people at the place to open the window and dilute the internal air with external air. This way, radon concentrations would be controlled only when it is necessary, with relevant savings in costs and energy consumptions of buildings.

Such an hypothesis would suggest precise features for the "smart" radon detector to be applied: *i*) the sensor should be an active detector; *ii*) accuracy and precision of measurements in "following" radon evolution vs. time (day–night cycles) should be demonstrated; *iii*) responses should be in real time e.g., frequency of measurements could be hourly, or each quarter, depending on device's detection limits.

At the Radiation Protection Laboratory of the Sapienza University of Rome, Department of Basic and Applied Sciences for Engineering, a photodiode-based detector for alpha particles has been developed and properly characterized. All the features of the device and results about the experimental campaign carried out can be found in Gugliermetti et al. [8]. The detector is open-source, it is called "Alphaino", and it can be suitable for different applications involving alpha particles detection and spectroscopy, so including measurements of radon and its progeny. The paper herein discusses how the "Alphaino" detector has been suitably modified to build a radon progeny monitor, called "Radonino".

At the Radiation Protection Laboratory, an experimental campaign with the "Radonino" detector, the AlphaGUARD [12] radon monitor, and the Tracerlab BWLM-PLUS-2S [10] PAEC monitor, has been carried out. Synchronized and contemporary sets of measurements have been registered for an entire month. Part of the data acquired have been used for the initial calibration of the "Radonino" detector. With calibration achieved, data have been used to verify the detector operation and demonstrate its capability in "following" radon and progeny evolution vs. time, as for as professional and expensive instrumentation.

This paper aims to show that reliable radon exposure measurements can be achieved even without expensive instrumentation, through a well-conceived detector. An effective and cheap radon monitor detector measuring directly the PAEC could be particularly suitable to provide continuous effective dose evaluation in radon-affected dwellings and workplaces, in order to warn people or initiate mitigation actions when truly needed.

2 DETECTOR DEVELOPMENT

The core of "Radonino" is the "Alphaino" detector discussed in Gugliermetti et al. [8]. Alphaino is an alpha-particle detector intended for applications where low cost and small dimension are key factors. It can be used both as an alpha particles counter, and for alpha spectroscopy. It can also be connected both to traditional nuclear spectroscopy electronics, and to cheap prototyping boards, e.g. Arduino [15]. The detector has been tested for low-rate alpha-particle counting and spectroscopy, demonstrating a maximum achievable count rate of $4 \cdot 10^3$ s^{-1}, with an energy resolution corresponding to a full width at half maximum of 160 keV over the entire energy range of measured alpha, namely 4 ÷ 6.5 MeV, the intrinsic efficiency being 100%. The detector is open-source: everyone can download the schematics, build the sensor and contribute to further development through the GITHUB repository for the project, https://github.com/bemxgm/Radon-Monitor.

The application of "Alphaino" to radon indoor concentration measurements comes from the fact that risks to people's health is given by the radon progeny, and not by radon gas itself. Detectors measuring radon (^{222}Rn or ^{220}Rn) gas directly can be fairly complex, and very expensive, due to the fact that radon isotopes are dispersed in air and radon is a noble gas, chemically inert almost. Detecting ^{222}Rn (or ^{220}Rn) decays means to "observe" a certain volume of the air to be monitored with the experimental capability to detect the decay event in such control volume: ion chambers or scintillation cells are used, typically. Professional instrumentation exploits this kind of sensitive element, usually.

Being ^{222}Rn (and the same ^{220}Rn) the initiator of an equilibrium chain with its Po, Pb, and Bi daughters, i.e., the progeny, detectors can rely on progeny measurements to infer the father concentration-in-air. The technical realization of the equilibrium among radon and its progeny is a quite interesting possibility to measure radon. As a matter of fact, the detection of the progeny is quite easier with respect to the measurement of the radon itself. Indeed, ^{222}Rn daughters, ^{218}Po, ^{214}Pb, ^{214}Bi, and ^{214}Po, are mainly produced by decay in form of positive ions [9]. Such atoms have a high probability to get attached among themselves and/or with dust particulate in air, forming "clusters" of different sizes [10], [11]. Such "clusters" are called "unattached" (with diameter ranging from 0.5 to 4 nm) when free or attached to vapor or gas in traces, and "attached" (with diameter from 0.1 to 0.4 μm) when attached to aerosol. The advantage in measuring radon's daughters is that the progeny can be easily collected in front of a detector for the measurement. Collection can be achieved by mechanical methods, e.g., forcing air to pass through a filter placed in front of the detector, or with electrostatic precipitation toward the sensor.

The "Alphaino" detector in Gugliermetti et al. [8] has been arranged in a suitable configuration to realize the mechanical collection of radon progeny on a paper filter by means of an air flow forced to pass throughout. The detector, Fig. 1, is called "Radonino". The photodiode sensor in front of the filter is capable in detecting alpha particles emerging from the filter itself. ^{218}Po, ^{214}Pb, ^{214}Bi, and ^{214}Po are determined by the analysis of the alpha particles' energy spectrum collected for one hour. By knowing the flow rate processed by the sensor, the Potential Alpha Energy Concentration, PAEC (MeV cm^{-3}), is calculated. Such energy, in form of a concentration of alpha particles decays in air, determines the risk to human health when air is breathed in the respiratory tract.

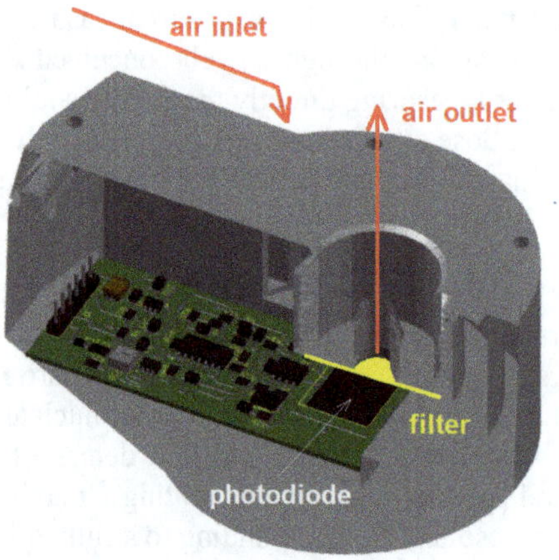

Figure 1: The "Radonino" detector. The chip is housed in a light-proof electromagnetic shielded ABS box sketched in gray color.

With the PAEC determined, it is possible to retrieve the corresponding radon concentration-in-air, being "action" or "reference" levels defined by legislation expressed in such units, or convert the PAEC in effective dose rate directly. Fig. 2 shows a comparison between the workflow with the "Radonino" detector here proposed and the classical workflow applied to professional radon detectors such as Bertin Instruments AlphaGUARD [12]. For the equilibrium factor, F, defined as the ratio between equivalent equilibrium concentration (EEC, i.e., the concentration corresponding to the PAEC) and the corresponding radon concentration-in-air, a nominal value of F=0.4 is assumed for indoor locations, conventionally. As to convert PAEC in effective dose, is assumed the effective dose coefficient per unit exposure by ICRP 115 [13], i.e. 0.465 μSv cm^3 h^{-1} MeV^{-1}.

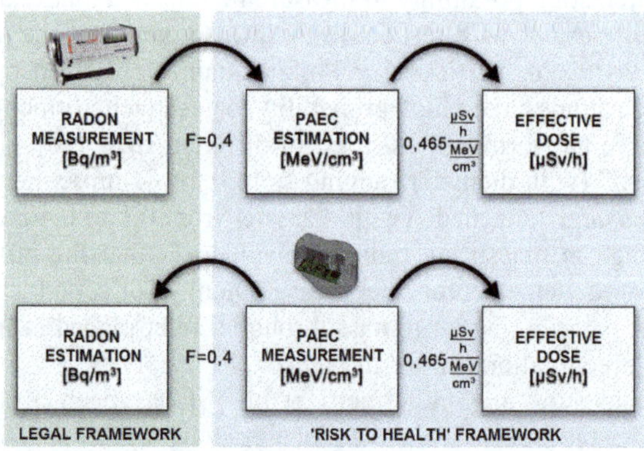

Figure 2: Workflow with professional radon monitors and with the "Radonino" detector here proposed.

The adjustment of "Alphaino" towards "Radonino" for realizing PAEC measurements asked for a new design of the casing with respect to Gugliermetti et al. [8]. The sensor is based on photodiode which is sensitive to light intrinsically, and a light-tight environment is needed for correct operations. For reducing the electronic noise, the casing has been covered with a conductive paint electrically connected to the ground, as to realize an electromagnetic shielding. The case has been produced by means of a 3D printer, and material selected is ABS, Acrilonitrile Butadiene Styrene, due to its excellent mechanical features. It allows the suction of air from the side and the outlet upright with respect to the photodiode sensitive element as depicted in Fig. 1. Prior to be expelled, the air is forced to pass throughout a suitable millipore paper filter, permitting the "dust" carrying the radon progeny to be settled, for alpha-particle detection.

The air flow is obtained by means of a volumetric micropump connected to the outlet of the detector. The flow rate realized corresponds to 0.3 L/s. Alpha particles energy spectra are recorded by means of a multi-channel analyzer ORTEC EASY-MCA-2k, and spectra are acquired with a frequency of one hour.

3 EXPERIMENTALS

The first test carried out was intended to demonstrate the feasibility of the project, i.e., verifying the effective capability of "Radonino" in detecting alpha particles from the radon progeny settled on the millipore paper filter.

Fig. 3 shows the first alpha energy spectrum measured with the "Radonino" detector, the acquisition lasting 24 h as to obtain statistically significant data to be compared with the professional detector Tracerlab BWLM-PLUS-2S. Peaks in the spectrum are due, in the order, to ^{218}Po, ^{214}Po, and ^{212}Po, the last one belonging to the ^{220}Rn decay chain.

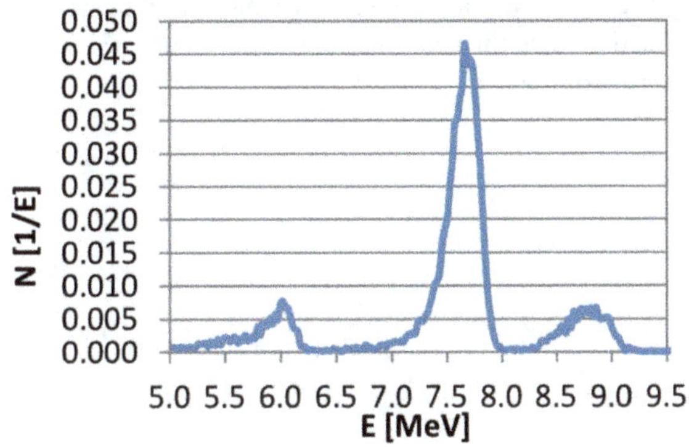

Figure 3: Alphaino recorded alpha spectrum of the progeny of ^{220}Rn-^{222}Rn. Peaks in the spectrum are due to ^{218}Po (6 MeV), ^{214}Po (7.7 MeV), ^{212}Po (8.8MeV).

The following experimental campaign was carried out for a quantitative characterization of "Radonino" aimed to calculating the efficiency of the system in detecting alpha particles vs. other working parameter selected, e.g. the air flow-rate. The calibration has been achieved by means of the Tracerlab BWLM-PLUS-2S. Alpha energy spectra, acquired with a frequency corresponding to 1 hour with both detectors synchronized in time, have been compared and the overall efficiency of "Radonino" in detecting alpha from the millipore filter was calculated.

Once ^{218}Po and ^{214}Po concentration-in-air are determined with the calibrated "Radonino", contributions to PAEC from ^{214}Pb and ^{214}Bi can be inferred by equilibrium considerations. The PAEC value [MeV/cm^3] vs. time is so determined for the following evaluations.

Three different experimental campaigns have been realized at the Radiation Protection Laboratory, involving the contemporary and synchronized use of:

- AlphaGUARD PQ2000 to measure indoor ^{222}Rn concentrations, one value per hour;
- Tracerlab BWLM-PLUS-2S to measure PAEC, one value per hour;
- The "Radonino" detector to measure PAEC, one value per hour.

The three experimental campaigns differ in duration and boundary conditions (open or closed windows, day–night cycle, build-up effect due to closed windows in the weekend, weather, etc.). Those variations were intended to analyze the real spectrum of situations for indoor conditions, as to include the dynamic evolution of radon at the building in dose evaluations and results. In particular:

- Experimental Campaign #1: working week, 88 hours, normal use for the analyzed room; weather: cloudy and rainy (low air pressure);
- Experimental Campaign #2: two working days, 45 hours, normal use for the analyzed room; weather: sunny (high air pressure);
- Experimental Campaign #3: weekend, 33 hours, room and building unused; weather: sunny (high air pressure).

As to discuss results, the dose approach in Fig. 2 (the "Risk to Health Framework") has been chosen to compare data. AlphaGUARD radon indoor concentrations measured values were converted in effective dose rate assuming F=0.4 and 0.465 μSv cm^3 h^{-1} MeV^{-1}. The same way, the 300 Bq m^{-3} "Reference Level" from 2013/59/EURATOM has been converted in the corresponding effective dose rate. The PAECs measured by Tracerlab and "Radonino" were converted to effective dose through the coefficient 0.465 μSv cm^3 h^{-1} MeV^{-1}. Figs 4, 5 and 6 shows the three sets of data.

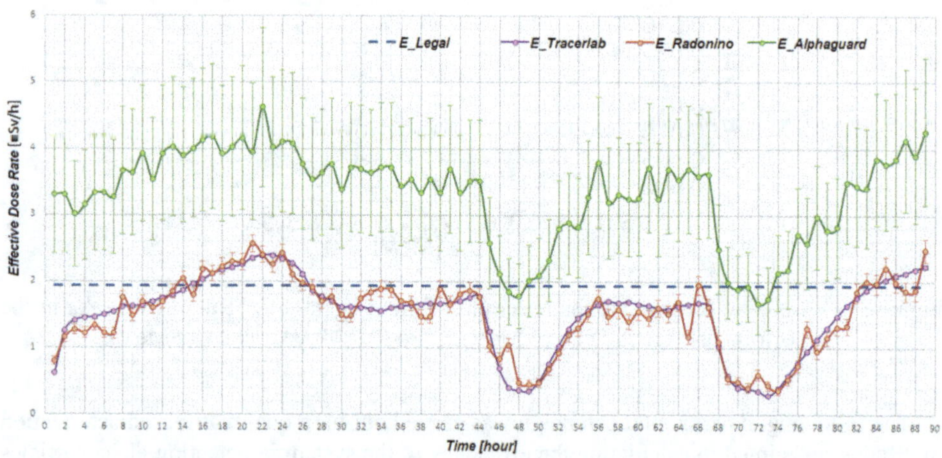

Figure 4: Results of the first experimental survey: comparison between effective dose rate computed from Radonino, Tracerlab BWLM-PLUS-2S, and AlphaGUARD measurements. 2013/59/EURATOM "Reference Level" converted to effective dose rate is provided as "E_Legal". This first set of measurements was performed during a whole week in a room regularly occupied by workers. The room is ventilated through windows manually operated according to occupants' arbitrary choices. F=0.18±24.3%, fp=0.13±15.1%.

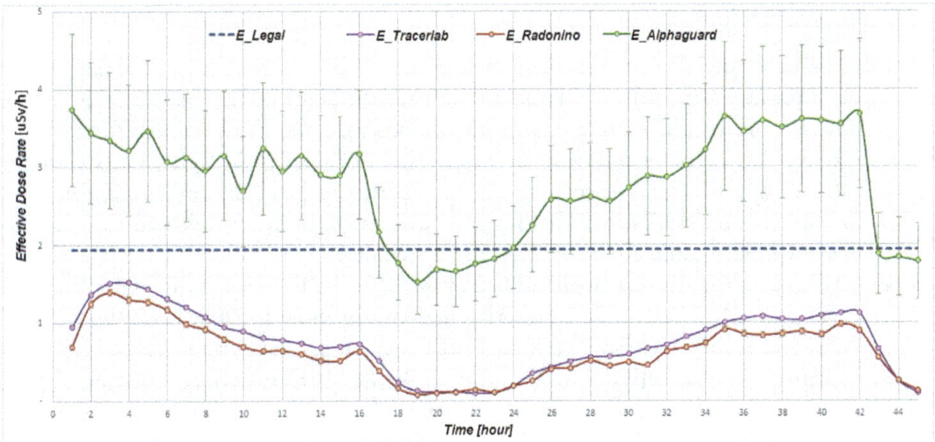

Figure 5: Results of the second experimental survey: comparison between effective dose rate computed from Radonino, Tracerlab BWLM-PLUS-2S, and AlphaGUARD measurements. 2013/59/EURATOM "Reference Level" converted to effective dose rate is provided as "E_Legal". This second set of measurements was performed in the same room as the first survey, but during the weekend. The room was closed and unattended during the whole period. F=0.21±29.8%, f_p=0.11±33.0%.

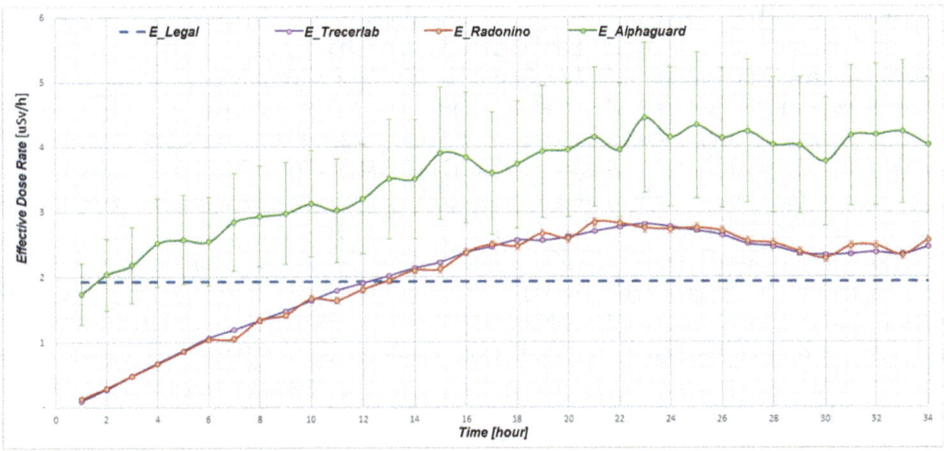

Figure 6: Results of the third experimental campaign: comparison between effective dose rate computed from Radonino, Tracerlab BWLM-PLUS-2S, and AlphaGUARD measurements. 2013/59/EURATOM "Reference Level" converted to effective dose rate is provided as "E_Legal". This third set of measurements was performed in the same room as the first two surveys. The observed period covered the transition from Sunday (weekend) to Monday (first day of working week). Due to the usual occupancy patterns, these hours are generally characterized by strong changes in all the climatic parameters (i.e., T, P, RH), as well as in Rn concentration, F, and f_p. F=0.10±43.8%, f_p=0.29±22.0%.

Some considerations follow:

- The calibrated "Radonino" is capable in determining PAEC values vs. time, as for the professional detector Tracerlab BWLM-PLUS-2S. The matching in PAEC means matching in effective dose rates values vs. time, as can be seen from Fig. 4.
- Indoor radon concentrations measured with AlphaGUARD PQ2000 (converted in PAEC by F=0.4 and effective dose rate later as happen for previous instruments) return overestimations of the effective dose rate due to the selected value F=0.4, assumed for indoor environments conventionally.
- At the room analyzed, the equilibrium factor F has been measured with Tracerlab BWLM-PLUS-2S vs. time. Experimental values show that F=0.4 can be assumed as a conservative assumption in this case, being the true value measured lower.
- Evaluate radon exposure starting from PAEC determination, or indoor radon concentration determination (Fig. 2), may lead to different results, as measurements and following calculations in Fig. 4 demonstrate; e.g. the 1st and 2nd experimental campaigns in Fig. 4 show that the exposure evaluation starting from radon measurement overcomes the legal 2013/59/EURATOM "threshold" (being F=0.4 an overestimation), while true values remain under such legal "threshold".

4 CONCLUSIONS AND FUTURE PERSPECTIVES

The paper discussed the application of the Alphaino sensor developed by the authors in Gugliermetti et al. [8] to radon measurements and the following effective dose evaluations.

The detector, named "Radonino" measured the Potential Alpha Energy Concentration in air, PAEC [MeV cm^{-3}], vs. time. It has been calibrated by means of professional instrumentation, Tracerlab BWLM-PLUS-2S and AlphaGUARD PQ2000.

Three experimental campaigns (variable in duration; boundary conditions, i.e., open or closed windows, day–night cycle, build-up effect due to closed windows in the weekend, weather; etc.) have been carried out with all instruments synchronized and with a 1-hour time resolution. Results showed that the inexpensive "Radonino" is capable in measuring PAEC vs. time as for the professional Tracerlab BWLM-PLUS-2S. PAEC values were converted to the corresponding radon concentration [Bq m^{-3}] by means of the equilibrium factor F, or to effective dose rate directly.

Experimental measurements here discussed demonstrate that "Radonino" could be suitable for future developments in terms of an inexpensive radon monitor with professional performances. Future works will include the coupling with inexpensive electronic board targeted for project developing, such as Arduino, as to register alpha energy spectra, to control the pump, to compute the detector response, and to start mitigating actions e.g. initiate mechanical ventilation, or warn the user to open the window and dilute indoor concentrations with external air.

REFERENCES

[1] EU Council, Council directive 2013/59/Euratom. *Official Journal of the European Union*, **17**(1), 2014.

[2] NRC, Health effects of exposure to Radon. BEIR VI Report, National Academy Press: Washington, DC, 1999.

[3] Tomasek, L., Rogel, A. & Tirmarche, M., Lung cancer in French and Czech uranium miners risk at low exposure rates and modifying effects of time since exposure and age at exposure. *Radiat. Res.*, **169**, pp. 125–137, 2008.

[4] Lecomte, J.-F. et al., Radiological protection against radon exposure. *Ann. ICRP*, **43**(3), 2014.

[5] International Commission on Radiological Protection (ICRP), *Human Respiratory Tract Model for Radiological Protection*, ICRP Publication 66, 1994.

[6] United Nations Scientific Committee on the Effects of Atomic Radiation, *Sources and Effects of Ionizing Radiation*, New York, 2008.

[7] World Health Organization, Indoor radon: A public health perspective. *International Journal of Environmental Studies*, **67**(1), 2009.

[8] Gugliermetti, L., Lepore, L., Remetti, R. & Colarieti Tosti, M., Alpha spectrometry with the inexpensive open-source detector Alphaino. *Nuclear Instruments and Methods in Physics Research Section A: Accelerators, Spectrometers, Detectors and Associated Equipment*, **928**, pp. 13–19, 2019.

[9] National Council on Radiation Protection and Measurements, Measurement of radon and radon daughters in air. Report no. 97, 1988.

[10] Porstendörfer, J., Properties and behaviour of radon and thoron and their decay products in the air. *J. Aerosol Sci.*, **25**(2), pp. 219–263, 1993.

[11] Porstendörfer, J., Pagelkopf, P. & Gründel, M., Fraction of the positive 218-Po and 214-Pb clusters in indoor air. *Radiation Protection Dosimetry*, **3**, pp. 342–351, 2005.

[12] Bertin Instruments. AlphaGUARD: Radon Monitor. www.bertin-instruments.com/product/radon-professional-monitoring/radon-alphaguard/. Accessed on: 9 May. 2019.

[13] Tirmarche, M., Harrison, J., Laurier, D. Paquet, F., Blanchardon, E. & Marsh, J., Lung cancer risk from radon and progeny and statement on radon. *Ann. ICRP*, **40**(1), 2010.

[14] Tracerlab. GmbH, 2018. www.tracerlab.com/cataloge/index.htm?http://www.tracer lab.com/cataloge/documents/wlmdesc.htm.

[15] Arduino, Arduino UNO. www.arduino.cc/en/Guide/ArduinoUno.

Author index

WIT*PRESS* ...for scientists by scientists

Urban Transport XXV

Edited by: ***G. PASSERINI****, Marche Polytechnic University, Italy and* ***C. BORREGO****, University of Aveiro, Portugal*

Containing research from the 25th edition of the Urban Transport conference, the papers included in this book address the need to solve important pollution problems associated with urban transport. There is also a growing need for integration with telecommunications systems and IT applications in order to improve safety, security and efficiency.

The need for better urban transport systems and for a healthier environment has resulted in a wide range of research originating from many different countries. These studies highlight the importance of innovative systems, new approaches and original ideas, which need to be thoroughly tested and critically evaluated before they can be implemented in practice.

The variety of topics covered in this volume reflects the complex interaction of the urban transport systems with their environment and the need to establish integrated strategies. The aim is to arrive at optimal socio-economic solutions while reducing the negative environmental impacts of current transportation systems.

ISBN: 978-1-78466-339-1 eISBN: 978-1-78466-340-7
Published 2019 / 204pp

www.ingramcontent.com/pod-product-compliance
Lightning Source LLC
LaVergne TN
LVHW060004031026
849161LV00025B/1374